Landscape Is...!

Gareth Doherty is an Associate Professor of Landscape Architecture and Affiliate Faculty in the Department of African and African American Studies at Harvard University.

Charles Waldheim is the John E. Irving Professor of Landscape Architecture and the Director of the Office for Urbanization at the Harvard University Graduate School of Design.

Edited by Gareth Doherty and Charles Waldheim

Landscape Is...!

Essays on the Meaning of Landscape

Routledge
Taylor & Francis Group

LONDON AND NEW YORK

Designed cover by Clara Waldheim

First published 2025
by Routledge
4 Park Square, Milton Park, Abingdon, Oxon OX14 4RN

and by Routledge
605 Third Avenue, New York, NY 10158

Routledge is an imprint of the Taylor & Francis Group, an informa business

British Library Cataloguing-in-Publication Data
A catalogue record for this book is available from the British Library

ISBN: 9780367708245 (hbk)
ISBN: 9780367708214 (pbk)
ISBN: 9781003148142 (ebk)

DOI: 10.4324/9781003148142

Typeset in Frutiger
by codeMantra

Contents

Preface

.

Landscape Is…! examines the implicit biases and received meanings of land-scape. Following on volume one in the series *Is Landscape…?* which examined the plural and promiscuous identities of the landscape idea, this second volume in the series reflects upon the diverse and multiple meanings of landscape as a discipline, profession, and medium. *Landscape Is…!* addresses various over-looked aspects of landscape that develop, disturb, and diversify received under-standings of the field. Framed as an inquiry into the relationship of landscape to the forms of human subjectivity it conceives and constructs, this publication fea-tures contributions from leading voices challenging the contemporary under-standings of the field in relation to capital and class, race and gender, power and politics, among other timely topics.

Acknowledgments

This book emerged from conversations with our wonderful former editors Louise Fox and Sadé Lee at Routledge about the possibility of an expanded second edition of our 2016 volume *Is Landscape…?* Rather than an augmented second edition of that volume, this volume was conceived as a critical reflection upon and response to that book. The original edition was intended to productively problematize landscape's plural and promiscuous identities, alternately understood as discipline or medium, profession or practice. That volume, which stemmed from an eponymous proseminar at the Harvard GSD, was organized as a series of chapters unpacking landscape's cognate identities. We have been gratified by the reception and critical response to that project and remain committed to the productive questioning of landscape's multiple alter egos. However, rather than repeat that effort in an expanded field, we took this opportunity to conceive of a counter-project that inverts the editorial lens to consider how and what landscape *means* in the world. This volume aspires to frame a series of critical questions examining landscape's embeddedness in power relations, subject formation, and various forms of world-building. These questions span a diverse array of sites and subjects from capital to class, race to gender, power to politics, and beyond. Our goal is not to negate nor abandon landscape or landscape architecture as we remain convinced of the field's intellectual, cultural, and environmental potential. Rather, our goal is to invite both discipline and profession to redefine themselves in relation to urgent societal change. This seems to us a project that is both long overdue and a timely antidote to the field's seemingly unqualified and too often uncritical ascendency over the past decades. We are indebted to our extraordinary roster of contributors who graciously shared their knowledge, insights, and time on these topics. We are equally indebted to our outstanding research assistants Elaine Stokes and Charles Gaillard, who contributed consequently to the project, and to Jake Starmer for his diligent copyediting. Finally, this book would not have been possible without the ongoing commitment and careful craft of our Routledge editorial team, including Selena Hostetler, Megha Patel, Kathryn Schell, and Sarah Rae.

Contributors

Ọlátúnjí Adéjùmọ̀ is a Professor of Landscape Architecture in the Department of Architecture, University of Lagos, Nigeria.

Kofi Boone is a University Faculty Scholar and Joseph D. Moore Distinguished Professor in the Department of Landscape Architecture and Environmental Planning in the College of Design at North Carolina State University.

Danielle Narae Choi is an Assistant Professor of Landscape Architecture at the Harvard University Graduate School of Design.

John E. Crowley is a Professor Emeritus in the Department of History, Dalhousie University in Halifax, Nova Scotia, Canada.

Gareth Doherty is an Associate Professor of Landscape Architecture at the Harvard University Graduate School of Design.

Sonja Dümpelmann is a Professor and Chair of Environmental Humanities and the Co-Director of the Rachel Carson Center at Ludwig-Maximilians-Universität in Munich, Germany.

Edward Eigen is a Senior Lecturer in the History of Landscape and Architecture at the Harvard University Graduate School of Design.

Alison B. Hirsch is an Associate Professor in the School of Architecture at the University of Southern California.

Tarna Klitzner is a Lecturer in Landscape Architecture at the University of Cape Town.

Jala Makhzoumi is an Adjunct Professor of Landscape Architecture at the American University of Beirut.

Douglas Spencer is the Director of Graduate Education and Pickard Chilton Professor in the Department of Architecture at Iowa State University.

Anne Whiston Spirn is the Cecil and Ida Green Distinguished Professor of Landscape Architecture and Planning in the School of Architecture and Planning at MIT.

Kate Thomas is the K. Laurence Stapleton Professor of Literatures in English at Bryn Mawr College.

António Tomás is a Program Convenor in the Graduate School of Architecture at the University of Johannesburg.

Burcu Yiğit-Turan is a Senior Lecturer in Planning and Cultural Environments in the Landscape Architecture Division of the Department of Urban and Rural Development at the Swedish University of Agricultural Sciences.

Charles Waldheim is the John E. Irving Professor of Landscape Architecture and the Director of the Office for Urbanization at the Harvard University Graduate School of Design.

Ed Wall is a Professor of Cities and Landscapes, Academic Portfolio Lead for Landscape Architecture and Urbanism at the University of Greenwich.

Thaïsa Way is the Director of Garden & Landscape Studies at Dumbarton Oaks Research Library and Collection, a Harvard University research institution in Washington, DC.

Ujijji Davis Williams is a Landscape Architect, Urban Planner, and Researcher based in Detroit where she is a founder and principal of JIMA Studio.

Foreword

Landscape Is History

Thaïsa Way

Landscape is history. It is history that is both human and more-than-human. This is where I begin. Landscape entails the ground under our feet, the air we breathe, the water we drink, and the rays of sunlight that touch us; it holds stories of the past and informs our futures. This collection of essays, following the 2016 volume *Is Landscape . . . ?*, re-engages with the multivalent ways that cultures shape their relationships to landscape and how that landscape shapes communities.

Landscapes are places in which land, water, air, and the more-than-human converge as communities. This interpretation offers more than a way of seeing. It is a way of being in and becoming within the more-than-human world. It is in the formation of these relationships that landscape is understood as so deeply powerful.

The essays in this publication focus on landscape as a discipline, profession, and medium, and they consider how landscape is shaped by inscriptions of power relations, cultural formations, place-relationships, and infinitely diverse forms of world-building. It presents a generous conversation inviting landscape to encompass a breadth of meanings and readings, rather than attempting to reduce it to some definable or even defensible territory of practice or thought. This volume draws from a rich tradition of landscape scholarship and practice while centering on the contemporary challenges of reading and writing race, gender, culture, capital, and property (among other topics) into landscape narratives and practices of making place. It poses critical questions we might ask if we think of landscape through the diverse lenses of human experience and cultural relationships.

Whether one reads landscapes as elitist or gendered, colonial or queer, this choice shapes the stories that we imagine and share. It determines whose stories are recognized and whose are suppressed. It reveals who is accountable and who has the potential to lead toward radical change. Reading landscape as capital suggests understanding it as a form of cultural and symbolic production that largely dictates how we steward the environment and respond to climate change. It connects capitalism to climate change. Labor in the landscape raises questions about what is material and what is not and where values lie.

The making and shaping of landscape, in all its forms, are present in all places, and yet we rarely acknowledge or honor that labor. Through this obfuscation of labor, scholars and practitioners not only devalue the communities of workers but dismiss the ways such places always inform our lives and our communities and cultures. Narratives of landscape, as inscribed by race, colonialism, and empire, challenge us to account for the dark side, the violence of dominating cultures. Writing these ideas into our readings of landscape expands discussions of both accountability and responsibility for pasts, presents, and futures.

The essays assembled here reveal the challenges of landscape and its histories and offer rich and productive ways of thinking and practice going forward. By reading landscape as indigenous, space opens up for alternative modes of thinking, acting, and being in relationships to the world. Considering the dynamics of public landscapes and commons against the advent of privatization suggests ways to reconfigure for and with whom places are made. Concepts of recognition, reconciliation, and collaboration offer avenues for rethinking how places are made and lived and what visions of collective futures might be nurtured. Within discourses of aesthetics and ideas of beauty, design might be reimagined in the move toward a more just landscape. These readings of landscape as a field, discipline, and place offer access to a rich collection of tools to uncover, reveal, understand, explore, and share narratives and counter-narratives. But only if one listens, observes, and reads carefully, struggling to identify and differentiate the particularities of places, cultures, and landscape.

Why is this significant now? Because landscape profoundly matters. It is no longer acceptable for it to be considered merely a medium of cultural production. Landscape is more than human. We live in and with landscape, comprising the places that we make through our inscriptions, constructions, and perceptions. Whether our cultures survive or thrive depends on our relationships to these landscapes. Thus, engaging in a rigorous discussion is essential for our present and our future. Our understanding of landscape will shape the future of the more-than-human, the Earth's living and nonliving worlds, and those futures will shape us. The discussion to which this volume contributes is thus not merely interesting but timely and critical.

Introduction

Gareth Doherty and Charles Waldheim

> Landscape is not a genre of art but a medium. . . . a medium of exchange
> between the human and the natural, the self and the other. As such, it
> is like money: good for nothing in itself, but expressive of a potentially
> limitless reserve of value.[1]
>
> —W.J.T. Mitchell, *Landscape and Power* (1994)

Landscape is a particular mode of aesthetic encounter between certain human subjects and their environments, imagined or perceived. As such, it is a medium that is inherently value laden. Landscape is a cultural construct with unique origins, histories, potentials, and predicaments. Rather than a putatively benign or restorative art, landscape is a way of being in the world and, therefore, a form of world-building through the work of imagination. This work of imagining is not neutral nor necessarily benevolent. Rather, it is enabled by and invested with various forms of power relations and vested interests concerning both the human subjects it forms and the worlds it imagines. Like language, landscape enables our perception of the world through the lenses of cultural construction while constraining the possible worlds we can imagine.

This volume is not the first to ask questions regarding the meaning of landscape as a mode of human subjectivity and power relations. Four decades ago, Denis Cosgrove's *Social Formation and Symbolic Landscape* opened the topic with now canonical authority.[2] A decade later, W.J.T. Mitchell's *Landscape and Power* framed these questions with precision and clarity.[3] In the first decade of the twenty-first century, Rachael Ziady DeLue and James Elkins's *Landscape Theory* renewed these questions in relation to art history and theory.[4] Their volume reprinted the Introduction to Cosgrove's *Social Formation and Symbolic Landscape* as the first essay in the collection following Ziady DeLue's own introduction.

Over the past decade, dozens of authors have contributed to the question of landscape's relations to subject formation from history, art history, history of science, architecture, landscape architecture, urban planning, geography, and more. These essays and chapters, edited volumes, and monographs bring to the surface timely topics in relation to landscape, including race and class, gender and sexuality, capital and labor, and empire and colony, among numerous other

DOI: 10.4324/9781003148142-1

topics. Many of those authors and their arguments are represented here in this volume.

Landscape Is...! aspires to contribute to those conversations as a critical response to our own 2016 volume *Is Landscape...?* The original edition was intended to productively problematize the plural and promiscuous *identities* of landscape and what landscape *is*. These identities were articulated across an array of cognate fields adjectively modified by landscape and construed as discipline or medium, profession or practice. That volume was organized as a series of chapters unpacking landscape's cognate identities, including landscape understood as literature, painting, photography, gardening, ecology, planning, urbanism, infrastructure, technology, history, theory, philosophy, and, yes, architecture. This publication was conceived as a counter-project that shifts the focus from landscape's identities (what it is) to questions concerning how and what landscape *means* in the world.

This volume is concerned with landscape's implication in human subject formation, power relations, and world-building. Our goal is not to negate nor minimize the intellectual, environmental, or cultural potential of landscape's renewal in recent decades. In fact, over the first decades of the twenty-first century, landscape has enjoyed growing name recognition, economic resources, and cultural capital around much of the world. Over those decades, landscape's renewal has been buoyed by neoliberal economies, globalized trade, and ongoing environmental crises. While the discipline and profession still have much to offer that world, the growth and prosperity of the field have neglected much of the "majority world" in favor of an expanded range of professional potentials.

This volume is conceived as an opportunity for ethical reflection on the conditions for practice as well as a timely reconsideration of how landscape lands on various sites for various human subjects. As a cultural product of Western Europe and the Renaissance, landscape has had a very particular role in conceiving power relations between states and their subjects, empires and their colonies, capitalists and laborers, and many others. This publication gathers contributions from eighteen extraordinary thinkers on these topics.

Thaïsa Way's generous Foreword: "Landscape is History" opens the publication by situating narratives and practices of landscape in relation to human and more-than-human experiences of place. Gareth Doherty asks, "Is Landscape Fieldwork?" Doherty's chapter considers the relationship between landscape architecture and its engagement with sites and subjects in the world. Douglas Spencer asks, "Is Landscape Capital?" Spencer's chapter rehearses a Marxian critique of landscape through the contemporary lenses of the climate crisis and the failure of neoliberal market-based solutions. Burcu Yigit Turan asks, "Is Landscape Colonial?" Turan's chapter considers landscape's role in coloniality, citizenship, and border politics as read through the case of the Superkilen Park in Copenhagen, Denmark. Jala Makhzoumi asks, "Is Landscape Empowerment?" Makhzoumi's chapter considers the extent to which landscape enables and embodies power relations through the case study of *Marj Bisri* in Lebanon.

Sonja Dümpelmann asks, "Is Landscape Gendered?" Dümpelmann's chapter considers the role of landscape in the social construction of gender identities as well as the gendered construction of landscape architecture and its imagined worlds. John E. Crowley asks, "Is Landscape Imperial?" Crowley's chapter examines the role of imaginative representation and landscape projection in the history of imperial art, arguing that projects of empire depended upon these representations prior to their use in nation-state construction. Olátúnjí Adéjùmò asks, "Is Landscape Indigenous?" Adéjùmò's chapter situates landscape in relation to indigeneity, colonial histories, and sacred space through the case of Ile-Obe, Nigeria. Alison Hirsch asks, "Is Landscape Elitist?" Hirsch's chapter returns to the origins of the landscape idea and its privileging of landscape literacy by and for certain classes of subjects. Hirsch advocates for a reconsideration of the landscape architecture practice and a renewed understanding of democratized landscape access and experience. Kofi Boone asks, "Is Landscape Just?" Boone's chapter considers landscape through the lens of social justice, asking to what extent landscape renders just public realms and for whom. Boone advocates for a form of spatialized justice through a framework of environmental justice. Danielle Narae Choi asks, "Is Landscape Labor?" Choi's chapter reconsiders labor in relation to landscape through Hannah Arendt's concepts of labor, work, and *vita activa*. Choi argues for the embeddedness of labor in and in relation to all built landscapes while advocating for a vision beyond our contemporary "Plantationocene" of extraction, consumption, and disembodied labor. Ed Wall asks, "Is Landscape Public?" Wall's chapter reconsiders landscapes' putative publicness through anthropogenic climate change lenses and sites for climate contestation and public protest. Kate Thomas asks, "Is Landscape Queer?" Thomas's chapter reconsiders a range of literary accounts of landscape read through a queer lens and in relation to queer theories. Tarna Klitzner asks, "Is Landscape Collaborative?" Klitzner's chapter considers the role of professional knowledge in relation to the lived experiences, collective identities, and agency of community in the construction of the public realm. Ujijji Davis Williams asks, "Is Landscape Reconciliation?" Williams's chapter tracks the re-emergence of the "land-back" movement in the United States and its extension beyond indigenous sovereignty to active reparations for Black Americans impacted by centuries of racist design and planning policies. António Tomás asks, "Is Landscape Insurgency?" Tomás's chapter surveys landscape in relation to state power and practices of resistance informed by local territorial knowledge. Edward Eigen asks, "Is Landscape Haunted?" Eigen's chapter turns the questions regarding landscape back upon themselves in a reflexive consideration of the field's relative capacity for self-awareness. Anne Whiston Spirn asks, "Is Landscape Language?" Spirn's chapter returns to the relations between landscape and language as culturally shaped determinates of human subjectivity and meaning. Charles Waldheim closes the volume by asking, "Is Landscape Human?" Waldheim's chapter returns to the question of landscape's origins and agency in relation to the construction of human subjectivity. The chapter rehearses various positions on the subject and the abandonment of techno-managerial assumptions of universal landscape literacy in the mid-twentieth century in favor of a more culturally specific under-standing.

NOTES

1 W.J.T. Mitchell, ed., "Imperial Landscape," in *Landscape and Power* (Chicago: University of Chicago Press, 1994), 5.
2 Denis Cosgrove, *Social Formation and Symbolic Landscape* (Kent: Croom Helm, 1984).
3 W.J.T. Mitchell, ed., "Introduction," in *Landscape and Power* (Chicago: University of Chicago Press, 1994), 1–4.
4 Rachael Ziady DeLue, "Introduction," in *Landscape Theory*, eds. Rachael Ziady DeLue and James Elkins (London: Routledge, 2008), 3–14.

Chapter 1: Is Landscape Fieldwork?

Gareth Doherty

Figure 1.1 Ian McHarg in the field. Courtesy: Bob Peterson/Getty Images.

This volume and its focus on the variable meanings of landscape offer an opportunity to reconsider the relations between landscape, landscape architecture, and the wider world (Figure 1.1). Landscape has often been defined as a medium of aesthetic reception between certain human subjects and their environments as imagined, perceived, or built.[1] Landscape architecture is a profession concerned with the realization of constructed landscapes, yet the relation of that practice to the sites and subjects in the world remains opaque. Many of the field's founders, such as Olmsted and Burle Marx, engaged in deep and meaningful forms of field-work. In its short history as a discipline, landscape architects have developed various methods and means of engaging with landscapes, including human subjects, and understanding their work through spending time in the field. Other disciplines have spent more time conceptualizing their methodological and epistemological approaches. This is to say that fieldwork is at the core of the practice of landscape architecture, and yet, as a discipline, we don't often recognize this. This chapter offers a consideration of how landscape architects might reconsider their work in relation to the fields in which they intervene and how this work might be

DOI: 10.4324/9781003148142-2

understood in relation to other disciplines. By addressing how a range of disciplines approach fieldwork on landscapes, this chapter asks, "Is landscape fieldwork?"

Fieldwork means different things to different fields. Fieldwork on a landscape, i.e., landscape fieldwork, demands a particular approach because landscapes are spatial, temporal, and sensual.[2] Various disciplines have developed their own fieldwork methods to understand, interpret, and intervene in a landscape. In this chapter, I trace some of the conceptual and methodological similarities and differences across a selection of fields, including the data-collection methods, goals, tools, and scales that are distinct for landscape fieldwork. In each case, we see how fieldwork has shaped the discipline. I examine the cognate fields alphabetically, beginning with what the anthropologist Tim Ingold terms the "four As"—anthropology, archaeology, architecture, art—and then move on to ecology, geography, history, and urban planning. This is not an exhaustive list but merely an illustrative sample. While landscape architecture, as a diverse and multifarious discipline, has often borrowed tools and techniques from these domains of knowledge and others, it can also benefit from a deeper understanding of and reflection upon the fieldwork pursued in other disciplines.

ANTHROPOLOGY

Anthropology, arguably more than any other field of study, relies heavily on insights obtained through fieldwork. At Harvard University's Department of Anthropology, the discipline is characterized as "the study of human diversity in the distant past and the present and teaches us to recognize the remarkable array of circumstances in which human beings live their lives and make meaning from them"[3] and emphasizes understanding the myriad ways in which humans live and derive meaning. To grasp these meanings, anthropologists often dedicate at least a year to immersive fieldwork. This involves living within a community, building trust, learning local languages and cultural codes, and meticulously documenting not just daily activities but also the intricate relationships between people, objects, the environment, and one another (Figure 1.2). "I did not know much about anthropology, but I lived it," remarked Laurie Olin, the renowned landscape architect[4], highlighting the immersive, embodied engagement the discipline demands. Olin further noted that the manner in which a landscape architect conducts fieldwork impacts the final constructed project.[5]

One of the founders of anthropology in the United States, Franz Boas, first trained as a physicist; his doctoral thesis attempted to understand the color of water and was titled *Contribution to the Understanding of the Color of Water*.[6] Boas realized it was impossible to appreciate the color of water from a purely scientific perspective. Having studied geography and physics, Boas writes:

> In the course of my investigation, I learned to recognize that there are domains of our experience in which the concepts of quantity, of measures that can be added or subtracted like those with which I was accustomed to operate, are not applicable.[7]

Figure 1.2 Margaret Mead doing fieldwork in Samoa. Still from Margaret Mead: Taking Note, 1981, by Ann Peck and Michael Ambrosino. Courtesy: Documentary Educational Resources.

In time, motivated by his dissatisfaction with purely scientific means, Boas found his way to what became anthropology.

This combination of qualitative and quantitative measures is often known as "thick description," a term coined by Gilbert Ryle and popularized by Clifford Geertz.[8] The term refers to the contextual understandings that anthropology ascribes to specific actions and objects, such as the difference between a twitch and a wink.[9] The anthropologist will look for contextual clues to help them decipher whether there is intentionality behind the blink of an eye or not. Usually, anthropology aims to understand various phenomena through their study in situ or through what's known as ethnographic fieldwork. As Robert M. Emerson, Rachel I. Fretz, and Linda L. Shaw explain, "Ethnographic field research involves the study of groups and people as they go about their everyday lives."[10] Through in situ fieldwork, anthropologists understand patterns and unearth relationships that might have gone unnoticed. The anthropologist aspires to do what Subhadra Mitra Channa calls "embodied engagement."[11] Channa describes embodiment as "the entire process of formation of the self and identity from within the bodily location of the consciousness that identifies itself."[12] In terms of fieldwork, one's embodied presence in the field influences the way others engage with you, as well as one's own gaze, which is "turned outside but within one's body."[13] Channa attributes the concept to both the Western philosophical discourse of phenomenology, following the philosophy of Edmund Husserl and others, and from Hindu philosophy, where mind and body are seen as unitary.[14]

The ethnographic method of fieldwork can be quite messy, but the final report or publication, known as the ethnography or ethnographic monograph, is usually carefully crafted.[15] Ethnography—which is comprised of the Greek "ethno" (for people) and "graphy" (for writing)—is the description of a group of people (or objects) based on spending time with them.[16,17] Ingold reminds us of an important distinction to keep in mind: anthropology and ethnography are not synonymous.[18] Anthropology, the larger study of the human condition, usually involves ethnography, but not always. Likewise, ethnography is not the sole domain of anthropologists; it is used by art historians, geographers, planners, and landscape architects, among others. Traditionally, anthropology has separated research in an emic/etic distinction, from the internal perspectives of the subject, known as emic, to the external perspective of the researcher, known as etic. Most anthropology is a mix of emic and etic.

Anthropologists have many tools at their disposal, including cameras, voice recorders, microphones, and video recorders. However, their primary instruments remain paper and pen, a smartphone, or another writing device for documenting and interpreting field notes. They meticulously maintain various types of notes—scratch notes, fieldnotes, and headnotes—which they repeatedly review as they develop their ethnography.[19] Boas's student, Margaret Mead, who became one of the most famous women in the world, is renowned for the nuances of her notetaking. Her attention to detail is attributed to the fact that Mead daily read and re-read fieldnotes, looking for connections between her observations. Aware of the possibilities of "observer bias," Mead preserved her fieldnotes for others to consult and interpret.[20] Mead is one of the few anthropologists to write about the future. Despite the sub-fields of anticipatory and design anthropology, anthropology is mainly concerned with the present rather than the future.

ARCHAEOLOGY

Archaeology, akin to anthropology, relies heavily on fieldwork, arguably even to a deeper extent since archaeologists frequently excavate field sites. Through these excavations, they uncover objects, artifacts, and physical remains, which they analyze to discern patterns of human history and, more commonly, prehistory.[21] Archaeologists seek out hidden structures in ancient fields or gardens to infer how people interacted with each other and their surroundings (Figure 1.3). Much like forensic scientists, archaeologists employ a form of thick description or understanding, known as situatedness, which acknowledges and describes the experiential relationships between individuals and their environments.[22]

Kathryn L. Gleason, Wilhemina Jashemski, and Amina-Aïcha Malek have researched ancient fields and gardens from the Roman era, such as Pompei and Herculaneum, uncovering not just the garden spaces but their contemporary social values too.[23] In a short text titled "What Do Archaeologists Look for When Looking for Gardens," Malek reports that Jashemski was first trained in literature and history and later studied the material form of Roman gardens via fieldwork. This fieldwork—combined with her knowledge of the literature and history and the support of an interdisciplinary team—gave Jashemski the basis for challenging fundamental assumptions about Roman-era gardens that, until then, were

Figure 1.3 Muhammad Zahir, Hazara University archaeologist, excavating at Bhamala. February 2013. Bhamala is one of the most important Buddhist archaeological sites in Pakistan.

derived primarily from historical texts.[24] Through evidence derived from field-work, Jashemski demonstrated that those who were too poor to have their own garden were allowed access to windows to private gardens so they could still benefit from them. This was not something previously discussed in the literature.[25,26]

Archaeologists don't just randomly stop and dig. They use remote sensing and sophisticated tools such as radar and geophysical investigations to read larger landscapes. Jason Ur, a specialist in early urbanism, mentions how the scale of landscape requires a more lateral focus than the focus of an intense dig, with the intensity being transferred from a 100 m² trench across a 200 km² region. Ur traced the development of landscapes in Mesopotamia using declassi-fied satellite photographs taken for surveillance.[27] These spy images are much higher resolution than those available through most civilian satellite imagery, allowing for a more nuanced reading of the landscape. Through a close analysis of CORONA spy images,[28] Ur deciphered "hollow ways," which are linear depressions formed by the passage of humans and animals across a landscape over a long period of time. Due to the subtlety of their topographic imprint, these sunken lanes are often not visible at eye level but are clearly distinguishable from satellite images. The nuances of light and shadow render them visible—and reinforce the importance of eye-level interaction with bird's-eye views.[29] To avoid anthropocentrism, the assumption that only humans have intrinsic value,[30] archaeologists, like anthropologists, need to be aware of the symbolic meanings of the objects and patterns they unearth, which demands broad contextual and historical knowledge.

The trowel is the archaeologist's most basic tool, since they engage in physical excavations. The trowel is complemented by shovels, brushes, and sieves which allow archaeologists to meticulously sift through the soil for remnants. For micro-excavations, some archaeologists use tiny blades and picks to uncover human or animal remains. In addition, they may employ a theodolite for site surveying, GIS, satellite imagery, and remote sensing for larger scale analyses. Similar to anthropologists, archaeologists meticulously document their observations with detailed fieldnotes, writing, sketching, drawing, and photographing. However, their work often requires more precise spatial and temporal measurements, necessitating the consultation of maps and plans.

ARCHITECTURE

Architectural fieldwork closely aligns with landscape fieldwork, as both disciplines are invested in understanding how we will inhabit spaces in the future. Complementing the archaeologist's study of past living conditions, and the anthropologist's examination of our present lifestyles, architecture shares a common projective dimension with landscape architecture and planning. Since architecture is generally concerned with the art of designing and building buildings rather than a landscape's living horizontal systems, it can also radically differ from landscape architecture.

A classic example of the fruits of architectural fieldwork is Robert Venturi, Denise Scott Brown, and Steven Izenour's *Learning from Las Vegas*. Through fieldwork, the trio and their Yale School of Architecture students challenged architectural conventions and, arguably, demonstrated that the study of the Las Vegas strip was as relevant to designers as the study of ancient Greece and Rome.[31] Consequently, the subsequent "Learning From" series of studios at Yale and later the University of Pennsylvania (where Steven Izenour was joined by Susan Nigra Snyder) was based on the concept that the study of everyday life can challenge perceptions of history and theory and offer new knowledge based on engagement with the field. The studios were based on a deep and meaningful engagement with not just form but human taste and values in how people went about their everyday lives. In some ways, they were seen as being more concerned with "low" culture than "high" culture. Still, this distinction is problematic as it implies a hierarchy of knowledge and privileged books over field-originated knowledge. Experiential knowledge from the field can complement and even transcend valuable knowledge gained from books (Figure 1.4 and 1.5).

"The more I work on it, the less I know," said Rem Koolhaas of his fieldwork in Lagos, for which he was heavily criticized for allegedly not leaving his limousine.[32] But even doing fieldwork from a limousine is a form of fieldwork. Although somewhat removed from the haptic site conditions, automotive

Figure 1.4 Robert Venturi and Denise Scott Brown are doing fieldwork on the Las Vegas Strip. Courtesy: The Architectural Archives, University of Pennsylvania, by the gift of Robert Venturi and Denise Scott Brown.

Figure 1.5 Plan of Las Vegas Strip with signs, Learning from Las Vegas. Courtesy: The Architectural Archives, University of Pennsylvania, by the gift of Robert Venturi and Denise Scott Brown.

fieldwork provides its own unique viewpoint, which is important to acknowledge. Koolhaas viewed Lagos more from the elites' point of view rather than the majority population. Alison Smithson's *AS in DS: An Eye on the Road* is a documentation of fieldwork from a car, a Citroën DS, which resulted in a book shaped like the car.[33] The 24-year-old Charles-Édouard Jeanneret, later Le Corbusier, famously undertook a voyage to Istanbul by train with his friend, Auguste Klipstein. This trip convinced Charles-Édouard to become an architect. The fieldnotes and drawings from that journey were later edited into his book *Le Voyage d'Orient*, or *Journey to the East*.[34] Perhaps, most importantly, this journey provided Le Corbusier with the ethic of keeping field notebooks, a practice he continued all his life. Le Corbusier was a fieldworker.

There is fieldwork on architects themselves. Albena Yaneva's ethnography of Koolhaas's Office for Metropolitan Architecture (OMA), based on two years of participant observation, shows that there is a design method even when the architects do not recognize it.[35] *Architects' People*, edited by Russell Ellis and Dana Cuff, both professors of architecture, speculate on the individuals who inhabit the imagination of architects, whether they are the users, other architects, or themselves.[36] There is also an activist dimension to architectural fieldwork. Taking a people-centered approach to live projects, Auburn University's

Rural Studio engages in design-build projects in the field, producing what they term "citizen architects" for their interlocutors who become part of the creative process.[37] Other firms, such as Berlin-based Raumlabor, conduct similar fieldwork through construction.[38]

Given their focus on space and form, architects naturally engage in drawing, sketching, painting, doodling, photographing, and notetaking. Often, their tools are quite simple: a notebook, pencil, pen, or another sketching implement. Their notes and drawings typically emphasize spatial relationships and projections. A notable, if unconventional, example of architectural drawing is the work of Kaijima Momoyo, co-founder of Atelier Bow-Wow, the Tokyo-based architecture firm. Kaijima, who teaches "behaviourology" at the ETH in Zürich, has devised a distinctive graphic style that portrays buildings based on their current and potential future uses. Unlike typical architectural drawings which show form before inhabitation, Atelier Bow-Wow envisions spaces in the context of everyday life. Kaijima explains their approach·as "exploring a method of observing and drawing architecture and urban space from the viewpoint of the people who use it, rather than the architects and planners involved in its construction."[39] The result is a form of representation that is both descriptive and projective; Kaijima describes it as architectural ethnography. In addition to sketching, drawing, painting, and writing, architects engage with photography, film, sound, and other media. And those architects engaged in an activist practice will work with chisels, saws, and hammers.

ART

Art encompasses many approaches and mediums that can be visual, sensual, phenomenal, conceptual, written, or all the above. Few writers are untouched by the influence of landscapes in various forms and subsequently convert those landscapes into their writing.[40] Let's start with the work of visual artists, particularly landscape painters. The painter may respond to a subject (i.e., a site) with their intuition, and this intuition will usually be informed by their own time-tested methods and technical training. The aim is rarely to depict physical reality but rather to make visible a landscape's latent qualities, be they transcendent, spatial, spiritual, or social.

David Hockney, the celebrated British artist, is known for his use of vibrant colors in depicting people and landscapes. When painting a landscape, Hockney sets up his easel and paints *en plein air*—literally "in the open air"—to capture the space of the outdoors, something he maintains a camera can never do. Hockney claims he paints landscapes because it's impossible to photograph them. He tells us, "The camera can't get the beauty of this. It just can't get the space, the thrilling space that I'm in."[41] Arguably, the camera flattens the image by reducing the multidimensional landscape to the two dimensions of the photograph. Painting offers more possibilities for dimensionalities through the act of interpretation (Figure 1.6).

Figure 1.6 Claude Monet Painting by the Edge of a Wood, c.1885, John Singer Sargent. Tate, Presented by Miss Emily Sargent and Mrs. Ormond through the Art Fund 1925. Photo: Tate.

The en plein air practice can be traced back to seventeenth- and eighteenth-century Rome and inspired the likes of Claude Lorraine, Nicholas Poussin, Paul Cézanne, and Claude Monet. In England, John Constable, a noted en plein air painter, was fascinated by meteorological details and studied the sky, light, and atmosphere with great precision. This knowledge informed his en plein air painting. The fundamental goal is for landscape painting to be based on observable facts as witnessed in the open air. In addition to the scientific literature, there are forms of knowledge we can only get from being in the field.

Landscape is more than visual; it is sensual, too. Sissel Tolaas, a scent artist, collects smells using specialized equipment and reconstructs them in her Berlin-based laboratory. Tolaas has a collection of over 7,000 scents in her lab. She is known for gathering and reproducing the smells of cities, exposing our inherent prejudices. Smell is a fundamental form of landscape knowledge that can help us read and understand a landscape.

ECOLOGY

The ecologist is concerned with the study of the relationships between organisms and their physical environment. Although organisms include human and non-human beings, ecology is generally understood as having an environmental, rather than human, focus. This is despite the field of human ecology[42] and the Guattarian view of The Three Ecologies, "ecosophy," which embraces environmental, social, and subjective ecologies.[43] The ecologist's fieldwork will usually involve systematic and quantitative surveys—and experiments—gathering data, though it can be quite sophisticated and objective. The first steps entail perceiving the problem and framing and testing hypotheses, which depend on the intuition and creative processes of the fieldworker. As James E. Brower, Jerold H. Zar, and Carl N. von Ende suggest in their foundational text, *Field and Laboratory Methods for General Ecology*, no matter how quantitative the methods, the fieldwork will succeed or fail based on the initial intuition and creativity of the fieldworker. This intuition, combined with creative hypotheses, drives modern ecology. While ecologists might, for instance, spend weeks methodically sampling deer droppings in fields, their initial inkling will be driven by their previous life experiences as much as by their scientific acumen and what they observe in the field. All other steps depend on the precision of this first intuitive step (Figure 1.7).[44]

In *Land Mosaics: The Ecology of Landscapes and Regions*, landscape ecologist Richard T.T. Forman reveals a world where the landscape is composed, both formally and operatively, as a series of matrices. Formally, Forman relates to mosaic patterns of the landscape as seen in aerial photography. This vantage enables him to operatively understand how the ecology works in unison as part of a complex whole.[45] For Forman, landscapes comprise patches, sometimes linked by corridors, and organized in a matrix.[46] These could be woodland patches linked by hedgerows and held in an agricultural landscape. Patches can be either isolated and known as islands or connected and called nodes. Nodes occur at the intersection of the corridors or linkages. Nodes can be a source of

Figure 1.7 Ecologist David Moreno Mateos digging a soil profile on terra preta in Novo Airao, Amazonas, Brazil. Charcoal at different levels of the profile helps in the study of pre-Colombian farmlands. Courtesy: David Moreno Mateos.

objects dispersing outward or inward.[47] Any node is a source or sink for a species. The edges, as the point of linkages, are essential. Linkages that comprise these landscapes need to be unraveled if we are to measure them. Forman's theory of landscape composition is based on careful field observations carried out over many years.[48]

Forman's maps of urban regions investigate the interaction between cities and their hinterlands. Forman charts 38 urban regions—areas of 1,000 km^2 or more—internationally, showing that cities interact with their regions and beyond. He tells us that urban ecology is the "study of the interactions of organisms, built structures, and the natural environment, where people are aggregated around city or town."[49] Forman's maps are designed to synthesize and facilitate a comparison between urban regions.[50] The maps are based on quantitative data and include more subjective coding, with owls representing biodiversity, strawberries, market gardening, and sound symbols representing noise from flying aircraft.[51]

According to ecological designer and planner Nina-Marie Lister, biodiversity is "the variety, distinctiveness, and abundance of life forms and processes" in a landscape.[52] The Nature Conservancy, the global environmental organization, developed a ten-step method to assess biodiversity that combines remote sensing with aerial imagery, field data collection, and spatial information visualization. These Rapid Ecological Assessments (REAs) are especially useful when time and financial resources are limited.[53] Field sampling is an essential part of the REA as it provides the necessary evidence to support conclusions that are generalizable from the aerial images. For example, Adisa Ogunfolakan et al. describe a REA for the Osun Sacred Grove in Osogbo, Nigeria, that identified "a rich diversity, consisting of two rock types, three minerals, eighty-eight plant species, 108 insect families, and four mammal species."[54] Having completed the REA, the ecologist might rightly ask what forces drive this habitat's maintenance or future change. Meanwhile, Allen and Lister point out that there can be no single method of measuring biodiversity due to the complexity of contexts and scales of investigation that lead to different biodiversity conclusions.[55]

Ecologists have access to a wide range of increasingly sophisticated tools, which vary depending on their area of specialization. These tools might include an altimeter, a clinometer for measuring angles and tree heights, a drone, environmental DNA (eDNA) for genetic testing, field guides, field glasses or binoculars, handbooks, image analysis (to measure spectral reflections from different trees), light meters for measuring light intensity, a meter stick, a Munsell Soil Color Chart Basic Set (used for chemical analysisin the lab), a pH test kit, and a planimeter for measuring distances on maps.[56]

GEOGRAPHY

Geographers writes about the land the way ethnographers write about people; however, this binary falls short since geography also incorportates humans.

Figure 1.8 Geography professor John Hadler seated on a mountaintop, reading a map, solar slide, ca. 1980. Courtesy: Harvard University Library.

As the human geographer Yi Fu Tuan articulates, geography is "the study of earth as the home of human beings."[57] Similarly, *National Geographic* describes it as "the study of places and the relationships between people and their environments."[58] There are two main approaches: those who take a physical approach and those who identify as human geographers, but both approaches are concerned with the study of the earth and humans' interaction with it. Geographers study places in the Certeauian sense, where human practice transforms a space into a place (Figure 1.8).

Because of their place-based focus, geography is often taught outdoors, especially at the secondary or high school level. In the United Kingdom, geography is an established part of the school curriculum. A former London Schools Commissioner, Tim Brighouse, emphasizing the benefits of taking schoolchildren on fieldwork, declared, "One lesson outdoors is worth seven inside."[59] There are many types of geography, including economic, feminist, and political.

The tools utilized by geographers encompass maps, globes, charts, data, GIS, and other various specialized instruments to study the Earth and explore the interactions between humans and the planet. Geographers often draw diagrams and maps to represent the complex data they derive from fieldwork. Alexander von Humboldt, the explorer, naturalist, and founder of physical geography, demonstrated the link between the physical earth and organic life. His maps of isotherms, for example, illustrate the relationship between climate and a range of factors, such as ocean currents and mountain ranges. Historian Susan Schulten praises Humboldt's ability to visualize complex data derived from fieldwork leading to discoveries; she describes him as a "visual thinker (Figure 1.9)."[60]

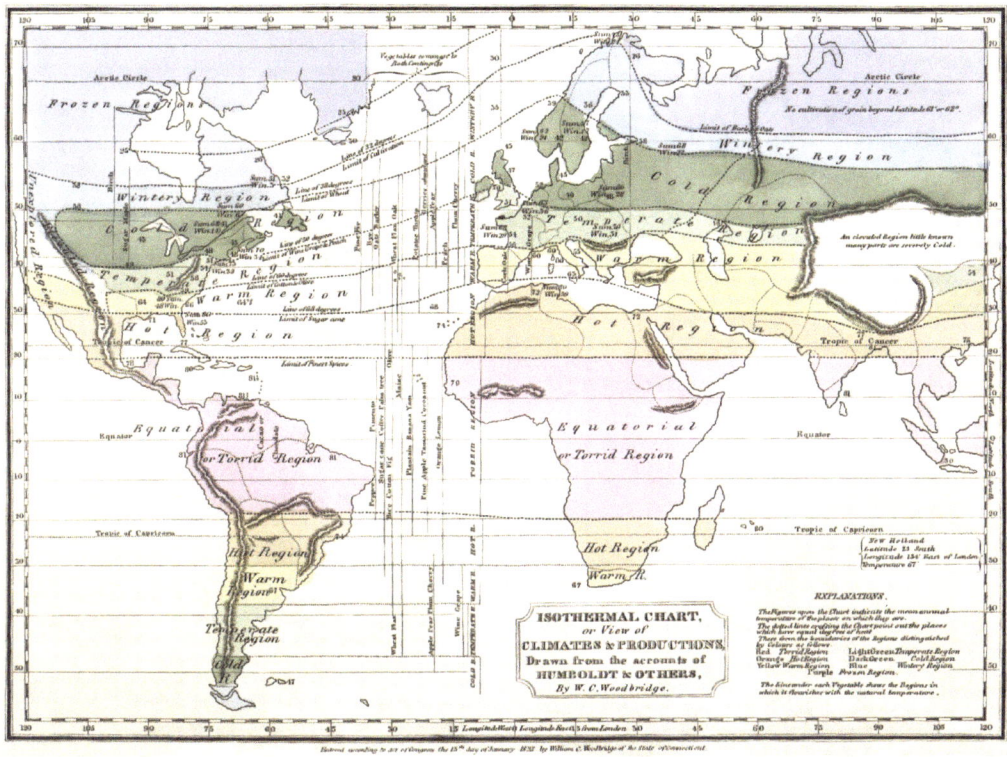

Figure 1.9 W.C.
Woodbridge's Isothermal
Chart, drawn from the
field accounts of
Humboldt and others.

HISTORY

The historian Ann Blair once remarked, "One of the great things about history is that everything has a history."[61] While historians have the breadth to explore any subject, this chapter will focus on garden histories. Garden and landscape historian, John Dixon Hunt, describes history as "the narration of past events."[62] Hunt elaborates that history requires both a narrator and an audience, and that "a narrative of the past implies or offers something for its present-day listeners, which, further, implies a future—by way of some guidance, instruction or emulation of the past for the years yet to be."[63]

Researching history can be a meticulous and tedious task. Like fieldworkers, historians do not necessarily know at the outset what the result will be: they need to be open to new information and leads. In *Crafting History: Archiving and the Quest for an Architectural Legacy*, Albena Yaneva unravels the complex process of forming design knowledge through the curation of archives. Yaneva asks what constitutes an architectural archive and what epistemologies it performs. Not only do the historians curate the information they find, but the documents they consult have already been curated by the archivists and the donors. In a reversal of power dynamics, we see that archives disempower the author of the archived work and set their work in comparison to the other work in an archive.[64]

One of the challenges for historians is when archives are limited, when documents, drawings, maps, photographs, or texts are unavailable or difficult to find. The historian might use the skills of the archaeologist to analyze objects or of the anthropologist to conduct interviews and oral histories. Chinese garden history, for example, comes with a particular set of challenges where gardens and garden history do not form distinct bibliographic categories. For references to historic gardens, the researcher must search under headings as diverse as "historical geography," "history of science," "horticulture," and "literary history." The researcher may consult biographies of the literati who owned the gardens. For images of gardens, they may search under "paintings" and "art."[65] As such, primary source research can be painstakingly slow. Stanislaus Fung has successfully navigated such challenges in the study of the historic Suzhou Gardens. Fung supplements his reading of the ancient gardens with field visits, photography, and 3-D scanning.

Together with a team of students, Fung conducted a visual analysis of two large Suzhou gardens—the Humble Administrator's Garden (*Zhuo Zheng Yuan*) and the Lingering Garden (*Liu Yuan*)—and contrasted them with the smaller

Figure 1.10 The Humble Administrator's Garden by Wen Zhengming (1470–1559), a Chinese painter, poet, and calligrapher of the Ming dynasty. Courtesy: Wen, Zhengming, *An Old Chinese Garden; a Three-Fold Masterpiece of Poetry, Calligraphy and Painting, by Wen Chên Ming*. Studies written by Kate Kerby; Translations by Mo Zung Chung (Shanghai China: Chung Hwa Book Company, 1922).

Figure 1.11 The Humble Administrator's Garden, Suzhou, 2019. Courtesy: The author.

Master of the Nets Garden (*Wang Shi Yuan*). Through this study, Fung challenged 200 years of Chinese garden history, which was based on Western readings of the spaces. In orthogonal drawings and plans, the sense of spatial depth is fixed, but Fung's fieldwork demonstrates that the spatial depth is unstable. The Suzhou Gardens play tricks with understanding Cartesian space and challenge the spatial understandings derived from plans and drawings. For example, Fung found that the occlusion of surfaces, whether ground or water, can make far-off objects seem closer, and when the ground surface is visible, the far-off object can seem farther away (Figures 1.10 and 1.11).[66]

Fully aware that different lighting and weather conditions may have different effects, Fung concluded that the gardens are based on what he terms the "instability of spatial depth." We see that the gardens are constructed with vertical layering—as in a Chinese painting—rather than the Western linear perspective as exemplified in Erwin Panofsky's *Perspective as Symbolic Form*.[67] In the case of instability of spatial depth, the body is "immersed" in the landscape rather than admiring it from afar. Fung describes his work as an "Attempt to realign the experience of traditional paintings in China with our experience of the gardens that we can still visit in Suzhou."[68] Fung likes to visit the gardens when they open shortly after 7:00 a.m. and to get most of the fieldwork completed before tourists arrive at 9:00 a.m. This necessitates staying at the only Suzhou hotel that serves breakfast as early as 6:00 a.m. Fung often does his fieldwork collaboratively, aided by his students, who take photographs that are subsequently used to test ideas.

Historians will employ a variety of resources including artifacts, documents, drawings, images, and photographs to piece together the intricate stories of a site and its surroundings. Often restricted to using only of a pencil and paper within archives and libraries, their primary tools are intellectual. Historians rely heavily on concepts. They seek out chronology, causation, evidence, perspective, and maintain a healthy amount of skepticism.[69] Their fieldwork may also involve measuring, photography, and the of use of 3-D scanners.

PLANNING

The planner's work is analytical, reflexive, imaginative, and projective; it is "the linkage between knowledge and organized action," according to John Friedmann and Barclay Hudson.[70] Planners work with communities in neighborhoods, cities, and regions to understand economic, environmental, legal, political, and social forces and their impact on the urban or rural landscape. "Planners take a broad view and look at how the pieces of a community—buildings, roads, and parks—fit together like pieces of a puzzle," according to the American Planning Association (APA).[71] To find that context and vision, planners analyze data, both quantitative and qualitative. Patsy Healey understands planning as "placemaking" and tells us that the plan provides the organizational framework for articulating how an area may change over the long term.[72] Planners prepare plans and policies that bring change to the built environment and facilitate and enable that change-making. Frederick R. "Fritz" Steiner emphasizes that "Plans require context and vision to provide the connection between what we know and what we want to do."[73] In other words, plans bridge the gap between our current situation and preferred futures.

The New York-based planning consultancy Interboro Partners develops participatory place-based approaches to find community consensus around complex urban issues. Interboro says their work involves "good listening, keen observation, and productive community engagement."[74] Community engagement can take many forms that exceed mandatory public meetings, which don't always attract a representative cross-section of society. To produce a more inclusive form of community engagement, Interboro worked with the City of Cambridge, Massachusetts, to develop a participatory strategy that included a mobile engagement station to encourage participatory mapping; a newspaper to keep the public updated on the plans; games to explain complex planning issues; and public exhibitions of the plans.[75]

Planners employ various analytic techniques, such as William H. (Holly) Whyte's time-lapse photography of small urban spaces in New York. This work led to his renowned film and book, *The Social Life of Small Urban Spaces*.[76] Whyte's analysis of human behavior patterns through photography and film has greatly influenced subsequent generations of designers and planners, including the London-based firm Space Syntax. Whyte consulted on many landscape architecture projects, including Hanna/Olin's Bryant Park renovation in New York City. Yasser Elsheshtawy uses similarly insightful observational techniques to examine the urban spaces inhabited by low-income migrant workers in Dubai and Abu Dhabi. Elsheshtawy's work is striking because it reveals the city's behind-the-scenes realities, showing how cities are inhabited, beyond the familiar formal depictions. Elsheshtawy's images expose the human dimension behind what Rahul Mehrotra terms the "landscape of impatient capital."[77]

In addition to documenting the results of time-lapse photography, planners also use spreadsheets, Geographic Information Systems (GIS), Building Information

Figure 1.12 Ray Lucas's flowchart of movement through the Tokyo Underground. Courtesy: Ray Lucas.

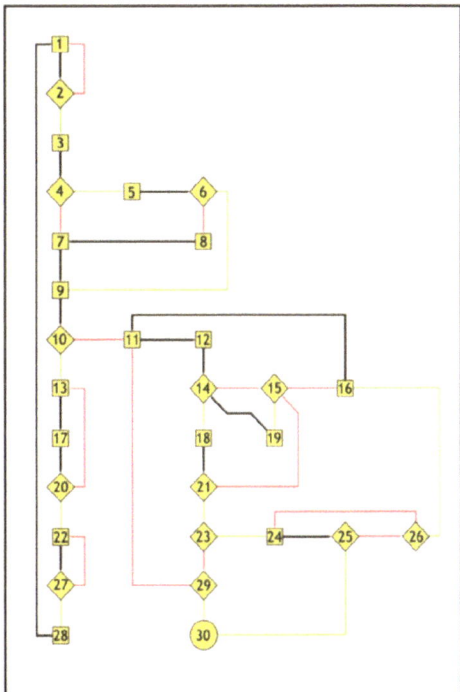

1. Stay on the train.
2. Is this your stop?
3. Stand up and wait at door.
4. Is someone else at the door?
5. Wait on them opening the door.
6. Do they open the door?
7. Press the door release.
8. Mutter, grumble and complain.
9. Exit train.
10. Are you changing lines?
11. Locate exit sign.
12. Move in the direction indicated.
13. Locate transfer sign.
14. Is there another sign?
15. Is there a transfer sign?
16. Find an open space.
17. Move in the direction indicated. (as 12)
18. Move in the direction indicated. (as 12)
19. Move away from that crowd.
20. Are you at the correct platform?
21. Is there another sign? (as 14)
22. Wait for the train to arrive.
23. Is there more than one sign?
24. Select an exit sign.
25. Does this exit lead out?
26. Have you tried other options?
27. Has the train arrived?
28. Board the train.
29. Does this exit lead out? (as 25)
30. Exit station.

Modeling (BIM), SketchUp, and Google Maps. Planners draw plans and write policy documents. There are city, ecological, environmental, regional, and urban planners. Given the plurality of approaches, planners use multiple site-specific tools and methods. Planners work with communities to understand their needs and concerns and use statistics and quantitative data to understand patterns of human interaction (Figure 1.12).[78]

FIELDWORK AS A SYNTHETIC ENDEAVOR

In this chapter, we have discussed some of the conceptual and methodological similarities and differences across a selection of disciplines from anthropology, archaeology, architecture, art, ecology, geography, history, and planning. Landscape architecture draws from all the above disciplines and more. Including art history, botany, engineering, and sociology in our list would be quite logical. While these disciplines offer valuable insights that can be applied to landscape architecture,[79] it is important to acknowledge that landscape architecture itself is inherently multidisciplinary. It is a hybrid field that encompasses a complex web of interconnected relationships. The forms of fieldwork listed above interact with one another. The disciplines themselves are hybrids and are further hybridizing into academic fields such as Situated Urban Political Ecology, for instance. Each of the forms of fieldwork above offers ways of understanding a landscape, and many of the tools are shareable. While the approaches have commonalities in techniques, tools, strategies, and even design intention, each example outlined above is based on a landscape. In the realm of landscape, disciplinary boundaries dissolve. But landscape is based on spatial, temporal, and sensual relationships; a defining challenge of landscape fieldwork is to understand space, time, and the senses (Figure 1.13).

Landscape *is* fieldwork. If landscape is the mutual shaping of people and place,[80] humans shape landscapes, and landscapes shape humans. Understanding the agency of landscapes requires ways and means of measuring their qualitative and quantitative dimensions. Across a range of disciplines, from anthropology to ecology and planning, we can see how fieldwork in landscapes creates new knowledge that complements and challenges existing norms. Fieldwork is central to the practice of landscape architecture, and yet, as a discipline, we don't often recognize this. If landscape architecture is to fully claim its agency to shape the landscapes and cities of the future, it will do well not to forget the landscape focus of allied fields, which all contribute to "landscape fieldwork."

Figure 1.13 Hydrosocial dynamics. Block 22 in Neighborhood 20, Buenos Aires by Brian Kohan, showing the landscape of the shantytown as it is inhabited, based on fieldwork in January 2024. Master in Landscape Architecture thesis, Harvard Graduate School of Design, 2024. Courtesy: Brian Kohan.

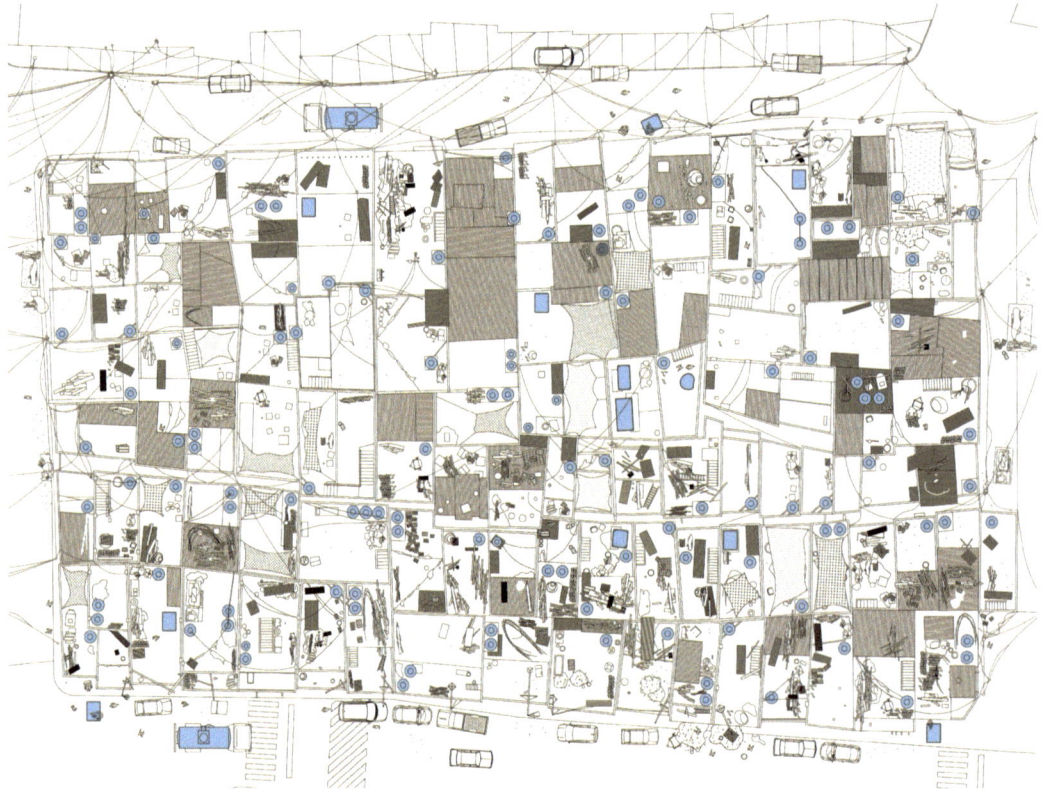

NOTES

1   See the introduction to this volume.
2   For more on landscape fieldwork, see my forthcoming book: Gareth Doherty, *Landscape Fieldwork: How Engaging the World Can Change Design* (Charlottesville: University of Virginia Press, forthcoming 2025) This chapter builds upon the introduction to that volume.
3   This is the definition given on the former website of the Department of Anthropology at Harvard University: https://anthropology.fas.harvard.edu/ (accessed April 1, 2020).
4   Laurie Olin, interview with author, 2022.
5   Laurie Olin, interview with author, 2022.
6   Franz Boas, *Beiträge zur Erkenntniss der Farbe des Wassers* (*Contribution to the Understanding of the Color of Water*) (Kiel, Germany: Druck von Schmidt & Klaunig, 1881).
7   Franz Boas, "An Anthropologist's Credo," *Nation* 147, no. 9 (1938): 201–204.
8   Clifford Geertz, "Thick Description: Toward an Interpretive Theory of Culture," in *The Interpretation of Cultures: Selected Essays* (New York: Basic Books, 1973), 3–30.
9   Clifford Geertz, "Thick Description: Toward an Interpretive Theory of Culture," in *The Interpretation of Cultures: Selected Essays* (New York: Basic Books, 1973), 3–30.
10 Robert M. Emerson, Rachel I. Fretz, and Linda L. Shaw, *Writing Ethnographic Fieldnotes*, 2nd edition (Chicago, IL and London: University of Chicago Press, 1995, 2011), 1.
11 Subhadra Mitra Channa, "Contextualizing Indian Feminist Scholarship within Local Patriarchy and Global Influences," Keynote lecture at the Federal University of Bahia, *What About Women in the History of Anthropology?* conference, July 25, 2018.
12 Subhadra Mitra Channa, "Embodied Engagement," personal correspondence with author, May 19, 2020.
13 Subhadra Mitra Channa, "Embodied Engagement," personal correspondence with author, May 19, 2020.
14 Subhadra Mitra Channa, "Embodied Engagement," personal correspondence with author, May 19, 2020.
15 Albert J. Mills, Gabrielle Durepos, and Elden Wiebe, *Encyclopedia of Case Study Research* (Los Angeles, CA and London: SAGE, 2010), 347.
16 For more on ethnography, see Rolf Malungo de Souza's excellent YouTube video, *O que é etnografia?* "What is Ethnography?" Souza identifies four basic principles of ethnography: (1) Participant observation; (2) Establishing rapport; (3) Listening deeply; and (4) Experimentation, stepping outside our comfort zones, https://www.youtube.com/watch?v=waWTIvMPmAM (accessed July 5, 2022).
17 With the emergence of multi-sited ethnography in the 1980s and 1990s, anthropologists have become less focused on a singular site. For more on this, see George E. Marcus, "Ethnography in/of the World System: The Emergence of Multi-Sited Ethnography," *Annual Review of Anthropology* 24, no. 1 (1995): 95–117.
18 Tim Ingold, "Anthropology is *Not* Ethnography," *Proceedings of the British Academy* 154 (2008): 69–92.
19 Roger Sanjek, *Ethnography in Today's World: Color Full before Color Blind* (Philadelphia: University of Pennsylvania Press, 2014), 69–71.
20 Margaret Mead, "Field Notes," Library of Congress; see https://www.loc.gov/exhibits/mead/mead-field.html# (accessed June 17, 2022).
21 In the United States, archaeology is often considered part of anthropology, one of the four fields alongside biological/physical anthropology, linguistic anthropology, and cultural anthropology.

22 See, for example, Anne Elizabeth Yentsch, "Introduction: Close Attention to Place–Landscape Studies by Historical Archaeologists," in *Landscape Archaeology: Reading and Interpreting the American Historical Landscape*, eds. Rebecca Yamin and Karen Bescherer Metheny (Knoxville: University of Tennessee Press, 1996), xxiii–xlii.

23 See, for example, Kathryn L. Gleason, *Gardens and Landscapes of the Past* (Philadelphia, PA: University Museum of the University of Pennsylvania, 1990); Wilhelmina F. Jashemski, *Gardens of the Roman Empire* (Cambridge; New York, NY: Cambridge University Press, 2018).

24 Amina-Aïcha Malek, "Preamble: What do Archaeologists Look for When Looking for Gardens?" in *Sourcebook for Garden Archaeology: Methods, Techniques, Interpretations and Field Examples*, eds. Amina-Aïcha Malek and Fondation des Parcs et Jardins de France (Bern: Peter Lang, 2013), 22–23.

25 Amina-Aïcha Malek, "Preamble: What do Archaeologists Look for When Looking for Gardens?," in *Sourcebook for Garden Archaeology: Methods, Techniques, Interpretations and Field Examples*, eds. Amina-Aïcha Malek and Fondation des Parcs et Jardins de France (Bern: Peter Lang, 2013), 22–23.

26 Anthropologists are turning their attention to more recent landscapes too. See Anne Elizabeth Yentsch, "Introduction: Close Attention to Place—Landscape Studies by Historical Archaeologists," in *Landscape Archaeology: Reading and Interpreting the American Historical Landscape*, eds. Rebecca Yamin and Karen Bescherer Metheny (Knoxville: University of Tennessee Press, 1996), xxiii–xlii; see also, William M. Kelso and Rachel Most, eds., *Earth Patterns: Essays in Landscape Archaeology* (Charlottesville: University of Virginia Press, 1990).

27 Jason Ur, "CORONA Satellite Photography and Ancient Road Networks: A Northern Mesopotamian Case Study," *Antiquity* 77, no. 295 (2003): 102.

28 CORONA are US spy satellite images used for surveillance between 1959 and 1972. They were replaced by KH-9 HEXAGON.

29 Jason Ur, "CORONA Satellite Photography and Ancient Road Networks: A Northern Mesopotamian Case Study," *Antiquity* 77, no. 295 (2003): 103.

30 See, for example, Brian Boyd, "Archaeology and Human–Animal Relations: Thinking Through Anthropocentrism," *Annual Review of Anthropology* 46 (2017): 301.

31 Robert Venturi, Denise Scott Brown, and Steven Izenour, *Learning from Las Vegas: The Forgotten Symbolism of Architectural Form* (Cambridge, MA: MIT Press, 1977), 87.

32 Suzanne Ewing, Jeremie Michael McGowan, Chris Speed, and Victoria Clare Bernie, eds., *Architecture and Field/Work* (London: Routledge), 1.

33 See: Alison Smithson, *AS in DS: An Eye on the Road* (Baden: Lars Müller Publishers, 2001).

34 See: Le Corbusier, *Journey to the East*, ed. Ivan Žaknić (Cambridge, MA: MIT Press, 2007).

35 Albena Yaneva, *Made by the Office for Metropolitan Architecture: An Ethnography of Design* (Rotterdam: 010 Publishers, 2009), 14.

36 Russell Ellis and Dana Cuff, eds., *Architects' People* (New York, Oxford: Oxford University Press, 1989).

37 "Educating Citizen Architects," http://ruralstudio.org/ (accessed March 22, 2021).

38 Raumlaborberlin, https://raumlabor.net/ (accessed March 21, 2021).

39 Momoyo Kaijima, "Learning from Architectural Ethnography," *Architectural Ethnography*, eds. Momoyo Kaijima, Laurent Stalder, Yu Iseki (Tokyo: Toro Kato, 2018), 8–14.

40 See for example: Gareth Doherty, "Is Landscape Literature?," in *Is Landscape...? Essays on the Identity of Landscape*, eds. Gareth Doherty and Charles Waldheim (London: Routledge, 2015), 13–43.

41 David Hockney, *David Hockney: A Bigger Picture*, 2017. https://vimeo.com/ondemand/116394, 1:11–1:25 (accessed July 2, 2021).

42 See Frederick R. Steiner, *Human Ecology: Following Nature's Lead* (Washington, DC: Island Press, 2002).

43 Félix Guattari, *The Three Ecologies* (London; New Brunswick, NJ; Somerset, NJ: Athlone Press; Distributed in the United States by Transaction Publishers, 2000).

44 James E. Brower, Jerrold H. Zar, and Carl von Ende, *Field and Laboratory Methods for General Ecology*, 3rd edition (Dubuque, IA: Wm.C. Brown Publishers, 1990), 1. Brower, Zar, and von Ende outline nine steps to an ecological study. They are "(1) Perceiving the problem; (2) Defining the entity of entities to be studied; (3) Designing the experiment or study; (4) Selecting a sampling (i.e. data-collection) procedure; (5) Obtaining representative samples; (6) Observing and measuring the samples to obtain data; (7) Objectively analyzing the data; (8) Interpreting and drawing conclusions from the data; and (9) Reporting the findings."

45 Ecology is the study of interrelated systems: "The science of the relationships between organisms and their environments. Also called bionomics." Dictionary.com, s.v. Ecology (n.) http://dictionary.reference.com/search?q=ecology (accessed May 31, 2024).

46 Richard T.T. Forman, *Land Mosaics: The Ecology of Landscapes and Regions* (Cambridge and New York: Cambridge University Press, 1995), 3–7.

47 Richard T.T. Forman, *Land Mosaics: The Ecology of Landscapes and Regions* (Cambridge and New York: Cambridge University Press, 1995), 253–262.

48 Richard T.T. Forman, *Land Mosaics: The Ecology of Landscapes and Regions* (Cambridge; New York: Cambridge University Press, 1995), 35–36.

49 Richard T.T. Forman, "Urban Ecology and the Arrangement of Nature in Urban Regions," *Ecological Urbanism,* eds. Mohsen Mostafavi with Gareth Doherty (Baden: Lars Müller Publishers, 2010), 312–321.

50 Richard T.T. Forman, *Urban Regions: Ecology and Planning Beyond the City* (Cambridge: Cambridge University Press, 2008), Color Figure 1.

51 Richard T.T. Forman, *Urban Regions: Ecology and Planning Beyond the City* (Cambridge: Cambridge University Press, 2008), Color Figures 2–39.

52 Roger Sayre, et al. *Nature in Focus: Rapid Ecological Assessment* (Washington, DC: Island Press, 2000), 84.

53 Roger Sayre, et al. *Nature in Focus: Rapid Ecological Assessment* (Washington, DC: Island Press, 2000), 84.

54 Adisa Ogunfolakan, Chinyere Nwokeocha, Ayodeji Olayemi, Moshood Olayiwola, Adebola Bamigboye, Adenike Olayungbo, Joan Ogiogwa, Oyeseyi Oyelade, and Olanipekun Oyebanjo. "Rapid Ecological and Environmental Assessment of Osun Sacred Forest Grove, Southwestern Nigeria," *Open Journal of Forestry* 6, no. 4 (2016): 243–258. https://doi.org/10.4236/ojf.2016.64020 (accessed May 31, 2024).

55 Nina-Marie Lister, "Bridging Science and Values: The Challenge of Biodiversity Conservation," *The Ecosystem Approach: Complexity, Uncertainty, and Managing for Sustainability*, eds. David Waltner-Toews, James Kay, and Nina-Marie Lister (New York: Columbia University Press, 2008), 86.

56 See "10 Things Aldo Leopold Used in the Field," https://onwisconsin.uwalumni.com/features/10-things-aldo-leopold-used-in-the-field/ (accessed June 20, 2022).

57 Yi Fu Tuan, "Foreword," *Geography and the Human Spirit*, ed. Anne Buttimer (Baltimore, MD and London: The Johns Hopkins Press, 1993), ix–xi.

58 "Geography," *National Geographic,* https://education.nationalgeographic.org/resource/geography-article/ (accessed June 5, 2024).

59 John Widdowson, "Fieldwork," *The Handbook of Secondary Geography*, ed. M. Jones (Sheffield: Geographical Association, 2017), 229.

60 Greg Miller, "The Pioneering Maps of Alexander von Humboldt," *Smithsonian Magazine*, October 15, 2019, https://www.smithsonianmag.com/history/pioneering-maps-alexander-von-humboldt-180973342/ (accessed May 31, 2024).

61 Ann Blair, https://history.fas.harvard.edu/about (accessed June 20, 2022).

62 John Dixon Hunt, "Is Landscape History," *Is Landscape...? Essays on the Identity of Landscape,* eds. Gareth Doherty and Charles Waldheim (London: Routledge, 2015), 247–260.

63 John Dixon Hunt, "Is Landscape History," *Is Landscape...? Essays on the Identity of Landscape,* eds. Gareth Doherty and Charles Waldheim (London: Routledge, 2015), 247–260.

64 Albena Yaneva, *Crafting History: Archiving and the Quest for an Architectural Legacy* (Ithaca, NY: Cornell University Press, 2020), 8–9.

65 Stanislaus Fung, "Guide to Secondary Sources on Chinese Gardens," *Studies in the History of Gardens & Designed Landscapes* 18, no. 3 (1998): 269.

66 Stanislaus Fung, "Recent Projects in Rural China," Harvard Graduate School of Design, November 16, 2018. YouTube video, 1:31:41, https://www.youtube.com/watch?v=ZjhM5slhK8k (accessed May 31, 2024).

67 Erwin Panofsky and Christopher S. Wood, *Perspective as Symbolic Form*, first paperback edition, 6th edition (New York: Zone Books, 2012). Panofsky's widely influential book translated from the German by Christopher Wood discusses the development of linear perspective in the European tradition.

68 Stanislaus Fung, "Recent Projects in Rural China," Harvard Graduate School of Design, November 16, 2018. YouTube video, 1:31:41, https://www.youtube.com/watch?v=ZjhM5slhK8k (accessed May 31, 2024).

69 See Lynn Hunt, "How to be an Historian: Five Tools," https://ilearn.laccd.edu/courses/85468/pages/how-to-be-an-historian-five-tools (accessed June 21, 2022).

70 John Friedmann and Barclay Hudson, "Knowledge and Action: A Guide to Planning Theory," *Journal of the American Institute of Planners* 40, no. 1 (1974): 2–16. https://doi.org/10.1080/01944367408977442 (accessed May 31, 2024).

71 American Planning Association, "What Is Planning?" https://www.planning.org/about-planning/ (accessed July 4, 2022).

72 Patsy Healey, "Collaborative Planning in a Stakeholder Society," *Town Planning Review* 69, no. 1 (1998): 1–21, https://doi.org/10.3828/tpr.69.1.h651u2327m86326p (accessed May 31, 2024).

73 Frederick R. Steiner, *Making Plans: How to Engage with Landscape, Design, and the Urban Environment* (Austin: University of Texas Press, 2018), 1.

74 Interboro Partners, https://www.interboropartners.com/ (accessed, July 5, 2022).

75 Interboro Partners, https://www.interboropartners.com/projects/envision-cambridge (accessed July 6, 2022).

76 See William Hollingsworth Whyte, *The Social Life of Small Urban Spaces* (Washington, DC: Conservation Foundation, 1980).

77 Rahul Mehrotra, *Architecture in India since 1990* (Mumbai; Ostfildern: Pictor Publishing Pvt Ltd; Hatje Cantz Verlag, 2011), 7.

78 Frederick R. Steiner, *Making Plans: How to Engage with Landscape, Design, and the Urban Environment* (Austin: University of Texas Press, 2018).

79 See Joseph Ragsdale, "Landscape Hybrids: Learning from Other Disciplines How to Read, Record, and Reveal the Landscape," in *Landscape Research Record* 7 (Blacksburg, VA: Council of Educators in Landscape Architecture (CELA) Conference, 2018).

80 Anne Whiston Spirn, *The Eye is a Door: Landscape, Photography, and the Art of Discovery* (Boston, MA: Wolf Tree Press, 2004), 1.

Chapter 2: Is Landscape Capital?

Douglas Spencer

As if further signs were needed, the recently published report by the International Panel on Climate Change signals our arrival at an epochal juncture.[1] The climate crisis is incontrovertibly caused by humans. Its effects are already irreversible, and we have only a short time to act in order to stave off the prospect of an all-but uninhabitable planet.

Given the ever-escalating and planetary-scale levels of destruction now being wrought by climate change, it's clear that the interests of capitalism are fundamentally and irredeemably at odds with those of human and other-than-human nature. Capitalism simply *is*, and always has been, a cause of numerous ongoing environmental crises. The question is not *if* or *how* capitalism will head off the worst impacts of climate change—since it plainly won't. Instead, it is *how* to exit this system that is inherently exploitative, extractive, and oppressive, in which life is cheap and the economy sacrosanct. This perspective has been argued most trenchantly by and on behalf of those who have most under capitalism. As The Red Nation writes in *The Red Deal: Indigenous Action to Save our Earth*:

> Market-based solutions must be abandoned. We have until 2050 to reach net-zero carbon emissions. . . The struggle for a carbon-free future can either lead to revolutionary transformation or much worse than what Marx and Engels imagined in 1848, when they forewarned that "the common ruin of the contending classes" was a likely scenario if the capitalist class was not overthrown.[2]

Likewise, for geographer Kathryn Yusoff, there is no escaping the environmental crisis without escaping capitalism:

> The Anthropocene is a project initiated and executed through anti-Blackness and inhuman subjective modes, from 1492 to the present, and it cannot have any resolution through individual liberal modes of subjectivity and subjugation. In short, that world must end for another relation to the earth to begin.[3]

For the collective authors of *The Tragedy of the Worker: Towards the Proletarocene*, common ruin and environmental destruction have always constituted the alpha and omega of capitalism. All life processes are subordinated to its

DOI: 10.4324/9781003148142-3

"homogenising frame of value production: M-C-M'. A regime of creative-destructive accumulation that is as inexhaustible as biospheric resources are finite."[4] Capitalism is, in fact, a "death cult."[5]

Our arrival at this juncture may also be discerned symptomatically from the intensification of efforts to persuade us otherwise. We are reassured that the existing state of things can be greened, guaranteed that there is no contradiction between capital and nature, and told that our relation to the Earth under capitalism is, in fact, already safely on the path of ecological redemption, promised that some ultimate unification of man and nature lies on the immediate horizon of capitalist development. Not coincidentally, the landscape has now returned to a position of cultural prominence equal to that enjoyed in the eighteenth and nineteenth centuries. Its contemporary uses are frequently aligned with such persuasive and reassuring messaging and are similarly symptomatic of our current conjuncture. Skyscrapers are dressed in lush foliage, botanical gardens housed in corporate atria, and tech campuses shrouded in verdant environments. Buildings are landforms, urban developments are landscaped, and the centuries-old aesthetics of the picturesque live on in the computer-rendered images through which these are publicized.[6]

To describe this as "greenwashing" is merely a dissimulating veneer designed to appease a generalized environmental consciousness and passes over much of what is at stake. Capitalism's capacity to naturalize its deep-seated, longstanding appropriation of human and non-human natures speaks to its death-cultish axiomatic and the accumulation of value as an end in itself. The charge of greenwashing is a contemporary variant of the most traditional practice of critical theory—the unveiling of ideology. This is of questionable and limited use in the practice of critique because landscape is as much creative and productive as it is dissimulating. Landscape is not simply or purely ideological but also lived. Landscape is a material condition of social existence. It shapes conduct, dispositions, and habits and creates perceptions of our relationship to capitalism, to nature, and to each other, all of which are, in turn, implicated in the reproduction of capitalism. A more substantial and wide-ranging critique of landscape—one reckoning with Marxian perspectives on the production of nature, the criticism of landscape as a genre of pictorial representation, and the ubiquity of landscape in the image worlds of contemporary development projects—is both possible and necessary.

THE CRITIQUE OF SYMBOLIC LEGITIMATION AND ITS LIMITS

A Marxian critique of landscape was developed in the 1970s and 1980s. It was practiced by art historians, literary critics, and cultural geographers such as Denis Cosgrove, John Barrell, and John Berger.[7] Cosgrove's *Social Formation and Symbolic Landscape*, first published in 1984, exemplifies this critique. As Cosgrove retrospectively underlined, this work was engaged with the question being addressed by many in the British New Left regarding the long transition from feudalism to capitalism. In his account of this transition, drawn from the perspective of cultural geography, the titular terms "social formation" and "symbolic landscape" correspond, more or less, to the notions of a societal "base" and an ideological "superstructure" as depicted by Marx in *A Contribution to the Critique of Political Economy* of 1859:

> In the social production of their existence, men inevitably enter into definite relations, which are independent of their will, namely relations of production appropriate to a given stage in the development of their material forces of production. The totality of these relations of production constitutes the economic structure of society, the real foundation, on which arises a legal and political superstructure and to which correspond definite forms of social consciousness.[8]

Cosgrove was concerned about underlining to his readers, however, that he was no "vulgar" Marxist. His argument was not that landscape—conceived, following Marx, as a superstructural expression of the "consciousness of men"—was merely determined by its economic base, but that landscape played an active role in legitimating and sustaining that economic base. Aligned with the broader currents and perspectives of the New Left, Cosgrove addressed landscape as a form of cultural and symbolic production that legitimized capitalism for the various social formations it had assumed during its historical ascendancy. Landscape, for Cosgrove, was an "idea" sustained by representational practices having social implications: "The landscape idea is a way of seeing—a way in which some Europeans have represented to themselves and others the world about them and their relationship with it, and through which they commented on social relations."[9] The history of this "way of seeing," as Cosgrove further expanded, could only be understood within the context of a "wider history of economy and society."[10]

While Cosgrove's handling of the relations between a "history of economy and society" and its forms of representation are sophisticated, landscape is nevertheless ultimately reduced to a concept: the "landscape idea." This concept is then further reduced to a "way of seeing." In this schema, the task consequently falls to the critic of correcting errors of vision so that the real workings of the economic base—the modes of production otherwise obscured by symbolic legitimation—can be made apparent. As Cosgrove wrote of his method,

> references to 'ideology' and 'hegemony' indicate attention to the social constitution and uses of symbolic discourse, and in fact the symbolic is generally treated theoretically as a veneer or veil, drawn with greater or lesser effect across material and social relations.[11]

Though restricted to the practice of ideological "unveiling," Cosgrove's method is to some degree apposite for the *longue durée* of the "landscape idea" with which his work, *Social Formation and Symbolic Landscape,* is concerned. From the Renaissance to the late nineteenth century, landscape was, as he writes, "the preserve of a privileged few—*cognoscenti* who were also politically and economically powerful."[12] But if landscape operates only within this rarefied class preserve, then the scope of its agency as an ideological force must be similarly constrained. Any symbolic legitimation of the mode of production can only be operating on those who own, govern, and profit from that mode of production. However, those whose land and labor have been appropriated over the long course of capitalist development are not themselves subject to the "landscape idea" or its "way of seeing." Landscape, in this account, is not a socially generalized practice of symbolic legitimation but a rarefied one in which the wealthy and the powerful practice a form of aestheticized self-deception as to the real basis of their wealth and power.

The objects of this critique of landscape are likewise rarefied, and the scope of their effectivity correspondingly confined: the poetry of Virgil and Pope, the painting of Bellini, Lorrain, and Bierstadt, and the designs of Palladio, Kent, and Lancelot Brown. Cosgrove and others practicing this kind of critique exemplify a further problematic feature of traditional critical theory in their near-exclusive fixation on the products of high culture. As Perry Anderson argued about this trait in his *Considerations on Western Marxism*, critical theory has always been more concerned with cultural expression than political economy.

> Marxist philosophy achieve[d] a general plateau of sophistication far beyond its median levels of the past; but the major exponents of Western Marxism also typically pioneered studies of *cultural* processes—in the higher ranges of the superstructures—as if in glittering compensation for their neglect of the structures and infrastructures of politics and economics.[13]

Anderson's point might be illustrated by John Berger's critique of Thomas Gainsborough's painting *Mr and Mrs Andrews* (circa 1750), depicting Lord and Lady Hardwicke in which the genres of portraiture and landscape are combined. In his *Ways of Seeing* in 1972, Berger takes issue with the disingenuousness of art historians, such as Kenneth Clark and Lawrence Gowing, for whom Gainsborough's painting ought only to be appreciated as, respectively, an "enchanting work" or a charming depiction of a couple engaged in the "philosophic enjoyment" of nature. But for Berger, the painting speaks of property relations and how the landed gentry—Lord and Lady Hardwicke—naturalized their proprietorial position by having their portraits painted in oils with the land they owned as a backdrop. Challenging orthodox art-historical aversions to overtly socioeconomic analysis, Berger situates Gainsborough's naturalization of property relations within the social formations of eighteenth-century England:

> Of course it is very possible that Mr and Mrs Andrews were engaged in the philosophic enjoyment of unperverted nature. But this in no way precludes them from being at the same time proud landowners. In most cases the possession of private land was the precondition for such enjoyment—which was not uncommon among the landed gentry.[14]

Berger, like Cosgrove, seeks to reveal the duplicities at work in representations of landscape with reference to the actual conditions of exploitation that would otherwise be allowed to appear as natural. His critical attention, also similar to Cosgrove's, focused on the "higher ranges of the superstructures." Gainsborough's *Mr and Mrs Andrews* was hardly a public work affecting ideological influence through mass dissemination, the painting being in the private possession of the Andrews family until 1960 and not even publicly exhibited at all until the 1930s. Ultimately, the contestation between Berger's interpretation and that of interlocutors such as Gowing was, then, one fought over art-historical interpretation rather than any broader or contemporary political significance.

SCENES OF PRODUCTION AND CONSUMPTION

Cosgrove wrote in the introduction to *Social Formation and Symbolic Landscape* that the importance of landscape as a form of cultural expression declined in the latter part of the nineteenth century. Landscape, he argued, "no longer carries the burden of social or moral significance attached to it during the time of its most active cultural evolution."[15] If true at the time of writing in 1984, it is scarcely so today. Landscape is now deeply implicated in envisioning green new worlds and sustainable futures; it is involved in a labor of naturalization through which capitalism seeks to persuade us of its ongoing viability. These landscaped future imaginaries, still bearing a family resemblance to their eighteenth- and nineteenth-century forbears, have now descended from the higher ranges of the superstructure to populate the everyday image worlds of the twenty-first century. The analysis of their role in the legitimation and reproduction of our relations to nature under capitalism is more pressing than ever.

To this end, the tools fashioned in the kind of analysis practiced by Cosgrove, Berger, and others remain valuable. The treatment of agricultural labor in landscape painting, as represented in Berger's writing on Jean-François Millet[16] or in John Barrell's *The Dark Side of the Landscape*, still significantly affects the analysis of how work is figured in contemporary landscapes. Similarly, Cosgrove's "landscape idea" continues to inform current imaginings of nature. Access to this idea is no longer the preserve of a privileged few. Nor is it confined now to the rarefied world of fine art or the design of private estates, and nor, consequently, is the "landscape idea" merely "a way of seeing" through which a ruling class practices upon itself the legitimation of its social standing. The expansion of the "landscape idea" to the worlds of the workplace, to "place-making," and to promotional strategies for sustaining capitalist development is one through which capitalism posits and realizes on a global scale its own ideal of the relations between humanity as a subject and nature as an object. At the same time, and through the same means, it posits and realizes its own conceptions of the "human" and the "natural." Landscape is, then, as much a site of the production of forms of subjectivity as it is one of their representations.

That capitalism produces the forms of subjectivity on which it depends for its reproduction is a fundamentally Marxian proposition. Production in capitalism, as Marx observed in the *Grundrisse*, "not only creates an object for the subject, but also a subject for the object."[17] In his essay "The Infinite Contradiction," Etienne Balibar expands on this often-overlooked form of production in capitalism in ways that significantly reformulate the "base and superstructure" model through which critical theory has tended to approach the relationship between ideology and economics. For Balibar, there are two mutually implicated "scenes" operative in capitalism: a "mode of subjection" and a "mode of production." "To name these different senses of the materiality of subjection and production," writes Balibar, "the traditional terms imaginary and reality suggest themselves."[18] In adopting these terms, Balibar urges that "one keep in mind that in any historical conjuncture, the effects of the imaginary can only appear through and by means of the real, and the effects of the real through and by means of the imaginary."[19]

In this reformulated schema, modes of production and modes of subjection—conventionally the economic and ideological, the base and the superstructure—are not deterministically but interdependently related. Each is required to take a detour through the "scene" of the other so as to realize itself. The real is produced through the detour of the imaginary, and the imaginary through the detour of the real. In this model, landscape would be understood not as an imaginary created to disguise the appearance of the real but as one through which the real is actually produced. And what the imaginary of landscape produces (among other things) is not so much an illusion to be dispelled through its unveiling, but the real and material formation of subjects; their conduct, dispositions, and habits, their perceptions and conceptions of their relationship to capitalism, nature, and each other.

The productive role of landscape design in subject formation might be approached via Manfredo Tafuri's *Architecture and Utopia: Design and Capitalist Development*, in which he engages with Marc-Antoine Laugier's notion of "the city as a forest."[20] The Marxian critique of landscape offered here by Tafuri is a distinct advance on the approaches of Cosgrove and Berger, for example, in that it engages with the legitimating power of design upon the crowd and the masses, not the connoisseur or the private patron of the arts. In dealing with urban planning and design, Tafuri is substantially engaged with the relationship between modes of production and modes of subjection, with the relation between infrastructure and superstructure in everyday experience, in a way that an analysis of traditional painting cannot be.

Tafuri describes in *Architecture and Utopia* how the aesthetics of the picturesque are employed, from the mid-eighteenth century onward, to naturalize capitalist development in respect of its impact upon the city and the country. Laugier's prescriptions for the design of the city reveal to Tafuri a "twofold inspiration": first, "that of reducing the city itself to a natural phenomenon," and second, "that of going beyond any a priori idea of urban organization by applying to the city the formal dimensions of the aesthetic of the picturesque."[21]

In the passage from *Observations sur l'Architecture* to which Tafuri refers, Laugier writes:

> Whoever knows how to design a park well will have no difficulty in tracing the plan of a city according to its given area and situation. There must be squares, crossroads, and streets. There must be regularity and fantasy, relationships and opposition, and casual, unexpected elements that vary the scene; get order in the details, confusion, uproar, and tumult in the whole.[22]

Tafuri employs Laugier's remarks to illustrate imperatives to landscape the city and naturalize its appearance through formal strategies already being deployed in garden design at a formative stage of capitalist development. For Tafuri, the city is thereby relegated to the "great sea of nature," and all attention focused, through the transplantation of an aesthetic derived originally from painting, "upon the suprastructural aspects of the city."[23] The aestheticization and naturalization of the city serve capitalist development by situating it beyond history, rendering its ruptures with tradition obscure. "Formal naturalism," writes Tafuri, was used to "consolidate and protect" the achievements of the emergent bourgeoisie "from any further transformation."[24] The landscaping of the city also served the development of capitalism through its negation of the disparities between city and country:

> Urban naturalism, the insertion of the picturesque into the city and into architecture, as the increased importance given to landscape in artistic ideology all tended to negate the now obvious dichotomy between urban reality and the reality of the countryside. They served to prove that there was no disparity between the value accredited to nature and the value accredited to the city as a productive mechanism of new forms of economic accumulation.[25]

Tafuri's larger purpose in *Architecture and Utopia* is to argue that design has long served as an ideological instrument of and for capitalist development even, perhaps especially, when it has imagined itself to be at the vanguard of some radical or utopian project. In this sense, Tafuri is visiting upon design the same kind of ideology critique that Theodor Adorno had brought to bear on the seemingly advanced music of Wagner or Stravinsky.[26] Tafuri sees the company towns and the city parks of the nineteenth century and the garden cities and urban planning of the twentieth century as heirs to Laugier's eighteenth-century notion of the "city as a forest." And yet, the project of naturalization to which the aesthetics of landscape are recruited in such instances is not only or merely an exercise in ideological masking. The idea of "city as a forest," alongside its contemporary variants, is also a lived and self-fulfilling proposition concerning the relations between man and nature that are realized through everyday conduct and experience. And, contra Tafuri, there is, beyond the register of their phenomenal appearance, no "obvious dichotomy between urban reality and the reality of the countryside"[27] to be ideologically masked because both are equally subject to the same extractive, improving, productive, and accumulative logics of capitalist development. Both are rendered as scenes of production in which the proposition of an unmediated relation between man and nature can be played out with such frequency and ubiquity as to actually become a kind of second nature.

BETWEEN LANDSCAPE AND INFRASTRUCTURE

In capitalism, the city and the countryside are both produced and sustained through infrastructural means that operate simultaneously and reciprocally in the register of the superstructural. For example, a Dezeen article from 2018 featured a video produced by Virgin Hyperloop One and BIG (Bjarke Ingels Group) purporting to demonstrate the travel experience it will present to its passengers.[28] "The high-speed transportation system's tubes will feature small circular windows that work in the style of a zoetrope—a device popular in the nineteenth century, which creates the illusion of motion."[29] The video opens with a porthole-like opening onto a landscape framed in black; the view is offered through the hyperloop tube, with its regularly spaced windows, from one of its travel pods. At the top of the screen, three counters register the journey's time, speed, and distance as it unfolds over the next 3 minutes and 34 seconds. As the train sets off on its journey, the window slides out of the frame to the right, and then another reappears from the left to glide past again and again, each time more rapidly, as the train accelerates. By the 30-second mark—by which point the train has reached 240 kilometers per hour—the intermittent appearance of the circular window has morphed into a strobing horizontal strip stretched across the width of the video frame. We catch glimpses of trees, green fields, telegraph poles, and isolated dwellings. At 1 minute 47 seconds, and having traveled 15 kilometers, the train reaches its full speed of 1,200 kilometers per hour. At this point, the frame disappears, and the hyperloop offers up an unimpeded widescreen experience of the landscape.

Ostensibly produced to assuage concerns of a claustrophobic travel experience, the hyperloop video also exemplifies a more longstanding dialectic of infrastructure and landscape in which each requires the other to enable its realization. For instance, this dialectic can also be seen at work in the American urban parkway systems of the late nineteenth and early twentieth centuries. Designed for an affluent class of car owners, these provided for a motorized experience of the picturesque. Drivers and passengers escaping the artifice of the metropolis along the landscaped corridors of the parkway were treated to a supposedly restorative contact with nature.[30] However, the desire to escape the industrial metropolis into some renewed connection with nature was also captured as a catalyst for further capitalist growth, automobile manufacturing, road building, fossil fuel extraction, and suburban real estate development. Objects are produced for the subject, and subjects are produced for the object.

Landscape is realized through infrastructural development, and infrastructural development is valorized through landscape. Each produces and sustains the other, forming and capturing the subject's tastes, habits, and sensibilities, acting as the fulcrum around which their relationship pivots. This is the dialectic of landscape and infrastructure in capitalism, the promise of an unmediated connection with nature subtended by an infrastructural apparatus staging its own disappearance. The work of Norman Foster exemplifies this dialectic; his oeuvre was founded on the premise that advanced building technology both sustains the environment and provides access to nature. Foster's techno-pastoralism is apparent in the Willis Faber and Dumas Headquarters of 1975. From the green-carpeted rooftop restaurant, employees can look out, through a seamlessly glazed curtain wall, onto scenes of colleagues resting and relaxing in the rooftop garden. At a grander scale, the architecture of Foster + Partner's Apple Park in Cupertino, California, situates the tech giant's employees in a landscape designed by Laurie Olin "with over six kilometers of walking and jogging trails. . . over 9,000 trees, orchards, as well as meadows, sports fields, terraces and a secluded pond." The exterior glazing of the park's ring-shaped building provides for an "uninterrupted connection to the landscape," and its eight atria create "social spaces that connect the park to the garden space within."

Corporate cultures of well-being are serviced through landscaping an environment whose ultimate purpose is to increase worker productivity, which is to be channeled toward the further growth of the multinational technology company. A lecture given by Foster in 2018 is especially revealing of his conception of the relationship between landscape and infrastructure.[31] Musing on the possibilities for a sustainable future for the planet, Foster imagines the countryside transformed so that the "dreams" of Frank Lloyd Wright's Broadacre City can at last become a reality. Likewise, Ebenezer Howard's vision of the Garden City "suddenly seems radically new"[32] to Foster, not least since his own Apple Park so strongly resembles Howard's formal diagram of ruralized urbanism. Foster looks forward to a future where "the infrastructure becomes invisible, autonomous," so we might "go to have a picnic in Battersea Park, and the picnic table disappears in the ground beneath us, and we pop up in San Francisco because of hyperloop."[33] As hard infrastructures disappear, the city becomes one great green and globally interconnected environment. "Smart technologies will connect entire cities, changing how we live and work. . . physical infrastructure will become obsolete,"[34] intones the voiceover from an accompanying video segment of the lecture while we watch a drone fly over a favela. The becoming invisible of infrastructure, on which Foster's green vision for the future is premised, is one where infrastructure services consumption, industries of travel, shopping, and leisure, an on-demand provision for individual needs and wants, the satisfaction of which is supposed to reconnect us with nature and to resolve for us, in the process, the environmental crisis. An infrastructure of neoliberalism, perhaps, but one whose historical specificity is disavowed in the claims to redeem the unfulfilled promise of Foster's forebears, Ebenezer Howard, Frank Lloyd Wright, and Buckminster Fuller.

The visions of the future promulgated by the billionaire class, by those personifying capital in its most hyper-concentrated form—Jeff Bezos, Elon Musk, Richard Branson—also share in Foster's retro-futurist and redemptive framing of the environmental question. Infrastructures of space travel, expanding the reach of capital out into the cosmos, will, it is argued, save the planet, allowing Earth to be salvaged as a haven for nature while space is transformed into a galactic reservoir of resource extraction. The mission statement for Bezos' Blue Origin company states:

> Blue Origin was founded by Jeff Bezos with the vision of enabling a future where millions of people are living and working in space to benefit Earth. In order to preserve Earth, Blue Origin believes that humanity will need to expand, explore, find new energy and material resources, and move industries that stress Earth into space.[35]

In a 2019 presentation titled "Going to Space to Benefit Earth," Bezos deployed four images to illustrate his vision for extra-planetary environments. One of these depicts Renaissance Florence in its fore- and middle-ground, extending into an alpine landscape, its skies dissolving at its outer edges to reveal the cylindrical frame of the spacecraft in which it is housed. Another presents an agricultural landscape, recalling Grant Woods' depictions of rural Iowa, traversed by a train speeding along an elevated rail line out of some more densely populated urban condition off in the distance. A third assumes a more sublime aesthetic. A lone stag stands on a rocky ledge. Waterfalls cascade into the valley, and mists cleave the mountainsides. The fourth image is equally theme park-worthy, a "Future-world" of greened urbanism and foliage-covered skyscrapers. These images echo and rework aesthetic devices long-established in landscape. Likewise, the "Blue Origin" project itself echoes and reworks the practices of colonization and exploitation of land for which landscape has been instrumental. Bezos, inter-viewed by NBC on returning to Earth from a short spaceflight, claimed: "We need to take all heavy industry, all polluting industry, and move it into space and keep Earth as this beautiful gem of a planet that it is."[36] But Earth, of course, can only appear gem-like from the distanced perspective of spaceflight. From up close and on the ground, Earth is marred by the sites of colonization and resource extraction within which certain of its inhabitants have been compelled to live and work and from which the concentrations of capital that Bezos person-ifies have been extracted. Turning space into a "sacrifice zone" for Earth only extends existing patterns of extraction and the labor they are necessarily based upon out into the cosmos and into a future that merely maintains existing forms of exploitative social and environmental relations.[37]

This centuries-long continuity in the uses of landscape is worth underlining, even where its imagery appears to evoke fantasies of a more recent vintage. Fred Scharmen, for example, has argued that Bezos' vision and accompanying images do little more than rehash a presentation on the future of space travel given to Congress in 1975 by the Princeton physicist Gerard O'Neill.[38] The so-called "O'Neill colonies" featured in this presentation, writes Scharmen, "were huge cylinders, spheres, and toruses with new surfaces for new kinds of civilizations inside."[39] Turning to the representation of these "colonies," Scharmen notes:

> The renderings produced for O'Neill's 1970s project were painted by Rick Guidice, who trained as an architect and graphic artist, and Don Davis, who had a background illustrating planetary science. Both men had roots in the counterculture, and they filled their space-habitat interiors with Buckminster Fuller's domes and Reyner Banham's architectural megastructures. In 1975, this was still what the future looked like.[40]

Scharmen persuasively argues that in appropriating O'Neill's original vision, all Bezos has to offer is a "watered-down version of nostalgia for yesterday's future." "Bezos's proposal," he continues, "is a version of O'Neill's project that somehow manages to look and feel less futuristic than its predecessor."[41] But it might also be added that this is a systemic rather than an individual failing, that a "nostalgia for the future" is all that the ruling class and their retinue of "imagineers" are capable now of dredging up from their near-exhausted imaginary. It might also be added that even "yesterday's future" traded in landscape imagery, standing not so much for the future as for reducing the future to a natural phenomenon. The O'Neill cylinder, as depicted by Rick Guidice and Don Davis, was a means to convey an essentially eighteenth-century aesthetic of the picturesque into outer space, landscape effecting its longstanding work of naturalization so that capitalism can continue to persuade us of its viability, consolidating and protecting its legacy. Infrastructure operates in these depictions, much as it does in those of Foster, as the vehicle for this payload. The visions of O'Neill and Bezos depict landscape as the subject and object of capital's unending expansion.

SPECIES, SUBJECTS, AND CRISIS

Earthbound imaginings of future landscapes extend this same proposition, offering up scenarios in which individuals are, through the interventions of design, placed at liberty to freely enjoy and directly experience nature: figures digging the ground on the floating islands of BIG's Oceanix or paddling kayaks in the "microhabitats" of Scape's designs for flood-resilience on the US Eastern Seaboard, for instance. In their depiction of the relationships between humans and their environments, these projective renderings repeat the representational devices of landscape painting. Figures isolated or depicted in small and typically familial groupings, the experience of nature idealized as an essentially private form of enjoyment, the collective and social labor through which nature is historically mediated is rendered obscure so that landscape can appear as given and immediate. To the extent that infrastructure appears at all, it does so as a mere prop within and for the landscape. But in the real conditions in which nature is currently experienced—conditions all too evidently mediated by capitalist development—immediacy and individualism assume other forms. For the migrant farmworkers digging the ground of regions experiencing extreme heat waves, immediate exposure to nature is life-threatening. The real-world version of the ubiquitous kayak is the makeshift raft, urgently fashioned by those forced by floods to find their own means of escape.

As Marx recognized, the condition of the human species is marked by an underlying alienation from nature. While the animal, in general, is "immediately one with its life activity," the human animal can and must engage in "free, conscious activity" in the production of its life activity as a social practice.[42] In this sense, our underlying alienation from nature is the basis from which we derive the potential to act upon nature socially, to conceive, plan, and produce nature in particular forms that reciprocally produce particular social formations. These are the species-specific capacities that have been hijacked by capitalism's "homogenizing frame of value production," by its inhuman and deathly drive to accumulate as an end in itself. Capitalism strips us of the collective capacity to consciously manage our social and environmental relations. It offers compensation for the false promise of individual communion with nature, a promise that landscape has long served to underwrite, both as a genre of pictorial representation and as a practice of design. Landscape and its promises, to recall Balibar's formulation, have been situated at a historical conjuncture where the effects of the imaginary have appeared through and by means of the real, and the effects of the real through and by means of the imaginary.

Attempting to break out of this closed and self-sustaining loop, many in the field of landscape education have begun to explore how their discipline might respond to the ongoing and increasingly catastrophic effects of global warming and environmental breakdown. This agenda cannot, though, be served by the very same landscape imaginary as has served capitalism, with its false and untenable promises of the private enjoyment of nature represented as an ahistorical given, as opposed to a social and historical product. Instead, the legacies of landscape and its uses for capitalism ought to be reckoned with in the process of repurposing its forms of knowledge and practice if these are to serve other ends.

Visualizing the kind of mega-scale and publicly financed infrastructural projects called for by, say, a Green, Black, or Red New Deal might be an opportunity to rethink the representational practices of landscape instead of packaging these in the modes of the pastoral and the picturesque. Where landscape has historically served to promote the private enjoyment of nature, it might instead explore how that enjoyment can be collectively experienced and represented. Where landscape has naturalized capitalism, it might turn to foreground the forms of artifice and labor that necessarily subtend the appearance and uses of what goes on in the name of nature. Rather than picturing seemingly natural environments, landscape might be a means to critically inform our cognitive comprehension of these and perhaps even to do so as a kind of mass participatory pedagogical project.

None of this suggests that landscape can or should simply refashion itself into some kind of anti- or post-capitalist practice, presenting itself as some spearhead of revolution. After all, these notions are unrealistic and only act as the self-aggrandizing and delusional fantasies through which the design disciplines sustain themselves while maintaining their client bases and positions of prestige. While facing up to its own ongoing uses for capitalism, landscape might challenge itself to forge tools of real use to other agendas and agents than those it has historically serviced.

NOTES

1 The Intergovernmental Panel on Climate Change, Sixth Assessment Report, *Climate Change 2021: The Physical Science Basis*, https://www.ipcc.ch/report/ar6/wg1/.

2 The Red Nation, *The Red Deal: Indigenous Action to Save our Earth: Indigenous Action to Save Our Earth* (Brooklyn, NY and Philadelphia, PA: Common Notions, 2021), 21–22.

3 Kathryn Yusoff, *A Billion Black Anthropocenes or None* (Minneapolis: University of Minnesota Press, 2018), 56.

4 The Salvage Collective, *The Tragedy of the Worker: Towards the Proletarocene* (London and New York: Verso, 2021), 18.

5 The Salvage Collective, *The Tragedy of the Worker: Towards the Proletarocene* (London and New York: Verso, 2021), 8, 10.

6 On architecture as "landform," see Stan Allen and Marc McQuade, eds., *Landform Building: Architecture's New Terrain* (Zurich: Lars Müller, 2011).

7 See Denis Cosgrove, *Social Formation and Symbolic Landscape* (Madison and London: University of Wisconsin Press, 1998); John Barrell, *The Dark Side of the Landscape: The Rural Poor in English Painting 1730–1840* (Cambridge: Cambridge University Press, 1980); John Berger, *Ways of Seeing* (London: BBC/Penguin, 1972).

8 Karl Marx, *A Contribution to the Critique of Political Economy*, trans. N. I. Stone (Chicago, IL: Charles H. Kerr & Company, 1904), 11–12.

9 Denis Cosgrove, *Social Formation and Symbolic Landscape* (Madison and London: University of Wisconsin Press, 1998), 1.

10 Denis Cosgrove, *Social Formation and Symbolic Landscape* (Madison and London: University of Wisconsin Press, 1998), 1.

11 Denis Cosgrove, *Social Formation and Symbolic Landscape* (Madison and London: University of Wisconsin Press, 1998), xxv.

12 Denis Cosgrove, *Social Formation and Symbolic Landscape* (Madison and London: University of Wisconsin Press, 1998), 263.

13 Perry Anderson, *Considerations on Western Marxism* (London: Verso, 1979), 17.

14 John Berger, *Ways of Seeing* (London: BBC/Penguin, 1972), 108.

15 Denis Cosgrove, *Social Formation and Symbolic Landscape* (Madison and London: University of Wisconsin Press, 1998), 2.

16 John Berger, "Millet and the Peasant," in *About Looking* (New York: Pantheon Books, 1980), 69–78.

17 Karl Marx, *Grundrisse: Foundations of the Critique of Political Economy*, trans. Martin Nicolaus (London and New York: Penguin, 1973), 92. Marx continues: "Thus production produces consumption (1) by creating the material for it; (2) by determining the manner of consumption; and (3) by creating the products, initially posited by it as objects, in the form of a need felt by the consumer. It thus produces the object of consumption, the manner of consumption and the motive of consumption. Consumption likewise produces the producer's inclination by beckoning to him as aim-determining need."

18 Etienne Balibar, "The Infinite Contradiction," trans. by Jean-Marc Poisson with Jacques Lezra, in *Depositions: Althusser, Balibar, Macherey, and the Labor of Reading* (Yale French Studies, no. 88), eds. Jacques Lezra (New Haven, CT: Yale University Press), 160.

19 Etienne Balibar, "The Infinite Contradiction," trans. by Jean-Marc Poisson with Jacques Lezra, in *Depositions: Althusser, Balibar, Macherey, and the Labor of Reading* (Yale French Studies, no. 88), eds. Jacques Lezra (New Haven, CT: Yale University Press), 160.

20 Manfredo Tafuri, *Architecture and Utopia: Design and Capitalist Development*, trans. Barbara Luigia La Penta (Cambridge: MIT Press, 1976).

21 Manfredo Tafuri, *Architecture and Utopia: Design and Capitalist Development*, trans. Barbara Luigia La Penta (Cambridge: MIT Press, 1976).

22 M.A. Laugier, *Observations sur l'Architecture* (The Hague: Desaint, 1765), cited in Manfredo Tafuri, *Architecture and Utopia: Design and Capitalist Development*, trans. Barbara Luigia La Penta (Cambridge: MIT Press, 1976), 4.

23 Manfredo Tafuri, *Architecture and Utopia: Design and Capitalist Development*, trans. Barbara Luigia La Penta (Cambridge: MIT Press, 1976), 8.

24 Manfredo Tafuri, *Architecture and Utopia: Design and Capitalist Development*, trans. Barbara Luigia La Penta (Cambridge: MIT Press, 1976), 7.

25 Manfredo Tafuri, *Architecture and Utopia: Design and Capitalist Development*, trans. Barbara Luigia La Penta (Cambridge: MIT Press, 1976), 8.

26 Theodor Adorno, *In Search of Wagner*, trans. Rodney Livingstone (London and New York: Verso, 2005); *Philosophy of New Music*, trans. Robert Hullot-Kentor (Minneapolis: University of Minnesota Press, 2006).

27 Manfredo Tafuri, *Architecture and Utopia: Design and Capitalist Development*, trans. Barbara Luigia La Penta (Cambridge: MIT Press, 1976), 8.

28 Rima Sabina Aouf, "Zoetrope Windows Turn Hyperloop Tubes 'Transparent,'" June 8 2018, https://www.dezeen.com/2018/06/08/zoetrope-windows-hyperloop-tube-transparent-big-virgin/ (accessed September 20, 2021).

29 Rima Sabina Aouf, "Zoetrope Windows Turn Hyperloop Tubes 'Transparent,'" June 8 2018, https://www.dezeen.com/2018/06/08/zoetrope-windows-hyperloop-tube-transparent-big-virgin/ (accessed September 20, 2021).

30 On the history of the parkway see, for example, Matthew Gandy, *Concrete and Clay: Reworking Nature in New York City* (Cambridge: MIT Press, 2003).

31 Architecture on Stage: Norman Foster, April 17, 2018, https://www.youtube.com/watch?v=V0Z5R69UWsw (accessed June 10, 2021).

32 Architecture on Stage: Norman Foster, April 17, 2018, https://www.youtube.com/watch?v=V0Z5R69UWsw (accessed June 10, 2021).

33 Architecture on Stage: Norman Foster, April 17, 2018, https://www.youtube.com/watch?v=V0Z5R69UWsw (accessed June 10, 2021).

34 Architecture on Stage: Norman Foster, April 17, 2018, https://www.youtube.com/watch?v=V0Z5R69UWsw (accessed June 10, 2021).

35 Blue Origin "vision" statement, https://www.blueorigin.com/about-blue.

36 See Denise Chow, "Jeff Bezos Says Spaceflight Reinforced His Commitment to Solving Climate Change," *NBC News*, July 20, 2021, https://www.nbcnews.com/science/science-news/jeff-bezos-says-spaceflight-reinforced-commitment-solving-climate-chan-rcna1467 (accessed December 27, 2022).

37 As the reporter Justine Calma noted of Bezos' proposals for extraterrestrial "sacrifice zones," "Long before rich, white men were catapulting themselves into space, they approached whatever was the 'frontier' at the time with dollar signs in their eyes and destruction in their wake." "Jeff Bezos Eyes Space as a New 'Sacrifice Zone,'" *The Verge*, July 21, 2021, https://www.theverge.com/2021/7/21/22587249/jeff-bezos-space-pollution-industry-sacrifice-zone-amazon-environmental-justice (accessed December 27, 2022).

38 Fred Scharmen, "Jeff Bezos Dreams of a 1970s Future," *Bloomberg*, May 13, 2019, https://www.bloomberg.com/news/articles/2019-05-13/why-jeff-bezos-s-space-habitats-already-feel-stale (accessed May 5, 2021).

39 Fred Scharmen, "Jeff Bezos Dreams of a 1970s Future," *Bloomberg*, May 13, 2019, https://www.bloomberg.com/news/articles/2019-05-13/why-jeff-bezos-s-space-habitats-already-feel-stale (accessed May 5, 2021).

40 Fred Scharmen, "Jeff Bezos Dreams of a 1970s Future," *Bloomberg*, May 13, 2019, https://www.bloomberg.com/news/articles/2019-05-13/why-jeff-bezos-s-space-habitats-already-feel-stale (accessed May 5, 2021).

41 Fred Scharmen, "Jeff Bezos Dreams of a 1970s Future," *Bloomberg*, May 13, 2019, https://www.bloomberg.com/news/articles/2019-05-13/why-jeff-bezos-s-space-habitats-already-feel-stale (accessed May 5, 2021).

42 Karl Marx, *Economic and Philosophic Manuscripts of 1844*, trans. Martin Milligan (Moscow: Progress Publishers, 1974).

Chapter 3: Is Landscape Colonial?

Burcu Yiğit-Turan

> Coloniality [is] the reverse and unavoidable side of "modernity"—its darker side, like the part of the moon we do not see when we observe it from Earth.
>
> — Walter Mignolo[1]

Sweden is internationally known for its labor movements, welfare state policies, democratic social life, anti-colonial and anti-racist discourses, a strong connection to nature, and an active outdoor lifestyle.[2] In the twentieth century, Swedish urban areas were said to produce "radically communal" public green spaces, including recreational infrastructures, to help the marginalized masses improve their physical and mental health.[3] Swedish landscape architecture is internationally branded and recognized for its supposedly moral virtues, deep natural knowledge, minimalist intervention, and contribution to democratic social life.[4] The *Folkhemmet* (people's homes) movement (Figure 3.1) advocated for high-quality working- and middle-class housing and generous public natural environments. Postwar modernist housing movements, including the Million Homes Programme,[5] also produced subsidized housing and green housing environments, which are defined as welfare landscapes.[6]

Nordic and Swedish vocabularies (e.g., art, architecture, and planning practices) might be used as seductive instruments within global urban development. They are said to be a trove of "eco-modernist knowledge" for environmentally and socially sustainable democratic planning and design within the Anthropocene. One of the most important pillars in Swedish city visions, "greening," lauds public landscapes as a basic right and societal equalizer. Such ideas have been integrated into perplexing future-city "utopias" at home and abroad (e.g., the City of Telosa[7]). Furthermore, they are used to expand the reach of architectural and planning offices in the global market.[8]

Under Olof Palme's Social Democrat government in the 1970s and 1980s, Sweden created an expansive welfare state and promoted anti-imperialist, anti-racist, pro-immigration ideas. From 1965 to 1975, the Social Democratic Party implemented the Million Homes Programme, which subsidized the construction of approximately one million new apartment units in response to housing shortages. This program produced high-rise suburban housing with

DOI: 10.4324/9781003148142-4

Figure 3.1 Bathing and playing children in Fredhälls Park in Stockholm, 1952. Photo: K W Gullres/Nordiska museet. Id på digitaltmuseum: 021016295939.

generous landscapes operated by the municipalities and was commonly associated with (often harsh, severe, and isolating[9]) architectural high modernism. White Swedes took advantage of economically advantageous policies and practices in the 1970s and 1980s to flee these units. This resulted in a concentration of non-white, non-European immigrants and refugees.[10]

Despite its image as anti-colonial and anti-racist, Sweden is deeply involved in the idea of race, white European supremacy, and the continuous subjugation of people of color, non-Europeans, and indigenous people.[11] Today, the global racialized hierarchy of labor has been fully reproduced within Sweden's discriminatory labor market.[12] Swedish society perpetuates racial segregation through white flight from non-white suburbs, racial discrimination in many domains (i.e., housing, school, health, labor market, and education), and racialized barriers to opportunities, power, and influence.[13] Contemporary Swedish cities are highly racialized—perhaps even the most segregated in Europe (in terms of race and class).[14]

The neoliberalization of state planning and housing, the deterioration of welfare policies, and the privatization of different sectors have exacerbated socio-spatial inequalities and polarization. Market-governed housing production that targets predominantly non-white neighborhoods for densification and urban renewal in the Million Programme (MP) areas is now a driving force in urban development.[15] Modernist welfare landscapes in those housing areas are taken away for development.[16] Inequality is distributed along racial and class fault lines in a polarized urban landscape. White residents move to new, low-density housing with ample natural surroundings, while non-white residents are disproportionately in high-density public housing. Such planning is characterized by "white socio-spatial epistemology"[17] that accumulates white privilege in the landscape.[18] The patterns of racialized green dispossession and green grabbing perpetuate internal colonial processes of racialized accumulation and dispossession in urban space.[19]

Against this backdrop, scholars in the field of urban landscape history often avoid the aspect of racism. This is particularly true in the construction of whiteness in the context of European national socialism and how it influences theories of urbanism and landscape architecture in the twentieth century. The avoidance obscures exclusionary sentiments in the public imaginary regarding the right to landscape in the European context.[20] The welfarist and modernist ethos of the 1930s through the 1970s—which promoted equality, solidarity, and socio-spatial justice—is offered as a potential discursive solution[21] against the current individualistic ideologies and market-oriented urban development projects that cause environmental injustice by discarding green spaces in so-called "immigrant-dense" neighborhoods.[22]

Despite the silence in architectural history, studies in social sciences have shed light on the "dark side"[23] of Swedish welfare and modernity. In a recent article, sociologists Diana Mulinari and Anders Neergaard argue that the Swedish welfare ethos was based on colonial racial ideology. The welfare state was disassembled in the context of growing anti-immigrant and racist political discourses that incorporated nostalgia for a past welfare society alongside a homogeneous white society project.[24] Similarly, Tobias Hubinette and Carin Lundström propose hegemonic whiteness as a shifting but persistent cultural imagination in the twentieth and twenty-first centuries that produces racial meanings and hierarchies in Swedish society. They identified three main phases of hegemonic whiteness in the societal imagination: "the white purity period (phase 1) between 1905–1968; the white solidarity period (phase 2) between 1968–2001; and the white melancholy period (phase 3) from 2001 and onwards."[25] If racism, white supremacy, and coloniality were so pervasive in society during the construction of welfarist and egalitarian ethos, how did they influence societal meanings of modernist welfare landscapes? How might landscape work as a colonial continuum of the overarching, racially driven exclusionary spatial imaginaries and practices of the past with today's Nordic context? Consequently, could the past ethos of Nordic welfare landscapes still be the inclusive, anti-racist instruments against landscape dispossessions in stigmatized non-white livelihoods?

In response to this tension, this chapter explores "the other side of modernity," the origins and components of how coloniality imposed race on the epistemologies underlying modernist welfare landscapes. Such insights are key to understanding the subjugating, dispossessing, and extractive ideological constructions in spatial thought that normalize non-white exclusions from the landscape today. The first sections of this chapter illuminate the spatial and environmental epistemologies of colonialism and landscape as a colonial instrument. I then examine the Swedish context to reveal specific elements of the colonial continuum in urban landscape production. Lastly, I outline key aspects of landscape coloniality in the Swedish context and call for further studies and practices that are attentive to how racism is perpetuated in planning and design.

THE SPATIAL EPISTEMOLOGIES OF COLONIAL LANDSCAPES

Case studies on European colonialism in architecture and planning usually focus on geographies beyond Europe. This can imply that colonialism only occurred elsewhere; however, as many scholars have argued, European colonialism not only originated in Europe but created a double movement inside and beyond.[26] Within Europe itself, the formation of a white European superior identity, oppressive territorialization, elimination, segregation, exploitation, othering, and racial categorization also occurred.[27] Autonomous political communities, their lands, and labor were violently territorialized and appropriated. Cartography and spatial science facilitated a collective forgetting of these appropriations.[28]

Potent landscape ideologies were mobilized as part of nation-building projects. In the nineteenth and twentieth centuries, early environmental studies dominated by sociobiological and social Darwinist theories were in a symbiotic relationship with Nazism.[29] Colonialism was justified by transposing sociobiological theories on space—for instance, Darwin's ideas about plants were put into the social context. Ideologies like *Heimat* (homeland) claimed that the survival of "a people" depended on their ability to expand territorially, adapt to new lands, and exploit. That colonial expansion was evidence of racial superiority. German architecture, green space, and urbanism were tools to enlarge the German people's *Lebensraum* ("living space").[30] This architectural approach that linked the land to "a people" advanced the colonial project, naturalized the idea of group superiority, and obscured indigenous claims to land.[31] During the early twentieth century, "a wide range of projects" occurred, "from landscape conservation and the design of garden suburbs to pioneering soil research and colonial planning" that applied "the methods of domestic and overseas colonization" for race enhancement (to boost the health and well-being of people deemed as the white, superior race). According to architectural historian and theoretician Kenny Cupers, such "Wilhelmine epistemolog[ies] of environment, race and colonization" endured political paradigm shifts over the Wilhelmine, Weimar, and Nazi periods. These first theories of environmental determinism (e.g., *Bodenständigkeit* and *Heimat*) silently shaped the development of modernist architecture and urbanism, even in the face of the Bauhaus' attempt to eliminate reactionary discourses and expand into socialist, communist, or capitalist modernist projects in the following decades.[32]

Sociobiological and racial theories of urbanism, community planning, housing, and landscape architecture spread across the Western European cities in the first half of the twentieth century in an ideological atmosphere of national socialism and in the postwar period, spreading further to other geographies.[33] The Western European cities were conceived as an "organic whole," or a "living organism," consisting of low-density satellite communities (seedling or garden city projects), which later became high-rise modernist estates,[34] imagined for white working and middle class and interwoven with networks of circulation and communal green open spaces surrounded by a green belt.[35]

COLONIALITY IN CONTEMPORARY URBAN SPACE AND LANDSCAPES

Western European city populations have shifted from "white purity" to diversity, notably after WWII, with an inflow of labor migrants and refugees from the near and far reaches of colonial empires. However, white supremacy remains intact. Global racial labor and land regimes are perpetuated in the city through structural discriminatory systems. Non-white populations have dominated the working class as they have been geographically sedimented in postwar modernist housing complexes previously envisaged for the white working class on the outskirts of cities. The stigmatization of non-white people and places, as well as exploitative tactics directed at them, has been reproduced in Western cities.[36] American Sociologist Charles Pinderhughes conceptualizes these socio-spatial processes as "internal colonialism," which he defines as:

> A geographically based pattern of subordination of a differentiated population, located within the dominant power or country [as] the outcome of systematic group inequality expressed in the policies and practices of a variety of societal institutions, including systems of education, public safety (police, courts, and prisons), health, employment, cultural production, and finance.[37]

Historians David Lloyd and Patrick Wolfe suggest that internal colonization occurred in two phases. The "first enclosure" of internal colonialism (state appropriation) established welfare state functions. The "second enclosure" in the second half of the twentieth century involved ongoing privatization by the modern liberal state to manage surplus populations generated by immigration.[38] The dismantling of welfare states, privatizing and enclosing of commons, symbolic and material dispossession, securitization, and political authoritarianization occurred incrementally in the industrialized Global North. The borders of the colonial modern state encompass tactics of elimination and seclusion that affect "Indigenous peoples of the fourth world" and a new surplus proletariat generated through large-scale immigration. In order to continue the colonial regime, neoliberal states must constitute new frontiers and create "other races" that belong to a "political outsider." These "new frontiers" facilitate new forms of dispossession and accumulation, stretching the settler colonial past into the present and the global into the local. Consequently, cities in colonial states are characterized by the spatial segregation of poorer immigrant populations from affluent groups.[39] Settler colonialism, racial capitalism, and shifting property regimes result in a "continually unfolding" structure of colonial domination through various temporalities.[40]

Postwar modernist housing estates inhabited by non-white groups are often disinvested and stigmatized as ghettos. Alternatively, such neighborhoods are revalorized through urban renewal projects, resulting in gentrification, dispossession, and displacement. While such projects claim to "mix" urban space and end "ghettos," they often end up further racializing people and urban space.[41] Landscape—as an added value and an instrument to impose the projection of an ideal population—contributes to gentrification by creating rent gaps.[42] This dynamic accumulation depends on "green grabbing"[43] and "green dispossession."[44]

> Accumulation by green dispossession still turns on some of the defining features of settler colonialism, e.g., private property as a civilizing mechanism on the frontier, the appropriation of collective land and resources, and the expendability of particular people and places. The production of this new urban frontier also depends, like any frontier, on erasure: the material and discursive work of presenting "empty" landscapes as in need of improvement by non-local actors.[45]

Settler colonial socio-spatial imaginations insidiously infiltrate contemporary green planning initiatives in urban development.[46] Non-white neighborhoods accumulate environmental degradation and symbolic and material dispossession; the value extracted from such places is transferred to white neighborhoods to protect and enhance the environment, property values, and living standards.[47] These spatial planning and design practices operate firmly within a neo-colonial, "white socio-spatial epistemology."[48] They distribute differentiated values, stigmatization, and erasure justified by analyses, representation, spatial projections, and interventions.[49]

COLONIALISM AND RACE IN THE SWEDISH CONTEXT

Despite its self-image of equality and tolerance, Sweden's national identity construction is heavily influenced by external and internal colonial processes and epistemologies that perpetuate discrimination based on background, physical appearance, language, and traditions[50]. For instance, the German romantic nationalist philosophy of *Heimat* was influential in defining the essence of the "Swedish people" through a common appearance, history, and culture believed to be intrinsically linked to the land as part of the geological past. "The Swede" dialectically evolved to stand for a white Westerner who is inherently good, kind, knowledgeable, and hard-working—a healthy person with good genes, a highly developed culture, and a belief in modern Western civilization. The formative years of national identity construction (between 1870 and 1940) contrasted Swedes with degenerated "others," such as Sami, Africans, Indians, "Orientals," and "savages."[51] Colonization on other continents shaped pseudo-scientific classifications and eventually "nations" in late nineteenth-century natural science.

Scientific racism, "race biology," and eugenics were recognized and institutionalized as legitimate scientific fields in Sweden. The first State Institute of Race Biology was founded in Uppsala in 1922 to promote ideas of race enhancement, including Nazi biology.[52] The Institute's data collection, analyses, and classifications propped up the "European race" and "Swedish race" as inherently superior and, therefore, worthy of survival. It also advised against race mixing since Europeans supposedly did not exhibit the defects of other races.[53] Native Sami people were portrayed as primitive and ignorant of the characteristics, material value, and possibilities of their land (unlike the Swedes with their purportedly higher intellect).[54] Furthermore, Swedes' emigration to the United States was supported based on the belief that it would help establish white supremacy.[55] Such ideas lay uncomfortably close to the Nazism that justified German colonization—a Social Darwinist logic that "better" races should be allowed to expand.[56]

In the 1930s, the Institution backed eugenics and genetic research. Eugenics was a central political issue in Swedish identity construction. Eugenics discourse constructs "the other" as an inferior or an "'extremely serious danger' for the nation and for 'the deterioration and destruction of the entire Swedish race.'"[57] Individuals were psychologically categorized based on their social usefulness in the imagined modern industrial society. "Negative eugenics" called for sterilizing psychologically inferior people who threatened Swedish folk and material culture, while "positive eugenics" supported reproduction among a population with "desirable genetic traits" for modern industrial society.[58] The parliament supported and passed a sterilization law in 1934.[59] The Swedish Society of Medicine, compromising its scientific ideals based on ideological pressures, also accepted eugenic social explanations for sterilization until 1947, when it shifted back to medical explanations.[60]

Between 1935 and 1975, a total of 62,888 individuals were sterilized in Sweden. About half of these sterilizations sought to improve the racial and genetic quality of the Swedish population by eliminating ethnic minorities.[61] As part of positive eugenics, the working-class population in poor living conditions was framed as harmful. State reforms (e.g., improving housing and living conditions and providing various child benefits and education) became part of a strategy to improve the quality of the population.[62] The Social Democrat's "modernist manifesto,"[63] *Crisis in the Population Question*,[64] written by the politicians Alva and Gunnar Myrdal,[65] advocated for eugenics and population management to modernize society and form a welfare state.[66] According to historians Broberg and Tyden,[67] the Myrdals rejected eugenics' racist discourses but accepted sterilization as a sociopolitical and race-hygienic/hereditary-hygienic method; they argued that modern society and technical development required certain intelligence and character. The political elite advanced racism with a more elaborate language and imagination to strip themselves of the "racist label" and more effectively enact the modernization of society.

COLONIALITY IN SWEDISH LANDSCAPES

German ideologies, philosophy, and environmental and spatial planning heavily influenced the racialization and modernization of epistemologies of landscape and space. During the eighteenth and nineteenth centuries, the Romantic nationalist movement and organicist ideas of "national spirit" and human dwelling formed in the philosophy of nature, arts, geology, and landscape painting.[68] An idealized imaginary of a national landscape and sense of belonging mystically located in the ancient layers of geology gave rise to an evolution of dialectics between people and their land(s).[69] The combination of romanticism and nationalism was paradoxical since the first carried a "nostalgia for permanence," while the second was "dynamic progressivism."[70] This paradox was reflected in state borders—the nationalist ideal eventually inhabited the abstraction of (internal) space. The social representation of space was constituted dialectically: inhabitants were never homogeneous, nor did they fit into the ideal imagined community. Abstraction meant representing Swedish geography through a common "folk identity" (i.e., selected folk tales, legends, and communitarian histories), representing space as "empty," and "eliminating" those who did not fit. Meanwhile, the natural environment and space (including their production through modernist spatial planning) became matters of scientific knowledge in the service of industrialization and modernization.[71]

According to geographer Kenneth Olwig,[72] the Swedish epistemology of spatial science and planning was heavily influenced by Nazi spatial scientists such as Walter Christaller[73] and Edgar Kant. They merged organicist environmental deterministic ideologies, a biologistic view of society, and Social Darwinist ideologies of *Lebensraum* to engineer Nazi spatial science and planning as "secret agents."[74] Chorographic mapping connected the Swedish race to the territory of the nation-state by imposing archaic, nostalgic, and romantic nationalistic imaginaries of landscape on absolute space. In doing so, they achieved racial territorialization and regionalization. This consolidation of national space within nation-state borders was based on the idea that societies with higher levels of economic and cultural evolution can make the best use of land.[75] The diabolical irrationality of reactionary nationalistic identarian landscape ideology and the ideology of *Lebensraum* merged with scientific and technical advancements in the service of modernism. The reactionary premodern and modern were never a binary but a "diabolic mixture" of "the modern" (cartographic abstract space) and "the reactionary" (scenic landscape).[76] These evil "subtle dialectics of reactionary modernism"[77] in socio-spatial epistemologies and imaginaries were unconsciously reproduced by figures in spatial science and planning long after

Nazism.[78] "Nature–culture and nationalism–regionalism"[79] were two deeply related discourses forming national unity and Swedish industrial capitalism within the abstracted national space.

The Swedish Tourist Association (Svenska Touristföreningen), established in 1885, encouraged people to explore the country's different regions and natural wonders. It mediated the experience of nature through patriotic sentiments—a shared love of nature, ideas about essential Swedishness and Swedish heritage, and the assimilation of minority groups such as the Sami and Finns.[80] The association's yearbooks reveal how Swedes became progressively represented as modern and highly cultured people, while the Sami were deemed as other, simple, exotic, and primitive. A common Swedish cultural heritage was constructed from natural areas through the 1970s (e.g., using "petroglyphs, menhirs, medieval churches, castles and courthouses"[81]). Later, these symbols became part of Sweden's international branding as an environmentally aware country.[82] Organic, unified classless national space was created around "Nature, love of nature, and the scientific knowledge of nature."[83] In 1909, "following the US. . . and inspired by German conservatism,"[84] Sweden created national parks. These parks, which were mainly planned on state-owned Sami lands, were chosen to represent an "archetypal Swedish landscape in [its] natural and primarily natural state" and the many "geomorphological classifications of Scandinavia."[85] Maps, texts, and images represented these areas as empty, geometrical spaces[86] that stripped "nature, landscape and the whole Scandinavia" of their historical and social complexity.[87] Meanwhile, the Sami were relegated to the passive background of untouched nature.

Political elites' ideology of "positive eugenics" in relation to housing and the environment came into realization in synergy with working-class movements. In the urban areas, *Folkets Parks* (People's Parks) emerged from the *Folkets Hem* (People's House) movement. Working-class agency and the democratic-socialist movement shaped both "the land and the folk, while also seizing and constructing the 'means of recreation.'"[88] *Folkets Parks* were created for physical well-being and to enable social conditions through an "ethos of solidarity and equality."[89] After the 1930s, the state strictly formalized the creation of parks and green spaces through a planning and design apparatus.[90] The city-scale functionalist park systems became an important part of the "progressive social program."[91] Open-air leisure in rural natural areas was encouraged and fashionable for urban populations, especially industrial workers.[92] *Allemansrätt* (the right of everyone's to roam in nature) was legitimized in 1940,[93] and exercise for bodily health among the (white)[94] population advanced the goal of building modern Sweden.

International exhibitions promoted racist, white supremacist architectural representations while idealizing Eurocentric notions of developmentalism and technological advancement, as well as establishing a stage for white Western nations to become visible to each other in order to compete and cooperate. At the Stockholm and Chicago architecture exhibitions in the 1930s, both the

Swedish and American sides equated the new functionalist style and modernist ideology through white bodies. In the American case, this meant cutting Native Americans and African Americans, while the Swedish cut the Sami and Roma.[95] In the 1930s and 1940s, many spatial-determinist housing and community development models were shared by the National Socialist "Third Reich" and the Social Democratic welfare state in Sweden (despite ideological differences between the two).[96] The Stockholm exhibition (1930) proved Swedish modernist architecture's success in "fusing the new functionalist aesthetics with nation building." Examples like the *Svea rike* (Swedish Kingdom) depicted "a radiantly hygienic, just, and economically efficient nation."[97] In the early 1940s, some Swedish planners readily connected with Nazi Germany to transfer community planning models to create racially homogenous, small, and integrated communities called *Volksgemeinschaft*.[98]

The "everyday" had become part of utopian social engineering. It was reflected in the creation of new human life and objects, spaces, and environments attached to the idealized ultra-modern human.[99] The landscapes of modernist architecture drew from modernist utopias, labor movements, and national conservative myths that allowed urbanized populations to reconnect with the rural.[100] Architectural practitioners actively contributed to the discursive and material production of modernity. For instance, the co-operative movement's architect office cited the Internationalist motto, "Workers in the countryside and in the cities, one day the earth shall be ours (or: let us henceforth claim the earth)," which combined the modernist, new style with an "atmosphere of a Swedish idyllic scenery"[101] (Figures 3.2 and 3.3).

NEW FRONTIERS OF LANDSCAPE DISPOSSESSION AND WHITE SOCIO-SPATIAL EPISTEMOLOGY IN CONTEMPORARY SWEDISH PLANNING

In the 1960s and 1970s, Sweden's Social Democrat government pursued the ambitious Million Homes Programme (1965–1974), a functionalist, modernist housing production project to address a severe housing shortage. It created decent, accessible subsidized housing on cheap land in the less connected peripheries of major cities.[102] However, the media claimed that the MP high-rise rental apartment segments with newly constructed or half-finished monotonous landscapes led to "anonymity, social deviation, and mental disorders."[103] The stigmatization of these built environments corresponded with the stigmatization of their mainly non-white inhabitants during a period of increasing international immigration in the 1970s.[104]

Today, many so-called "immigrant-dense" MP areas are designated as "vulnerable areas" by the Swedish Police.[105] They are also frontiers for property development and disproportionately selected for densification projects under the pretext of "social sustainability," which supposedly aims to break segregation through "social mixing," "crime prevention," and "job creation." Yet, a systematic review of these Swedish densification planning practices[106] revealed no evidence that the

Figure 3.2 KF Architects entry One day the Earth shall be ours (En dag skall Jorden bliva vår), exterior, painting by Arvid Fougstedt, gouache, 1932. Photo: ArkDes, Stockholm (ARKM.1999-37-01). © Arvid Fougstedt/BUS.

densification in these areas would solve current problems. This illustrates how the socio-spatial epistemology of planning and design is associated with "whiteness" that analyzes, abstracts, surveils, categorizes, and hovers over non-white neighborhoods to essentialize societal problems and implement large-scale extraction projects; meanwhile, white neighborhoods are idealized and conserved.[107]

The high-rise modernist MP estates are now experiencing "green dispossession." As the MP areas mature, leafy landscapes are destined to be dispossessed for densification projects and defined as "empty," even dangerous, prone to criminal activities, or without heritage, including social and ecological values

Figure 3.3 KF Architects entry One day the Earth shall be ours (En dag skall Jorden bliva vår), interior, painting by Arvid Fougstedt, gouache, 1932. Photo: ArkDes, Stockholm (ARKM.1999-37-02). © Arvid Fougstedt/BUS.

(Figures 3.4 and 3.5). They are said to obstruct urban social life and are trans-formed into "abstract space" in planning documents.[108] Furthermore, landscape analysis legitimizes these acts by invoking a "green spinner" that balances the distribution of urban public green space relative to the amount of surface covered by building blocks. This two-dimensional approach ignores the amount of construction, the height and number of people inside the blocks, and the dis-tribution of different material assets and power relations. The historical eco-nomic, social, political, and spatial relations and systems of oppression and segregation remain obscured.

Green space conveys positive connotations of equality, well-being, climate crisis mitigation, public meeting facilitation, and access to nature. Contemporary spatial and landscape planning is dominated by sustainable development dis-courses that declare public landscapes the right of "all"; such areas are said to equalize elements and remedy social polarization by creating "meeting places." However, the material distribution of the landscape in "sustainable" urban devel-opment is far from equalizing. In the face of a severe housing shortage, market-led housing and urban development projects have transformed urban landscapes along racial lines. The supposedly neutral, objective, and scientific landscape-analysis tools mask how racialization is inscribed into space. Land-scape architectural projects stabilize dominant social meanings ascribed to

Figure 3.4 Views from the central rental housing areas in the Gottsunda Million Programme Neighborhood, Uppsala, 2022. Photo courtesy: The author.

different areas. For instance, low-density single-family housing projects located near nature areas are considered important cultural landscapes that recreate middle- and upper-middle-class white-only enclaves. The low density is "justi-fied," as it does not put heavy pressure on culturally and ecologically valuable landscapes. "High-tech green solutions" are promised to decrease the negative impact of sprawl.[109] Sustainable urban development narratives shield against cri-tique and displace political contestation, claiming there is no alternative to suita-bly address climate change, economic development, environmental conservation, and abstract notions of democracy, human rights, and responsibility.[110]

CONCLUSION

Sweden's dark past has given rise to the realities of spatial racialization and socio-spatial inequality. The racialized landscape created through *Heimat* exists in the abstract space of modernist planning and architecture that produces func-tional communities, housing environments, parks, and gardens to ensure the reproduction of (white) industrial society. Coloniality and modernity are "two sides of a single coin."[111] The egalitarian welfare state and its modernist plan-ning and architecture emerged in synergy with a race enhancement project to

Figure 3.5 Views from the central rental housing areas in the Gottsunda Million Programme Neighborhood, Uppsala, 2022. Photo courtesy: The author.

enable the (white) population to take its position atop the global colonial power matrix.

Today, racialized minorities' senses of belonging and socio-spatial agency in the modernist estates that have emerged from the cracks of the white spatial imaginary have been repeatedly recolonized by urban development. Contemporary practices insidiously uphold the colonial gaze to territorialize, dispossess, eliminate, and racialize diversified populations and their socio-spatial practices and histories. Such practices disassemble landscapes in postwar modernist high-rise housing areas and deny racialized minorities' rights to the landscape. In a context of increasing hostility, non-white Swedes, "immigrants," and refugees are deemed undeserving "political outsiders."[112]

Today, the welfare landscapes and modernist housing estates are assets uplifting the living conditions of the inhabitants, as they were aimed in the past. They are full of histories and everyday practices. The problem is not the modernist architecture and landscape as material artifacts but the constant actions of macro- and micro-erasures, disinvestment, and material and symbolic dispossessions and displacements in the landscapes. In symbolic displacement and dispossession, demonizing simplistic, socio-spatially deterministic narration of modernist architecture in mass housing areas in architectural and urban planning history and theory plays an important role, as seen in the instance of the critique of the Pruitt-Igoe housing scheme in the United States, which ignored the value of the housing scheme in providing decent living conditions and creating a community for historically marginalized black communities, and why it was deteriorated.[113]

Most often today demands for spatial justice are articulated through the narrative of histories of modernist landscapes, which are often geographically disjointed and lack colonial and racial context.[114] These arguments rightfully emerge in defense of today's welfare society, which is under attack from a conservative right that seeks to dismantle its collective public assets.[115] However, while defending welfare society today, remembering that bottom-up working-class movements are essential for the democratization of landscapes, the dark side of social democratic welfare in Sweden should not be overlooked. Such "colorblind" defenses unintentionally generate the impression that modernist welfare architecture and landscapes were never under the influence of wider exclusionary of white colonial spatial imaginaries (or that they were even anti-racist by proxy). Current urban spatial production and racialization do not seem to have a history. Yet, this perspective ignores welfare's dialectical nature. The Swedish concepts of *welfare*, *all*, *solidarity,* and *equality* are historically laden with white supremacist connotations and reflect on the contemporary notion of "white melancholia," a longing for the myths of an ethnically homogenous white society.[116] Elements like *folk hemmet* (people's house) and *Allemansrätt* (the right of everyone's to roam in nature) were imagined for a homogenous white society at the expense of "other" people and geographies. Diabolical relational functions continue insidiously in the spatial epistemology perpetuated by today's market-oriented planning and design practices.

By highlighting the coloniality and racialization of space and landscape in the Nordic context, my intention in this chapter is not to denigrate the qualities of modernist architecture and welfare landscapes as they stand today for the lives of the non-white working class or to undermine the value of grassroots labor, or anti-imperialist, anti-racist movements, but to historicize essentialized racialized imaginations in order to know what to dismantle for genuinely sustainable, democratic, anti-racist, and egalitarian planetary welfare futures.

ACKNOWLEDGMENTS

I thank the editors of the volume and Sued Ferreira Da Silva for their comments on earlier versions of the chapter.

NOTES

1   Walter D. Mignolo, *Local Histories/Global Designs: Coloniality, Subaltern Knowledges, and Border Thinking* (Princeton, NJ: Princeton University Press, 2012), 22.
2   Catharina Nolin, "Public Parks in Gothenburg and Jönköping: Secluded Idylls for Swedish Townsfolk," *Garden History* 32, no. 2 (2004): 197–212; https://doi.org/10.2307/4150381 (accessed May 30, 2024).

3 Johan Pries, Erik Jönsson, and Don Mitchell, "Parks and Houses for the People," *Places Journal* (May 2020); https://doi.org/10.22269/200512 (accessed January 8, 2022).

4 For an example of such national branding see Thorbjörn Andersson, "Landscape Architecture Used as a Society Service: The Swedish Example," *Green Visions: Green Space Planning and Design in Nordic Cities*, eds. Kjell Nilsson, Ryan Weber, and Lisa Rohrer (Stockholm: Arvinius + Orfeus Publishing, 2021).

5 For more on the Million Housing Program in Sweden see Thomas Hall and Sonja Vidén, "The Million Homes Programme: A Review of the Great Swedish Planning Project," *Planning Perspectives* 20, no. 3 (2005): 301–328.

6 Johan Pries and Mattias Qviström "The Patchwork Planning of a Welfare Landscape: Reappraising the Role of Leisure Planning in the Swedish Welfare State," *Planning Perspectives* 36, no. 5 (January 22, 2021), 923–948.

7 See more on the project via https://cityoftelosa.com (accessed January 27, 2022).

8 Anna Hult, "The Circulation of Swedish Urban Sustainability Practices: To China and Back," *Environment and Planning A: Economy and Space* 47, no. 3 (March 1, 2015): 537–553.

9 Allan Pred, "Somebody Else, Somewhere Else: Racisms, Racialized Spaces and the Popular Geographical Imagination in Sweden," *Antipode* 29, no. 4 (1997): 383–416; https://doi.org/10.1111/1467–8330.00053 (accessed May 30, 2024).

10 Allan Pred, "Somebody Else, Somewhere Else: Racisms, Racialized Spaces and the Popular Geographical Imagination in Sweden," *Antipode* 29, no. 4 (1997): 383–416; https://doi.org/10.1111/1467–8330.00053 (accessed May 30, 2024).

11 Michael McEachrane, "Universal Human Rights and the Coloniality of Race in Sweden," *Human Rights Review* 19, no. 4 (2018): 471–493; https://doi.org/10.1007/s12142-018-0510-x (accessed May 30, 2024).

12 Karl Gauffin, "Precariousness on the Swedish Labour Market: A Theoretical and Empirical Account," *The Economic and Labour Relations Review* 31, no. 2 (2020): 279–298.

13 Michael McEachrane, "Universal Human Rights and the Coloniality of Race in Sweden," *Human Rights Review* 19, no. 4 (2018): 471–493; https://doi.org/10.1007/s12142-018-0510-x (accessed May 30, 2024).

14 Catharina Thörn and Håkan Thörn, "Swedish Cities Now Belong to the Most Segregated in Europe," *Sociologisk Forskning* 4 (2017): 293–296.

15 Lina Olsson, "The Neo-liberalization of Municipal Land Policy in Sweden," *International Journal of Urban and Regional Research* 42, no. 4 (2018): 633–650.

16 Alva Zalar and Johan Pries, "Unmapping Green Space: Discursive Dispossession of the Right to Green Space by a Compact City Planning Epistemology," *City* 26, no. 1 (January 2, 2022): 51–73; https://doi.org/10.1080/13604813.2021.2018860 (accessed May 30, 2024)

17 Owen J. Dwyer and John Paul Jones III, "White Socio-Spatial Epistemology," *Social & Cultural Geography* 1, no. 2 (January 1, 2000): 209–222; https://doi.org/10.1080/14649360020010211 (accessed May 30, 2024).

18 Burcu Yiğit-Turan and Mia Ågren, "Segregation and Landscape Injustice in the Shadows of White Planning and Green Exceptionalism in Sweden," *Urban Matters Journal*, Issue: Dislocating Urban Studies (May 11, 2022); https://urbanmattersjournal.com/segregation-and-landscape-injustice-in-the-shadows-of-white-planning-and-green-exceptionalism-in-sweden/ (accessed May 30, 2024).

19 Cf. Garcia-Lamarca et al., "Urban Green Boosterism and City Affordability: For Whom Is the 'Branded' Green City?" *Urban Studies* 58, no. 1 (January 2021): 90–112; https://doi.org/10.1177/0042098019885330 (accessed May 30, 2024); Sara Safransky, "Greening the Urban Frontier: Race, Property, and Resettlement in Detroit," *Geoforum* 56 (September 1, 2014): 237–248; https://doi.org/10.1016/j.geoforum.2014.06.003 (accessed May 30, 2024).

20 Cf. Koenraad Danneels, "Nature's Offensive: The Sociobiological Theory and Practice of Louis Van der Swaelmen," *Journal of Landscape Architecture* 14, no. 3 (2019): 52–61; https://doi.org/10.1080/18626033.2019.1705581 (accessed May 30, 2024).

21 Cf. Johan Pries, Erik Jönsson, and Don Mitchell, "Parks and Houses for the People," *Places Journal* (May 2020); https://doi.org/10.22269/200512 (accessed January 8, 2022).

22 Alva Zalar and Johan Pries, "Unmapping Green Space: Discursive Dispossession of the Right to Green Space by a Compact City Planning Epistemology," *City* 26, no. 1 (January 2, 2022): 51–73; https://doi.org/10.1080/13604813.2021.2018860 (accessed May 30, 2024).

23 Walter D. Mignolo, *Local Histories/Global Designs: Coloniality, Subaltern Knowledges, and Border Thinking* (Princeton, NJ: Princeton University Press, 2012)

24 Diana Mulinari and Anders Neergaard, "The Swedish Racial Welfare Regime in Transition," in *Racism in and for the Welfare State*, ed. Fabio Perocco (Cham: Springer International Publishing, 2022), 91–116.

25 Tobias Hübinette and Catrin Lundström, "Three Phases of Hegemonic Whiteness: Understanding Racial Temporalities in Sweden," *Social Identities* 20, no. 6 (2014): 423–437; https://doi.org/10.1080/13504630.2015.1004827 (accessed May 30, 2024).

26 See Kenneth Robert Olwig, Landscape, Nature and the Body Politic: From Britain's *Renaissance to America's New World* (Madison: University of Wisconsin Press, 2002); Claudio Minca, "Humboldt's Compromise, or the Forgotten Geographies of Landscape," *Progress in Human Geography* 31, no. 2 (April 1, 2007): 179–193; https://doi.org/10.1177/0309132507075368 (accessed May 30, 2024); Kenny Cupers, "Bodenständigkeit: The Environmental Epistemology of Modernism," *The Journal of Architecture* 21, no. 8 (November 16, 2016): 1226–1252; https://doi.org/10.1080/13602365.2016.1254271 (accessed May 30, 2024).

27 Amber Murray, "Colonialism," *The International Encyclopedia of Human Geography: People, the Earth, and Technology,* 2nd edition, eds. D. Richardson, N. Castree, M. F. Goodchild, A. L. Kobayashi, W. Liu, and R. Marston (Amsterdam: Elsevier, 2020), 315–326.

28 Claudio Minca, "Humboldt's Compromise, or the Forgotten Geographies of Landscape," *Progress in Human Geography* 31, no. 2 (April 1, 2007): 179–193; https://doi.org/10.1177/0309132507075368 (accessed May 30, 2024); Kenneth Robert Olwig, Landscape, *Nature and the Body Politic: From Britain's Renaissance to America's New World* (Madison: University of Wisconsin Press, 2002); Denis Cosgrove, "Prospect, Perspective and the Evolution of the Landscape Idea," *Transactions of the Institute of British Geographers* 10, no. 1 (1985): 45–62; https://doi.org/10.2307/622249 (accessed May 30, 2024).

29 Pierre Cornut and Erik Swyngedouw, "Approaching the Society-nature Dialectic: A Plea for a Geographical Study of the Environment," *Belgeo* 1-2-3-4 | 2000, Special Issue 29th International Geographical Congress, July 12, 2015; https://doi.org/10.4000/belgeo.13873 (accessed December 10, 2020).

30 Kenny Cupers, "Bodenständigkeit: The Environmental Epistemology of Modernism," *The Journal of Architecture* 21, no. 8 (November 16, 2016): 1226–1252; https://doi.org/10.1080/13602365.2016.1254271 (accessed May 30, 2024). *Lebensraum*—living space—is defined as a concept that "served as a critical component in the Nazi worldview that drove both its military conquests and racial policy," see *Holocaust Encyclopedia* https://encyclopedia.ushmm.org/content/en/article/lebensraum (accessed July 16, 2023).

31 Kenny Cupers, "Bodenständigkeit: The Environmental Epistemology of Modernism," *The Journal of Architecture* 21, no. 8 (November 16, 2016): 1226–1252; https://doi.org/10.1080/13602365.2016.1254271 (accessed May 30, 2024). *Lebensraum*—living

space—is defined as a concept that "served as a critical component in the Nazi world-view that drove both its military conquests and racial policy," see *Holocaust Encyclopedia* https://encyclopedia.ushmm.org/content/en/article/lebensraum (accessed July 16, 2023), 1245.

32 Kenny Cupers, "Bodenständigkeit: The Environmental Epistemology of Modernism," *The Journal of Architecture* 21, no. 8 (November 16, 2016): 1226–1252; https://doi.org/10.1080/13602365.2016.1254271 (accessed May 30, 2024). *Lebensraum—* living space—is defined as a concept that "served as a critical component in the Nazi worldview that drove both its military conquests and racial policy," see *Holocaust Encyclopedia* https://encyclopedia.ushmm.org/content/en/article/lebensraum (accessed July 16, 2023).

33 David H. Haney, *When Modern was Green: Life and Work of Landscape Architect Leberecht Migge* (Oxon and New York: Routledge, 2010). For more about the adaptation of self-sufficient Siedlung planning by the Zionist movement in Palestine see p. 156; for racial planning and architecture in Latin America see Fabiola Lopez-Duran, *Eugenics in the Garden: Transatlantic Architecture and the Crafting of Modernity* (Austin: University of Texas Press, 2018); https://doi.org/10.7560/314951 (accessed May 30, 2024).

34 Kenny Cupers, "The Social Project: On the Complex," *Places Journal* (April 2014; https://doi.org/10.22269/140402 (accessed December 30, 2023).

35 David H. Haney, *When Modern was Green: Life and Work of Landscape Architect Leberecht Migge* (Oxon and New York: Routledge, 2010).

36 Stefan Kipfer, "Neocolonial Urbanism? La Rénovation Urbaine in Paris," *Antipode* 48, no. 3 (2016): 603–625; https://doi.org/10.1111/anti.12193 (accessed May 30, 2024).

37 Charles Pinderhughes, "Toward a New Theory of Internal Colonialism," *Socialism and Democracy* 25, no. 1 (March 2011): 235–256; https://doi.org/10.1080/08854300.2011.559702.

38 David Lloyd and Patrick Wolfe, "Settler Colonial Logics and the Neoliberal Regime," *Settler Colonial Studies* 6, no. 2 (April 2, 2016): 109–118, https://doi.org/10.1080/2201473X.2015.1035361 (accessed May 30, 2024).

39 David Lloyd and Patrick Wolfe, "Settler Colonial Logics and the Neoliberal Regime," *Settler Colonial Studies* 6, no. 2 (April 2, 2016): 109–118, https://doi.org/10.1080/2201473X.2015.1035361 (accessed May 30, 2024).

40 See Brenna Bhandar, *Colonial Lives of Property: Law, Land and Racial Regimes of Ownership* (Durham and London: Duke University Press, 2019), 23.

41 Stefan Kipfer, "Neocolonial Urbanism? La Rénovation Urbaine in Paris," *Antipode* 48, no. 3 (2016): 603–625; https://doi.org/10.1111/anti.12193 (accessed May 30, 2024); Stefan Kipfer and Jason Petrunia, "'Recolonization' and Public Housing: A Toronto Case Study," *Studies in Political Economy* 83, no. 1 (March 2009): 111–139; https://doi.org/10.1080/19187033.2009.11675058 (accessed May 30, 2024).

42 Melissa Garcia-Lamarca et al., "Urban Green Boosterism and City Affordability: For Whom Is the 'Branded' Green City?," *Urban Studies* 58, no. 1 (January 1, 2021): 90–112; https://doi.org/10.1177/0042098019885330 (accessed May 30, 2024).

43 Isabelle Anguelovski et al., "Expanding the Boundaries of Justice in Urban Greening Scholarship: Toward an Emancipatory, Antisubordination, Intersectional, and Relational Approach," *Annals of the American Association of Geographers* 110, no. 6 (November 1, 2020): 1743–1769; https://doi.org/10.1080/24694452.2020.1740579 (accessed May 30, 2024).

44 Sara Safransky, "Greening the Urban Frontier: Race, Property, and Resettlement in Detroit," *Geoforum* 56 (September 1, 2014): 237–248; https://doi.org/10.1016/j.geoforum.2014.06.003 (accessed May 30, 2024).

45 Sara Safransky, "Greening the Urban Frontier: Race, Property, and Resettlement in Detroit," *Geoforum* 56 (September 1, 2014): 237–248; https://doi.org/10.1016/j.geoforum.2014.06.003 (accessed May 30, 2024), 237.

46 Sara Safransky, "Land Justice as a Historical Diagnostic: Thinking with Detroit," *Annals of the American Association of Geographers* 108, no. 2 (March 4, 2018): 499–512; https://doi.org/10.1080/24694452.2017.1385380 (accessed May 30, 2024).

47 Danielle M. Purifoy and Louise Seamster, "Creative Extraction: Black Towns in White Space," *Environment and Planning D: Society and Space* 39, no. 1 (February 1, 2021): 47–66; https://doi.org/10.1177/0263775820968563 (accessed May 30, 2024); George Lipsitz, *How Racism Takes Place* (Philadelphia, PA: Temple University Press, 2011).

48 Owen J. Dwyer and John Paul Jones III, "White Socio-Spatial Epistemology," *Social & Cultural Geography* 1, no. 2 (January 1, 2000): 209–222; https://doi.org/10.1080/14649360020010211 (accessed May 30, 2024).

49 Edward G. Goetz, Rashad A. Williams, and Anthony Damiano, "Whiteness and Urban Planning," *Journal of the American Planning Association* 86, no. 2 (April 2, 2020): 142–156; https://doi.org/10.1080/01944363.2019.1693907; George Lipsitz, *How Racism Takes Place* (Philadelphia, PA: Temple University Press, 2011).

50 Karin Idevall Hagren, "Othering in Discursive Constructions of Swedish National Identity, 1870–1940," *Critical Discourse Studies* 19, no. 4 (2021): 384–400; Allan Pred, *Even in Sweden: Racisms, Racialized Spaces, and the Popular Geographical Imagination* (Berkeley: University of California Press, 2000).

51 Karin Idevall Hagren, "Othering in Discursive Constructions of Swedish National Identity, 1870–1940," *Critical Discourse Studies* 19, no. 4 (2021): 384–400.

52 Niels Lynöe, "Race Enhancement Through Sterilization: Swedish Experiences," *International Journal of Mental Health* 36, no. 1 (2007): 17–25.

53 Karin Idevall Hagren, "Othering in Discursive Constructions of Swedish National Identity, 1870–1940," *Critical Discourse Studies* 19, no. 4 (2021): 384–400.

54 Karin Idevall Hagren, "Othering in Discursive Constructions of Swedish National Identity, 1870–1940," *Critical Discourse Studies Critical Discourse Studies* 19, no. 4 (2021): 384–400.

55 Sayaka Osanami Törngren, "Does Race Matter in Sweden? Challenging Colorblindness in Sweden," *Sophia Journal of European Studies*, no 7 (2015): 125–137, ISSN 1883–5635.

56 Kenny Cupers, "Bodenständigkeit: The Environmental Epistemology of Modernism," *The Journal of Architecture* 21, no. 8 (November 16, 2016): 1226–1252; https://doi.org/10.1080/13602365.2016.1254271 (accessed May 30, 2024).

57 Karin Idevall Hagren, "Othering in Discursive Constructions of Swedish National Identity, 1870–1940," *Critical Discourse Studies* 19, no. 4 (2021): 384–400, 13.

58 Karin Idevall Hagren, "Othering in Discursive Constructions of Swedish National Identity, 1870–1940," *Critical Discourse Studies* 19, no. 4 (2021): 384–400, 12–13.

59 Alberto Spektorowski and Elisabet Mizrachi, "Eugenics and the Welfare State in Sweden: The Politics of Social Margins and the Idea of a Productive Society," *Journal of Contemporary History* 39, no. 3 (2004): 333–352.

60 Niels Lynöe, "Race Enhancement Through Sterilization: Swedish Experiences," *International Journal of Mental Health* 36, no. 1 (2007): 17–25.

61 See Lynöe, 2007, for exact numbers, if the sterilization happened with personal initiative or without a consent, and the reasoning behind sterilization.

62 Eva Rudberg, "Building a Utopia of the Everyday," *Swedish Modernism: Architecture, Consumption, and the Welfare State*, eds. Helena Mattson and Sven-Olov Wallenstein (London: Black Dog, 2010), 152–159.

63 Hedvig Ekerwald, "The Modernist Manifesto of Alva and Gunnar Myrdal: Moderniza-tion of Sweden in the Thirties and the Question of Sterilization," *International Journal of Politics, Culture, and Society* 14, no. 3 (2001): 539–561.

64 Alva Myrdal and Gunnar Myrdal, *Kris i befolkningsfragan (Crisis in the Population Question)* (Stockholm: Albert Bonniers förlag 1934, third, revised and extended edition, 1935).

65 Their modernist villa and its garden are featured in an article, "To Erase the Garden: Modernity in the Swedish Garden and Landscape" Thorbjörn Andersson (2002), who celebrates Sweden as an international role model with reference to the "humanist" aspects of the welfare state. See Thorbjörn Andersson, "To Erase the Garden: Modernity in the Swedish Garden and Landscape," *The Architecture of Landscape 1940–1960*, ed. Marc Treib (Philadelphia: University of Pennsylvania Press, 2002), 2–27.

66 Karin Idevall Hagren, "Othering in Discursive Constructions of Swedish National Iden-tity, 1870–1940," *Critical Discourse Studies* 19, no. 4 (2021): 384–400; Allan Pred, *Even in Sweden: Racisms, Racialized Spaces, and the Popular Geographical Imagina-tion* (Berkeley: University of California Press, 2000).

67 "Sterilization debate," September 13, 1981, Hedvig Ekerwald, "The Modernist Mani-festo of Alva and Gunnar Myrdal: Modernization of Sweden in the Thirties and the Question of Sterilization," *International Journal of Politics, Culture, and Society* 14, no. 3 (2001): 539–561, 544.

68 See Michelle Facos, *Nationalism and the Nordic Imagination: Swedish Art of the 1890s* (Berkeley: University of California Press, 1998).

69 Jakob Stougaard-Nielsen, "Nordic Nature: From Romantic Nationalism to the Anthro-pocene," *Introduction to Nordic Cultures*, ed. Jakob Stougaard-Nielsen and Annika Lindskog (London: UCL Press, 2020), 165–180; https://doi.org/10.2307/j.ctv13x-prms.17.

70 Joep Leerssen, "Notes toward a Definition of Romantic Nationalism," *Romantik: Journal for the Study of Romanticisms* 2, no. 1 (February 3, 2013): 9–35; https://doi.org/10.7146/rom.v2i1.20191 (accessed May 30, 2024).

71 Kenneth R. Olwig, "Geese, Elves, and the Duplicitous, 'Diabolical' Landscaped Space of Reactionary Modernism: The Case of Holgersson, Hägerstrand, and Lorenz," *GeoHumanities* 3, no. 1 (2017): 41–64; https://doi.org/10.1080/2373566X.2016.1245108 (accessed May 30, 2024).

72 Kenneth R. Olwig, "Transcendent Space, Reactionary-Modernism and the 'Diabolic' Sublime· Walter Christaller, Edgar Kant, and Geography's Origins as a Modern Spatial Science," *GeoHumanities* 4, no. 1 (January 2, 2018): 1–25; https://doi.org/10.1080/2373566X.2017.1291310 (accessed May 30, 2024).

73 Walter Christaller's spatial models facilitated the development of "racially" purified organized areas in the newly conquered eastern territories of Nazi Germany. See Kenneth R. Olwig, "Transcendent Space, Reactionary-Modernism and the 'Diabolic' Sublime: Walter Christaller, Edgar Kant, and Geography's Origins as a Modern Spatial Science," *GeoHumanities* 4, no. 1 (January 2, 2018): 1–25; https://doi.org/10.1080/2373566X.2017.1291310 (accessed May 30, 2024).

74 Kenneth R. Olwig, "Transcendent Space, Reactionary-Modernism and the 'Diabolic' Sublime: Walter Christaller, Edgar Kant, and Geography's Origins as a Modern Spatial Science," *GeoHumanities* 4, no. 1 (January 2, 2018): 1–25; https://doi.org/10.1080/2373566X.2017.1291310 (accessed May 30, 2024), 183.

75 Kenneth R. Olwig, "Transcendent Space, Reactionary-Modernism and the 'Diabolic' Sublime: Walter Christaller, Edgar Kant, and Geography's Origins as a Modern Spatial Science," *GeoHumanities* 4, no. 1 (January 2, 2018): 1–25; https://doi.org/10.1080/2373566X.2017.1291310 (accessed May 30, 2024), 194.

76 Kenneth R. Olwig, "Transcendent Space, Reactionary-Modernism and the 'Diabolic' Sublime: Walter Christaller, Edgar Kant, and Geography's Origins as a Modern Spatial Science," *GeoHumanities* 4, no. 1 (January 2, 2018): 1–25; https://doi.org/10.1080/2373566X.2017.1291310 (accessed May 30, 2024), 173.

77 Kenneth R. Olwig, "Transcendent Space, Reactionary-Modernism and the 'Diabolic' Sublime: Walter Christaller, Edgar Kant, and Geography's Origins as a Modern Spatial Science," *GeoHumanities* 4, no. 1 (January 2, 2018): 1–25; https://doi.org/10.1080/2373566X.2017.1291310 (accessed May 30, 2024), 211.

78 Kenneth R. Olwig, "Transcendent Space, Reactionary-Modernism and the 'Diabolic' Sublime: Walter Christaller, Edgar Kant, and Geography's Origins as a Modern Spatial Science," *GeoHumanities* 4, no. 1 (January 2, 2018): 1–25; https://doi.org/10.1080/2373566X.2017.1291310 (accessed May 30, 2024).

79 Tom Mels, "Nature, Home, and Scenery: The Official Spatialities of Swedish National Parks," *Environment and Planning D: Society and Space* 20, no. 2 (April 1, 2002): 135–154; https://doi.org/10.1068/d14s (accessed May 30, 2024).

80 Karin Idevall Hagren, "Nature, Modernity, and Diversity: Swedish National Identity in a Touring Association's Yearbooks 1886–2013," *National Identities*, 23 no. 5 (2020), 1–18; https://doi.org/10.1080/14608944.2020.1803819 (accessed May 30, 2024).

81 Karin Idevall Hagren, "Nature, Modernity, and Diversity: Swedish National Identity in a Touring Association's Yearbooks 1886–2013," *National Identities*, 23 no. 5 (2020), 1–18; https://doi.org/10.1080/14608944.2020.1803819 (accessed May 30, 2024), 10.

82 Karin Idevall Hagren, "Nature, Modernity, and Diversity: Swedish National Identity in a Touring Association's Yearbooks 1886–2013," *National Identities*, 23 no. 5 (2020), 1–18; https://doi.org/10.1080/14608944.2020.1803819 (accessed May 30, 2024).

83 Ella Ödmann, Eivor Bucht, Maria Nordström, *Vildmarken och välfärden: om natursky-ddslagstiftningens tillkomst* (Wilderness and welfare: the origins of nature conservation legislation) (Stockholm: Liber Förlag, 1982), 154; cited in Mels, "Nature, Home, and Scenery," 138.

84 Tom Mels, "Nature, Home, and Scenery: The Official Spatialities of Swedish National Parks," *Environment and Planning D: Society and Space* 20, no. 2 (April 1, 2002): 135–154; https://doi.org/10.1068/d14s (accessed May 30, 2024) 138.

85 Tom Mels, "Nature, Home, and Scenery: The Official Spatialities of Swedish National Parks," *Environment and Planning D: Society and Space* 20, no. 2 (April 1, 2002): 135–154; https://doi.org/10.1068/d14s (accessed May 30, 2024), 144.

86 Tom Mels, "Nature, Home, and Scenery: The Official Spatialities of Swedish National Parks," *Environment and Planning D: Society and Space* 20, no. 2 (April 1, 2002): 135–154; https://doi.org/10.1068/d14s (accessed May 30, 2024), 142–143.

87 Tom Mels, "Nature, Home, and Scenery: The Official Spatialities of Swedish National Parks," *Environment and Planning D: Society and Space* 20, no. 2 (April 1, 2002): 135–154; https://doi.org/10.1068/d14s (accessed May 30, 2024), 144.

88 Cf. Don Mitchell, Erik Jönsson, and Johan Pries, "Making the People's Landscape: Landscape Ideals, Collective Labour, and the People's Parks (Folkets Parker) Movement in Sweden, 1891–Present," *Journal of Historical Geography* 72 (April 1, 2021): 23–39; https://doi.org/10.1016/j.jhg.2020.11.002 (accessed May 30, 2024).

89 Don Mitchell, Erik Jönsson, and Johan Pries, "Making the People's Landscape: Landscape Ideals, Collective Labour, and the People's Parks (Folkets Parker) Movement in Sweden, 1891–Present," *Journal of Historical Geography* 72 (April 1, 2021): 23–39; https://doi.org/10.1016/j.jhg.2020.11.002 (accessed May 30, 2024), 39.

90 Don Mitchell, Erik Jönsson, and Johan Pries, "Making the People's Landscape: Landscape Ideals, Collective Labour, and the People's Parks (Folkets Parker) Movement in

Sweden, 1891–Present," *Journal of Historical Geography* 72 (April 1, 2021): 23–39; https://doi.org/10.1016/j.jhg.2020.11.002 (accessed May 30, 2024).

91 Thorbjörn Anderson, "Erik Glemme and the Stockholm Park System," *Modern Landscape Architecture: A Critical Review*, ed. Marc Treib (Cambridge: MIT Press, 1993), 114–133.

92 Mattias Qviström, "Shadows of Planning: On Landscape/Planning History and Inherited Landscape Ambiguities at the Urban Fringe," *Geografiska Annaler: Series B, Human Geography* 92, no. 3 (September 2010): 219–235; https://doi.org/10.1111/j.1468-0467.2010.00349.x (accessed May 30, 2024).

93 Mattias Qviström, "Shadows of Planning: On Landscape/Planning History and Inherited Landscape Ambiguities at the Urban Fringe," *Geografiska Annaler: Series B, Human Geography* 92, no. 3 (September 2010): 219–235; https://doi.org/10.1111/j.1468-0467.2010.00349.x (accessed May 30, 2024).

94 Carl Marklund and Peter Stadius, "Acceptance and Conformity: Merging Modernity with Nationalism in the Stockholm Exhibition in 1930," *Culture Unbound* 2, no. 5 (2010): 609–634; https://doi.org/10.3384/cu.2000.1525.10235609 (accessed May 30, 2024).

95 Carl Marklund and Peter Stadius, "Acceptance and Conformity: Merging Modernity with Nationalism in the Stockholm Exhibition in 1930," *Culture Unbound* 2, no. 5 (2010): 609–634; https://doi.org/10.3384/cu.2000.1525.10235609 (accessed May 30, 2024).

96 David Kuchenbuch, "Architecture and Urban Planning as Social Engineering: Selective Transfers between Germany and Sweden in the 1930s and 1940s," *Journal of Contemporary History* 51, no. 1 (2016): 22–39.

97 David Kuchenbuch, "Architecture and Urban Planning as Social Engineering: Selective Transfers between Germany and Sweden in the 1930s and 1940s," *Journal of Contemporary History* 51, no. 1 (2016): 22–39, 29.

98 David Kuchenbuch, "Architecture and Urban Planning as Social Engineering: Selective Transfers between Germany and Sweden in the 1930s and 1940s," *Journal of Contemporary History* 51, no. 1 (2016): 22–39, 35.

99 Eva Rudberg, "Building a Utopia of the Everyday," *Swedish Modernism: Architecture, Consumption, and the Welfare State*, eds. Helena Mattson and Sven-Olov Wallenstein (London: Black Dog, 2010), 152–159.

100 Eva Rudberg, "Building a Utopia of the Everyday," *Swedish Modernism: Architecture, Consumption, and the Welfare State*, eds. Helena Mattson and Sven-Olov Wallenstein (London: Black Dog, 2010), 152–159..

101 Eva Rudberg, "Building a Utopia of the Everyday," *Swedish Modernism: Architecture, Consumption, and the Welfare State*, eds. Helena Mattson and Sven-Olov Wallenstein (London: Black Dog, 2010), 152.

102 Thomas Hall and Sonja Vidén, "The Million Homes Programme: A Review of the Great Swedish Planning Project," *Planning Perspectives* 20, no. 3 (2005): 301–328.

103 Irene Molina, "Planning for Patriarchy? Gender Equality in the Swedish Modern Built Environment," *The Routledge Companion to Modernity, Space and Gender*, 1st edition, ed. Alexandra Staub (New York: Routledge, 2018), 33; https://doi.org/10.1201/9781315180472-4 (accessed May 30, 2024).

104 Irene Molina, "Planning for Patriarchy? Gender Equality in the Swedish Modern Built Environment," *The Routledge Companion to Modernity, Space and Gender*, 1st edition, ed. Alexandra Staub (New York: Routledge, 2018), 33; https://doi.org/10.1201/9781315180472-4 (accessed May 30, 2024).

105 Polisen, Utsatta områden – sociala risker, kollektiv förmåga och oönskade händelser. Polismyndigheten, nationella operativa avdelningen, 2015; https://polisen.se/siteassets/

dokument/ovriga_rapporter/utsatta-omraden-social-ordning-kriminell-struktur-och-utmaningar-for-polisen-2017.pdf (accessed April 19, 2023).

106 Per A. Haupt, Meta Y. Berghauser Pont, Victoria Alstäde, and Per G. Berg, "A Systematic Review of Motives for Densification in Swedish Planning Practice," *IOP Conference Series: Earth and Environmental Science* 588 (November 21, 2020): 052030; https://doi.org/10.1088/1755-1315/588/5/052030 (accessed May 30, 2024).

107 Owen J. Dwyer and John Paul Jones III, "White Socio-Spatial Epistemology," *Social & Cultural Geography* 1, no. 2 (2000): 209–222; https://doi.org/10.1080/1464936 0020010211 (accessed May 30, 2024).

108 Burcu Yiğit-Turan and Mia Ågren, "White landscapes: Tracing Socio-spatial Epistemologies of Contemporary Swedish Planning," presented at the Dislocating Urban Studies: Rethinking Theory, Shifting Practice Digital Workshop Series organized by Malmö University, Dept. N. 1. *A Non-Occidentalist West: Learning from Theories Outside the Canon,* 2021.

109 Burcu Yiğit-Turan and Mia Ågren, "White landscapes: Tracing Socio-spatial Epistemologies of Contemporary Swedish Planning," presented at the Dislocating Urban Studies: Rethinking Theory, Shifting Practice Digital Workshop Series organized by Malmö University, Dept. N. 1. *A Non-Occidentalist West: Learning from Theories Outside the Canon,* 2021.

110 See Toni Adscheid and Peter Schmitt, "Mobilising Post-Political Environments: Tracing the Selective Geographies of Swedish Sustainable Urban Development," *Urban Research & Practice* 14, no. 2 (March 15, 2021): 117–137; https://doi.org/10.1080/17535069.2019.1589564 (accessed May 30, 2024).

111 Ramón Grosfoguel, "The Epistemic Decolonial Turn: Beyond Political-Economy Paradigms," *Globalization and the Decolonial Option*, eds. Walter Mignolo and Arturo Escobar (London: Routledge, 2011), 72; For the delineation of "the colonial and racial origins of the welfare state" see, Gurminder K. Bhambra and John Holmwood, "Colonialism, Postcolonialism and the Liberal Welfare State," *New Political Economy* 23, no. 5 (September 3, 2018): 574–587; https://doi.org/10.1080/13563467.2017.1417369 (accessed May 30, 2024); and Diana Mulinari and Anders Neergaard, "The Swedish Racial Welfare Regime in Transition," *Racism in and for the Welfare State*, ed. Fabio Perocco (Marx, Engels, and Marxisms; Cham: Springer International Publishing, 2022), 91–116; https://doi.org/10.1007/978-3-031–06071-7_4 (accessed May 30, 2024).

112 According to Bhambra and Holmwood, "exclusive solidarities" have played a role in reducing welfare rights in the several Western countries, including Sweden; see Gurminder K. Bhambra and John Holmwood, "Colonialism, Postcolonialism and the Liberal Welfare State," *New Political Economy* 23, no. 5 (September 3, 2018): 574–587; https://doi.org/10.1080/13563467.2017.1417369 (accessed May 30, 2024), 584. Mulinari and Neergard drew similar conclusions regarding the Swedish context, see Diana Mulinari and Anders Neergaard, "The Swedish Racial Welfare Regime in Transition," *Racism in and for the Welfare State*, ed. Fabio Perocco (Marx, Engels, and Marxisms; Cham: Springer International Publishing, 2022), 91–116; https://doi.org/10.1007/978-3-031–06071-7_4 (accessed May 30, 2024).

113 In his book *The Language of Postmodern Architecture* (1984), North American architectural historian and theorist Charles Jencks praised the demolition of Pruitt-Igoe as the end of modern architecture and the beginning of a presumed more pluralistic field.

114 cf. Ellen Braae, Svava Riesto, Henriette Steiner, and Anne Tietjen, "European Mass-Housing Welfare Landscapes," *Landscape Research* 46, no. 4 (May 19, 2021): 451–455; https://doi.org/10.1080/01426397.2021.1914568 (accessed May 30, 2024).

115 cf. Don Mitchell, Erik Jönsson, and Johan Pries, "Making the People's Landscape: Landscape Ideals, Collective Labour, and the People's Parks (Folkets Parker) Movement in Sweden, 1891–Present," *Journal of Historical Geography* 72 (April 1, 2021): 23–39; https://doi.org/10.1016/j.jhg.2020.11.002 (accessed May 30, 2024).

116 Tobias Hübinette and Catrin Lundström, "Three Phases of Hegemonic Whiteness: Understanding Racial Temporalities in Sweden," *Social Identities* 20, no. 6 (November 2, 2014): 423–437; https://doi.org/10.1080/13504630.2015.1004827 (accessed May 30, 2024).

Chapter 4: Is Landscape Empowerment?

Jala Makhzoumi

"The landscape is never inert,"[1] argues Barbra Bender, "people engage with it, re-work it, appropriate and contest it" in the process of creating and disputing individual, group, and state identities.[2] History abounds with examples of landscapes that were shaped by powerful rulers and feudal lords, just as in today's globalizing world, market forces manifest their power in the landscape of cities and the lands beyond. In this chapter, I argue that landscape is both the tool of centralized authorities, a staging and scenography for the manifestation of power, and the lived and mutually constitutive construct which, by its very nature, empowers everyone who is given, or insists upon, their right to it. Drawing on the Bisri Valley in Lebanon, I explore the discourse of the state and its plan to construct a dam on the Bisri River with funding from the World Bank, and the opposing discourse of communities whose livelihoods and very identities are threatened by the dam. Opposition to the project is suppressed through intimidation and political coercion. However, it was once again revived with the mass uprising in Lebanon on October 17, 2019, against state failures and corruption. The fight against the Bisri Dam gathers national support, as the valley landscape becomes a symbol in the fight for social and environmental justice and the protection of cultural heritage in Lebanon. I reflect on the valley's culturally meaningful and emotionally charged landscape to understand why it gained national support and how it enabled disempowered rural communities to contest and reclaim the landscape they value.

LANDSCAPE, POWER, EMPOWERMENT

Landscape is a complex idea and a word with many meanings. The earliest meaning, corresponding to the German word *Landschaft* implying "a sheaf, a patch of cultivated ground, something small-scale that corresponded to a peasant's perception," was replaced by words that correspond to the "larger political spaces of those with power—*territoire*, *pays*, domain."[3] The meaning changed again to gain its present scenic meaning, implying a particular "way of seeing" that was human-centered, domineering of nature, and the less privileged. Olwig distinguishes between two central meanings: (1) landscape as "scenic" space, something that can be quantified and planned by those in power, the domain of landscape

DOI: 10.4324/9781003148142-5

architects, architects, and urban planners; and (2) landscape as place, polity, and community, a social, political and legal phenomenon, the domain of landscape historians, archaeologists, and geographers.[4] The two meanings, the earth-bound original meaning and the human-centered, detached scenic one, survive today. The choice of meaning depends on how "landscape" is framed and by whom.

For the largest part of its relatively short professional history, landscape architecture has adhered to the earlier meaning of landscape as a product, land that is managed to meet the needs of a client. Here, a landscape architect's role is to package the "product" of the landscape they design. In the last couple of decades, professional practice has evolved toward greater sensitivity to place and local culture and responsiveness to the perceptions, values, and valuation of those inhabiting the landscape. "The broadening of the professional scope," explains Waterman,[5] "is the outcome of ongoing enrichment by the increasingly post-disciplinary field of landscape studies, which deals with landscape relationships—urban, rural, and everything in between—across the widest set of scales from the local to is the continental."[6]

The European Landscape Convention (ELC)—the first international treaty devoted to all aspects of the European landscape, urban and rural, ordinary and outstanding, degraded and pristine—defines landscape as "an area perceived by people whose character is the result of the action and interaction of natural and/ or human factors."[7] The emphasis on "perception" is significant in several ways. It is an admission that there is no "universal" definition, that landscape is context-specific and culturally rooted. "The way in which people—anywhere and everywhere—understand and engage with their worlds will depend upon the specific time and place and historical conditions."[8] If we accept that there are variations in the perception of landscape within the same culture—for example, the powerful and the subordinate—it becomes clear how and why one group's ideas take precedence over another. Here, landscape becomes a medium for the dispute and contestation of power.

The ELC builds on the overlap with "nature," "nation," and "culture" with "landscape" that is integral to the "hidden" discourse, underwriting the legitimacy of those who exercise power in society.[9] This has encouraged new political interpretations of "landscape," for example, in framing human rights.[10] Landscape combines tangible physicality, anchored in the land, with subjective, intangible perceptions and emotional attachments as the basis for survival and the kind of spiritual and emotional well-being that are essential to human existence. Here, the prefix *land* is similarly of value "as the meaning of the place, region, or country of the people of the body politic," while the suffix, *scape*, embodies not only a collective "sense of creative shaping and carving," but also imbues the concrete with a more abstract character, "the sense of a condition as in friendship,"[11] of community and social collective. The duality of landscape as a subjective place of emotions, a collective sense of belonging and identity, and as an objective, rational space for the enactment of power renders the idea of landscape "elusive" but with a potential that is "all-encompassing."[12]

Having elaborated on the complexity of landscape and the versatility of landscape framing, I turn to the meaning of "power" and its derivative "empowerment." Power implies possession of the ability to wield force, authority, or influence[13] either to affect change or to ward off a threat. Power drives everything to realize itself with increasing intensity and extensity,[14] and while "intensity" implies strength, "extensity" is the attribute of sensation from which the perception of spatial extension is developed. If we accept a basic definition of "empowerment" as "the process of becoming stronger and more confident, especially in controlling one's life and claiming one's right,"[15] then landscapes empower through the shared sense of kinship and belonging, attachment and inherited values that support/sustain the ability of individuals and communities to act to affect change or to resist change. The idea of landscape as empowerment is relevant as human-induced change transforms the global ecology and undermines life on the planet. The dialogic discourse of landscape empowerment is an opportunity to adapt to radical landscape change, to debate those things that should change (exploitative use of energy, water, land) and those that should resist change (natural and cultural heritage) because they are too valuable, and to find ways to adapt to change gracefully. The case study from Lebanon exemplifies the intertwining of "landscape" and "power." The landscape of the Bisri Valley is the setting for a power struggle between powerful state authorities' plan to construct a dam in the Bisri Valley and local inhabitants of villages in the watershed, who oppose the project that has come to threaten their livelihoods, undermine the values they uphold, and impend their very existence.

LANDSCAPE AS THE MANIFESTATION OF POWER

As one of Lebanon's five coastal rivers, the Awali River[16] has its source in Mount Lebanon. It is from the mountains that the Barouk and Array rivers emerge and join to form the Bisri River. The wide Bisri Valley (Arabic: *Marj Bisri*) narrows as it approaches the coast and changes its name to the Awali River before discharging into the Mediterranean Sea north of Sidon. The Roum seismic fault accounts for the valley's complex morphology. Successive collapse of the valley walls accounts for an exceptionally wide valley through which the river meanders freely (Figure 4.1). The south-facing valley wall is steep and forested, and the limestone cliffs of the northside are sparsely vegetated. Landform, climatic sheltering, and rich alluvial valley soils sustain a diverse landscape mosaic that includes pine forests, native Mediterranean scrublands, orchards, and arable farming. The landscape sustains communities inhabiting the Bisri watershed—the valley has been inhabited since the Paleolithic era (circa 8300 BCE). Archaeological finds include remains from the Bronze Age, Babylonian/Persian, Hellenistic, Roman, Byzantine, Arab, and Ottoman times, a testimony to the importance of the valley, which served to connect the Mediterranean across the mountains to Damascus and the desert beyond.[17] Inhabitants of the 14 villages north and south of the valley[18] were, throughout history, custodians of the river, its lands, and resources.

Figure 4.1 The meandering Awali River in the cultivated Bisri Valley. Courtesy: Nadim Asfar.

The mutually constitutive relationship that binds people to their land accounts for today's exceptionally beautiful and productive landscape and their strong attachment to Marj Bisri.

In 2015, the Lebanese parliament approved the construction of a dam in the town of Bisri. The project triggered a power struggle at the heart of the landscape of the Awali/Bisri Valley. The French Mandate was the first to propose dams that were seen as a tool to "modernize" colonized territories. The Bisri Dam project was developed years later by the US Bureau of Reclamation, which was assisting the Lebanese state water management plan[19] (Figure 4.2). In the decades since, there has been growing awareness of the adverse impact of large dams on ecosystems and biodiversity,[20] the disruption of local livelihoods, and the displacement of resident communities.[21] While countries in the West are demolishing dams constructed in the first half of the twentieth century, dams continue to be favored by countries in the Global South.[22] This is partly because large infrastructural engineering projects, dams, and highways are symbolic, the physical expressions of state power and embodiment of engineering science and skills associated with an outdated understanding of "progress" and "development."[23] Donors are similarly supportive of large infrastructural projects because, as tangible "products," they show off their investment. It was the promise of a USD 474 million loan from the World Bank that enabled the Lebanese Council for Development and Reconstruction (CDR) to revive and approve plans for the construction of the Bisri Dam. The World Bank is known, and often criticized, for

Figure 4.2 Dams built and proposed on coastal rivers in Lebanon, among which is the dam on the Awali River that would inundate the Bisri Valley, 2020. Courtesy: Jude Mabsout.

its funding of "mega-dam projects in developing countries,"[24] let alone that it would give a multi-million-dollar loan to a bankrupt country like Lebanon.[25]

The four-kilometer-long reservoir created by the dam will inundate the lower parts of the Barouk and Array tributaries and approximately 570 hectares of the Marj Bisri downstream. The project, as such, threatens to disrupt the lives and livelihoods of rural communities in the watershed, destroy a landscape of outstanding natural and cultural significance, and interrupt the historical and ecological link between the mountains and the sea. Quantified, the reservoir will destroy 82 hectares of pine forests, 150 hectares of agricultural fields, 131 hectares of orchards, 77 surveyed archaeological sites, remains of a large Roman temple, a Byzantine church, the Saint Sofia monastery, necropolis, and rock-cut tombs and Ottoman bridges.[26] Beyond the valley, the repercussions will damage the entire watershed ecology. The state defended its position, citing the urgent need to meet the growing need for potable water for the 1.5 million residents of the capital city, Beirut, which sits 35 kilometers northwest of the valley. However, the facts continue to stack up against the project, as they do against other dams in Lebanon. Foremost is the present and imminent risk of locating the Bisri Dam on a geologically complex and tectonically active Roum fault.[27]

Lebanese and foreign experts agree that water scarcity is the result of corruption and mismanagement and that the water needs of the capital city can be met through efficient management and tapping into aquifers close to Beirut.[28] Riachi argues that large dams can be replaced by cost-efficient small- to medium-sized water collection ponds with little ecological impact.[29] Questions were raised on the Environmental Impact Assessment commissioned in 2016 by the CDR. Not only was the assessment incomplete and outdated but it was also commissioned by a consulting firm that is a stakeholder in the project, clearly ignoring this "conflict of interest."[30] The project ignores the Ministry of Environment's designation of the entire Awali River as a protected natural site (decree No. 131/1, 1/9/1998). The Bisri Dam project similarly disregards the National Physical Master Plan for the Lebanese Territory (NPMPLT),[31] which identifies the Marj Bisri as one of six "exceptional natural landscapes" that must be protected, a status that can't be readily ignored.

Why would the Lebanese state persevere with plans for the project despite all the arguments against it? Riachi argues that the answer lies in Lebanon's political-economic history, which is "based on neoliberal thought that enshrines concessions for private interests in the water sector."[32] Lebanon's political elite favors dams because large-scale infrastructural engineering projects empower politicians by presenting opportunities "to claim that they are achieving something concrete. . . [which] can quickly be turned into opportunities for political self-advertising and tools to maintain clientelist networks."[33] Alongside political empowerment is the promise of financial gain. Lebanese politicians favor market-driven approaches to services because they maximize water production and profits gained, explicit and illicit.[34] CDR, the key player and "uber-state agency that oversees the Bisri Dam project," and the bulk of Lebanon's public tenders is "the epitome of patronage, [disbursing] funding for public projects to enrich Lebanon's sectarian leaders."[35]

THE THREAT OF DISEMPOWERMENT

The promise of political acumen and financial gain that empowered state authorities and politicians spelled disaster for the watershed's inhabitants. Water from the reservoir will displace residents of three villages and destroy agricultural and pastoral livelihoods for inhabitants of the remaining 11 villages. In this section, I draw on the extensive coverage by the Legal Agenda (2020) research team to

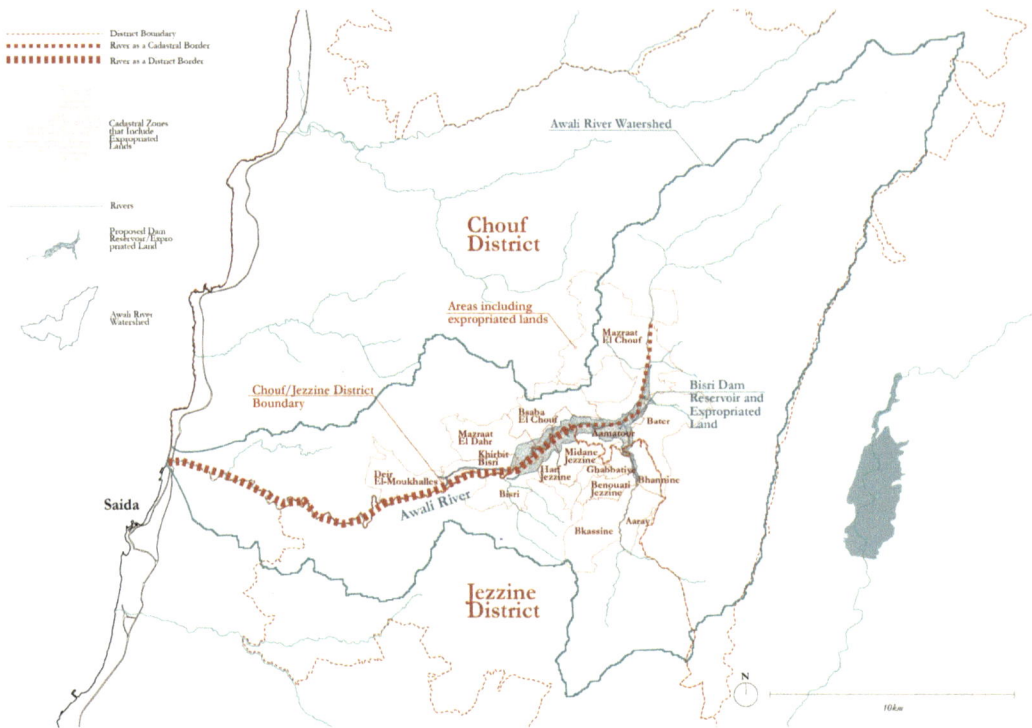

narrate the fears and grievances of the 14 villages affected by the project (Figure 4.3).

Figure 4.3 The Awali River demarcates the administrative boundary between the two districts of Chouf and Jezzine. Location of the proposed dam reservoir that will submerge expropriated lands in the Bisri Valley, 2020. Courtesy: Jude Mabsout.

Ma'moun Abu Shakra, Amatour village, owns one of the largest farms in Marj Bisri. He explains that the people of Jezzine and Chouf, respectively, south and north of the river, invest in their lands either directly or by leasing the land to other farmers,

> the economic return from the *marj* from agriculture is no less than USD 6 million annually for one season, more seeing the people of *marj* enjoy two planting seasons, spring and late summer. . . in addition to cattle that graze in the pastoral lands of the *marj*, sheltering there in the winter.[36]

The sentiments are echoed by Kamal Kiwan, a teacher, who objects to the belittling of the role of agriculture, "*people owning lands in the Marj* in Bater [his village] cultivate their lands to meet living expenses for their families including those employed or have alternative source of income."[37]

Herding in the valley is another source of income. A cattle owner, Najib Bu Kurum from Mazraat al Chouf, laments the loss of the diverse microclimate in the valley, "in the lowest part of the valley we keep our goats, next to it the cows and sheep. . . the weather there is warm in the winter."[38] Livelihoods from summer tourism in the valley are similarly threatened. Tony Habib, a resident of Mazraat al Mathana, fears the closure of restaurants along the river that support tourism in the *Marj*. Another resident of the village, Mishal Tawfiq, praises the fish in Bisri River, "my father raised a large family, educated [his children] from fishing in Bisri."[39]

Alongside the concern for livelihoods was anger at the unfair compensation of private property in the valley—the state expropriated a total of 966 plots purchased at below-market price from residents of the 14 villages in preparation for the project. The injustice of expropriation of land that had been inherited over generations is echoed by Yahya Abu Al Kurum, mayor of Mazraat al Chouf; his village is poor in water resources and grazing lands and thus fully reliant on the *Marj Bisri*. Abu Al Kurum complains that state estimation of property values for the purpose of expropriation "does a great injustice to those families whose livelihoods relies on the land."[40]

Municipal councils in Jezzine and Chouf had little choice but to give in to the powerful state. Nor could they oppose their confessional leaders, the zu'ama, respectively, the Free Patriotic Movement in Maronite villages of Jezzine and the Druz followers of the Progressive Socialist Party and its leader, Jumblatt. The exception is Midan village, Jezzine. Its mayor, Norma Feghali, consistently opposed the dam. No amount of pressure by the state or coercion by the zu'ama could force ordinary citizens to accept what was perceived as a blatant injustice. Abu Al Kurum ridicules the shift from objecting to supporting the project when their za'em, Jumblatt, sent word to his subjects to support the project. Thereafter, he says, "Everyone was talking of the water needs of the capital city."[41]

Damage to the environment resulting from the construction of the dam is high on the list of concerns. Inhabitants in the watershed are well aware of the environmental assets of the Marj Bisri, and that it is considered one of the last "unspoiled" landscapes in Lebanon, which is a matter of pride. Inhabitants are also aware of the problematic edaphic conditions in the valley and the ever-present threat of earthquakes. "Our forefathers built their houses on a sand tell following an earthquake that happened a thousand years ago," explains Jebran Mtanos Habib, an inhabitant of Mazraat, "the soil is fluid, we lose land when the waters of Bisri rise and continue to construct stone walls to protect our property."[42] Survivors of the 1956 earthquake top the list of those opposing the dam. The thunderous sound of the earthquake is alive in the memory of Bahiya Tabet and images of half of her village, Bater, buried under the debris. "I don't think of myself," declares Shafiq Boulos, now in his eighties, "but I must think of my children and grandchildren, I don't want to bequeath to them a legacy of displacement and earthquakes, I have to resist this project to my last breath."[43] Raouf Mohammad Darwich, mukhtar of Bater, contrasts the aspirations of the village community with those of the state, "We strive to enhance agriculture and safeguard the health of people and environment [in Bater] whereas the dam threatens with destruction should an earthquake strike."[44] Emile Habib, Mazraat al Mathana, is worried by the prospect of landslides, recalling when the river washed away 200 olive trees. "Our house was at a distance of 1000 meters from the river," explains Claudette, "slowly the land is sliding, and year after year, the distance separating us from the river grows smaller. We fear that one day we will find ourselves under the water of the river."[45]

Another concern is the threat to the environment that comes from the water reservoir. Abu Shakra, Amatour village, speaks of expected changes to the local climate with the introduction of 125 million cubic meters of water,

> my village and agricultural lands will be affected by the reservoir. . . if it does not explode from pressure on fragile rocks of the area or an earthquake caused by [building] the dam on a Faultline. . . the high humidity from evaporation will destroy our crops. . . kill the grapes and figs, pomegranate and other trees that can't withstand the humidity.[46]

The threat of increased humidity similarly worries villages in Jezzine because they threaten pine forests covering the southern slopes of the valley. Habib Khoury, a member of the municipal council of Bkessine, points out the economic significance of pine forests. The estimated 120,000 mature pine trees, *Pinus pinea,* extend for over 220 hectares, the largest productive pine forest in the Middle East. He explains that the production of pine nuts makes up half the municipal revenues in Bkessine. The waters of the reservoir will cause landslides, and the increased air humidity will be detrimental to the growth of pine trees. "We protect the pine forest because it is the 'mother' of Bkessine," he explains, the "protective talisman of our village alongside the saint Mar Takla."[47]

There is also the political discourse in Marj Bisri, which is historically rooted. After the withdrawal of the Syrian army occupation in Lebanon in 2005, Bisri came to mark the boundary between warring Druz communities of the Chouf and Christian Maronite political warlords, respectively, north and south of the Bisri River. Memories of the conflict, buried under everyday rural life in the *Marj*, surface with the Bisri Dam. On the one hand, the project would not have proceeded without the endorsement of the zu'ama, and their coercion that had five Chouf villages and four from Jezzine acquiesce, and their municipalities formally consent to the construction of the dam. These villages understand stripping them of livelihoods and property is not only crippling economically but disempowering. "Self-sufficiency," argues Ma'moun Abu Shakra, the village of Amatour, is a key to "liberate citizens, farmers and village communities, from the servitude to the zu'ama throughout Lebanon."[48] On the other hand, displacement and economic destitution are compounded by physical separation. The reservoir will impede the reconciliation and healing of communities north and south of the valley. Yahya Abu Kurum, mayor of Mazraat al Chouf, laments that "the dam constitutes the final blow to return and resettlement" of the Christian population that constituted a third of his village, Mazraat al Chouf.[49]

To summarize, landscape is at the heart of the local discourse opposing the Bisri Dam. Even if the word "landscape" is absent, it is implicit in the perception of "nature" and "environment" and as resources that sustain agricultural, pastoral, and silvicultural livelihoods; the discourse on cultural heritage, be it tangible land, orchards and fields, woodlands and pastoral lands, handed over generations, and intangible practices and values shared by villages across the political and religious divide. Beyond land and heritage, landscape embraces the intimate relationship that binds people to the Marj Bisri, defines their collective identities and what they hold dear, and underlies the rights and entitlements they are willing to fight for.

LANDSCAPE AS EMPOWERMENT

The first wave of objections to the Bisri Dam was by local authority representatives, mayors, and mukhtars to the 14 villages directly affected by the Bisri Dam. Their opposition to the project was voiced during meetings in 2012 and 2013 organized by the CDR—consulting with local stakeholders is bound by the Environmental Impact Assessment (EIA) law, Ministry of Environment. Meetings were followed by petitions to the CDR signed by Mazraat al Chouf, Bisri, and Mazraat al Mathana village municipalities, refusing to consent to the project. Political zu'ama succeeded where intimidation and false promises by the CDR failed. All 13 villages withdrew their objections except for the village of Midan, Jezzine, which consistently opposed the project. Midan municipality was dissolved. Undeterred, Norma Feghali, the deposed mayor, followed with legal action invalidating the project because it doesn't constitute "public benefit"[50] as claimed by the CDR. Legal objections were ignored, overridden by the presidential decree endorsing the project, authorizing the CDR to proceed with implementing the Bisri Dam. The *Marj* was fenced and guarded by manned security by the contracting firm commissioned with construction. Just as the local communities reluctantly admitted to their failure, news of their struggle with the authorities was slowly coming to the attention of environmental activists elsewhere in Lebanon. Lebanese citizens could see through false claims by the CDR and the World Bank—namely, the provision of potable water benefiting the public good was yet another cover for financial gains by the authorities and political zu'ama. The Bisri Dam came to embody all that was wrong with the confessional political system in Lebanon and its injustice to people and the environment.

A key figure mobilizing action against the Bisri Dam is Roland Nassour, an architect and environmental activist affiliated with the Lebanon Eco Movement, an umbrella for concerned citizens, activists, and NGOs. Nassour explains that his dedication to the fight against the Bisri Dam builds on past grievances regarding failed dam projects elsewhere in Lebanon.[51] In 2014, Nassour actively opposed the Jannah Dam north of Beirut, which was smaller than Bisri but similarly aimed to supply the capital with potable water.[52] Failure of the completed dam at Jenna was a living example of the disaster in store should the Bisri Dam be implemented.[53] Opposition gathered momentum as the public became better informed of sustainable alternatives to water resource management without damaging the environment and destroying rural landscapes. The "National Campaign to Protect the Bisri Valley"[54] in 2018 was led by Nassour, who mobilized social media to secure the support of communities from all over Lebanon. The national campaign followed two strategies: the first was to petition[55] the various

Lebanese Ministries to stop the project and conduct an updated Environmental Impact Assessment[56]; the second targeted the World Bank Inspection Panel, expressing concern about the negative impact of the dam and the absence of comprehensive analysis and lack of public participation. The complaints were investigated but not taken seriously.

Riding the wave of the October 17th *Intifada* mass uprising that spread across the entire country, the fight to protect Marj Bisri turned into embodied resistance on November 9, 2020. More than 1,400 demonstrators poured into the valley (Figure 4.4). Joined by the local community, protestors "smashed through the barricades and tied themselves to the remaining trees [camped in] tents, vowing to thwart any work on the site."[57] The contractor for the Bisri Dam was given 48 hours to remove all machinery from the construction site, and a permanent base was established to ensure that the contractor didn't return to the site. Villagers hosted the protestors and sold traditional foods, incredulous and grateful for the support. Residents and protestors alike demanded that the state comply with the NMPLT designating the Bisri Valley as a national park accessible to all, one of five in Lebanon. Alongside embodied resistance in the valley, protestors camped outside government buildings and the World Bank headquarters in the capital, Beirut. Public debates were organized, and testimonials from independent experts were secured. Geologists explained the threat of earthquakes. Hydrologists proposed alternatives to

Figure 4.4 Campaigners against the Bisri Dam celebrate after breaking through fences constructed by the contractor in preparation for construction of the dam. Courtesy: Archives of the Save the Bisri Valley Campaign.

Figure 4.5
Demonstrations in downtown Beirut against the Bisri Dam continued until the World Bank announced that it was suspending its loan to the Lebanese State. Courtesy: Archives of the Save the Bisri Valley Campaign.

meet the capital's need for potable water that was ecologically sound and incurred minimum cost (Figure 4.5). The discourse of activists and ordinary citizens fighting to stop the dam and protect Marj Bisri rests on four arguments: "The seismic threat, threat to agriculture and rural livelihoods, destruction of forests and the obstructing the NMPLT law designating the valley as a national park."[58]

The masses were slowly swaying the stance of politicians and parliamentarians. Some politicians called for new legislation to stop the dam for the same reasons cited by protestors during the Intifada. Jumblatt called for talks to thrash the pros and cons but refused to voice his objections to the project, to the disappointment of his followers in the Chouf. The most comprehensive petition against dam construction was by the International Council on Monuments and Sites (ICOMOS-Lebanon) in April 2020. ICOMOS-Lebanon compiled a dossier that included arguments against the dam with testimonials from experts and parliament members.[59] For the first time, Marj Bisri was framed as a cultural and natural heritage, a cultural landscape of significance. The World Bank could no longer ignore mounting objections to the project in Lebanon. In June 2020, the World Bank gave in and suspended the loan to the Lebanese State. Funding would resume on the condition that the government address community fears in "reaching an agreement on the operation and maintenance of the dam with the concerned parties."[60]

CONCLUSION

Marj Bisri's discourse on power is complex and layered. For those in power, landscape is reduced to a material setting, a "terra nullius," ready to be colonized for profit. State authorities and the zu'ama manifest their power through "language and image to conceptualize and naturalize a particular, and in this case, deeply unequal, way of relating to the land and other people."[61] In cities across the Lebanese countryside, the totalizing landscape of power is exploitative in its use of environmental resources, oblivious to public rights to the land and indifferent to the social injustice caused. For the disempowered, "landscape" is the lived, a mutually constitutive place that meets their needs for survival and defines who they are. However, because organized action to claim communal and public rights is "context-specific, it unfolds in a tangible space,"[62] the landscape becomes the source of empowerment. By severing the relationship that binds people and their landscape, state plans for the Bisri Dam creates a grave injustice, which follows a pattern that has reoccurred throughout history. Waterman cites one such example, the historic act of *enclosure* in England, whereby "land shared in common has been enclosed privatized, thereby weakening a common foundation for community life and identity."[63] So, too, the threat of "enclosure," albeit by the dam's water reservoir, provoked and empowered inhabitants of the Marj Bisri—then eventually the entire nation—enabling them to mobilize and contest the project with direct action and embodied resistance. Equally problematic is the focus on capital cities, which are part and parcel of colonial rule in the region. In Lebanon, the capital city, Beirut, receives the largest share of funding for development at the expense of denying rural communities their share of betterment. Whether intentionally or inadvertently, those in power worked *against* the practices necessary for a good life in the peripheries, underestimating that being grounded in the landscape, literally and metaphorically, is equally empowering.

Attending a discussion organized by the National Campaign to Protect the Bisri Valley in the summer of 2018, I was surprised that the word landscape was not mentioned in the presentations nor the following discussions. The absence is understandable, considering that the Western concept of "landscape" is new to the local culture and is generally associated with the public realm in cities. The fact that no Arabic word embraces the complexity of the English term is another reason landscape is absent from the discourse on rights. There is, however, a layered conception of landscape in rural Lebanon, even if there is no single, totalizing word to express this conception.[64]

Does it matter that the word landscape was missing from the discourses of Marj Bisri? Bender argues that it does, "Words define experience, experience takes its meaning from time and place. Words empower action (or inaction),"[65] only by understanding how their meanings have been constructed can we begin to democratize them. I believe the absence of "landscape" from Bisri's discourse is significant and needs to be addressed. On the one hand, no other concept bridges *tangible space*, land, and natural resources with *intangible* community values, issues of identity, and emotional attachment. As such, a landscape framing is more likely to consider not only objective, scientific arguments opposing the Bisri Dam, the riparian ecology, and sustainable use of environmental resources but also subjective, cultural perceptions and shared values of those that inhabit the landscape. On the other hand is the political dimension, where landscape serves as a platform to discourse individual and collective rights. This is the case in the rural landscapes as well as in cities,[66]

> where larger entities such as governments or corporations, extract resources or (re)claim land for the purpose of economic gain, while those who live in, depend upon or identify with the environment are weaker with regard to their political and economic resource base.[67]

Here, landscape contextualizes universal and moral standards, such as human rights, that aspire to "guarantee both concrete needs for survival and the spiritual/emotional/psychological needs that are quintessential to the human experiences."[68]

The potential of landscape as a platform to discourse the rights of the inhabitants of Bisri Valley brings us back to the question raised at the beginning of this chapter, "Is Landscape Empowerment." Whether valued as unspoiled "nature" and a healthy environment or as the foundation for livelihoods, the landscape was a source of pride and heritage for local communities and the basis for their identity. The people of Bisri were not ready to relinquish their entitlement to the land nor disregard their attachment to the landscape. This, in turn, empowered them to take action, resist the all-powerful state and the zu'ama, and fight for social and environmental justice.

There is no closure to the political narrative from Bisri. Those in power are persistent in their plans to construct the dam—the profits they would gain are too big for them to ignore the project. They will ride out the public uproar, waiting for the opportunity to resume their plans and construct the dam. For the people of Bisri, the landscape has acquired yet an added value, that of community solidarity, an action that bridged the religious divide, a sense of empowerment that has become yet another layer in the heritage of the Bisri Valley landscape.

ACKNOWLEDGMENTS

I would like to thank Roland Nassour, for meeting with me, answering my questions and sharing his experience as an activist relentlessly protesting the construction of dams in Lebanon; Antoine Atallah and Souheir Mabsout for sharing background material on the Bisri Project; Nadim Asfar, for allowing me to use his photo of the Marj Bisri; and Jude Mabsout for permitting me to use the beautiful maps she prepared for the Bisri Landscape.

NOTES

1 Barbara Bender, "Introduction: Landscape—Meaning and Action," in *Landscape: Politics and Perspectives*, ed. Barbra Bender (Oxford: Berg, 1995), 1–17.

2 Barbara Bender, "Introduction: Landscape—Meaning and Action," in *Landscape: Politics and Perspectives*, ed. Barbra Bender (Oxford: Berg, 1995), 3.

3 Barbara Bender, "Introduction: Landscape—Meaning and Action," in *Landscape: Politics and Perspectives*, ed. Barbra Bender (Oxford: Berg, 1995), 2.

4 Kenneth Olwig, "Foreword," in *Defining Landscape Democracy: Perspectives on Spatial Justice*, eds. Shelley Egoz, Karsten Jorgensen and Demi Ruggeri (London: Edgar Publishers, 2018).

5 Tim Waterman, "Introduction," in *Landscape Citizenships*, eds. Tim Waterman, Jane Wolff, and Ed Wall (London: Routledge, 2021).

6 Realizing the expanding role of the profession scope, the international Federation of Landscape Architects (IFLA) felt the need to change the definition and scope of landscape architecture for the International Labor Organization (ILO). The definition, not yet approved by the ILO, appears on the IFLA website, available on https://www.iflaworld.com/the-profession (accessed February 07, 2021).

7 Council of Europe, https://rm.coe.int/1680080621 (accessed June 02, 2023).

8 Barbara Bender, "Introduction: Landscape—Meaning and Action," in *Landscape: Politics and Perspectives*, ed. Barbra Bender (Oxford: Berg, 1995), 2.

9 Kenneth Olwig, "Sexual Cosmology: Nation and Landscape at the Conceptual Interstices of Nature and Culture or What Does Landscape Really Mean?," in *Landscape Politics and Perspectives*, ed. Barbara Bender (Oxford: Berg, 1995), 307.

10 Shelley Egoz, Jala Makhzoumi, and Gloria Pungetti, eds. *The Right to Landscape: Contesting Landscape and Human Rights* (London: Ashgate, 2011).

11 Kenneth Olwig, *The Meaning of Landscape. Essays on Place, Spaces, Environment and Justice* (London: Routledge, 2019), 130.

12 Howard Morphy, "Colonialism, History and the Construction of Place: The Politics of Landscape in Northern Australia," in *Landscape Politics and Perspectives*, ed. Barbara Bender, (Oxford: Berg, 1995), 205.

13 Merriam Webster Dictionary, https://www.merriam-webster.com/dictionary/power (accessed August 02, 2023).

14 Adam Kahane quoting Paul Tilich, 2010, 2.

15 See https://www.lexico.com/definition/empowerment.

16 The total length of the Awali River is 48 km, water discharge estimated at 10.16 cubic meters per second. The total area of the Awali basin is 29,000 hectares, the Bisri Valley, the focus of the article, an estimated hectare.

17 Wissam Khalil, "Routes Et Fortifications Dans Le Chouf Libanais," in *Dossier d'Archéologie* no. 350 (Mars/Avril 2012), 70–75.

18 The Bisri Valley is the boundary that separates two governorates. The district of Chouf with six villages north of the Bisri River (Khirbet Bisri, Mazraat el-Daher, Bsaba, Mazraat el-Chouf, Aamatour and Bater) and the district of Jezzine, to the south with eight villages south of the valley (Bisri, Harf Jezzine, Midane, Ghabbatiyeh, Benouati, Bkassine, Array, and Bhanine).

19 More than 18 dams were planned, some implemented others awaiting funding for implementation (Interview with Riachi, Legal Agenda, 2020).

20 Thomas New and Zonggiang Xie, "Impact of Large Dams on Riparian Vegetation: Applying Global Experience to the Case of China's Three Gorges Dam," in *Biodiversity and Conservation* 17, no. 13 (2008): 3149–3163; Thayer Scudder, *Large Dams: Long Term Impacts on Riverine Communities and Free Flowing Rivers* (Singapore: Springer, 2019).

21 Vinodh Jaichand and Alexandre Sampaio, "Dam and be Damned: The Adverse Impact of Belo Monte on Indigenous People in Brazil," in *Human Rights Quarterly* 35, no. 2 (2013): 408–447.

22 Alexander Matthews, "The Largest Dam-removal in US History," *BBC Future Planet*, November 10, 2020.

23 The shortcomings of the post WWII model of development are discussed by Esteva et al., who argue that "development" continues to be a banner for "enterprises that destroys both physical environments, and human culture all over the world." Gustavo Esteva, Salvatore Babones, and Phillip Babcicky, *The Future of Development: A Radical Manifesto* (Bristol: Policy Press, 2013), viii.

24 Dalal Mawad, "World Bank-funded Dam in Lebanon Mirrors Governance Crisis," *The Washington Post* website, September 09, 2020.

25 Lebanon continues to endure an unprecedented economic crisis, having defaulted on its foreign bond commitments, unprecedented inflation and collapse of the currency has left the entire population in abject poverty. See Vivian Yee and Hwaida Saad, "To Make Sense of Lebanon's Protests, Follow the Garbage," *The New York Times*, December 3, 2019.

26 Compiled report and testimonials by the International Council on Monuments and Sites of Lebanon (ICOMOS-Lebanon), August 2020, addressed to the World Bank to protest construction of the Bisri Dam.

27 Tony Nemer, "The Bisri Dam Project: A Dam on the Seismogenic Roum Fault, Lebanon," in *Engineering Geology* 261 (2019): 1–11.

28 Roland Riachi, "Water Policies and Politics in Lebanon: Where is Groundwater?," in International Water Management Institute, *Groundwater Governance in the Arab World*. Taking stock and addressing the challenges, Report No. 9, 2016.

29 Ibid.

30 Nassour, quoted in: Dalal Mawad, "World Bank-funded Dam in Lebanon Mirrors Governance Crisis," on *The Washington Post* website, September 9, 2020.

31 N.P.M.P.P.T., 2005, The N.P.M.P.L.T. is the strategic reference that overrides all documents concerning regional and local urban development planning and guides public investment.

32 Roland Riachi, "Water Policies and Politics in Lebanon: Where is Groundwater?," in International Water Management Institute, *Groundwater Governance in the Arab World*. Taking stock and addressing the challenges, Report No. 9, 2016.

33 Joey Ayoub and Maroun Christophe, "Stopping the Bisri Dam: From local to National Contestation," in *Arab Reform Initiative*, April 2020.

34 Joey Ayoub and Maroun Christophe, "Stopping the Bisri Dam: From local to National Contestation," in *Arab Reform Initiative*, April 2020.

35 Nabih Bulos and Marcus Yam, "Climate Change and Corruption Endanger an Ancient Valley in Lebanon," in the *Los Angeles Times*, February 21, 2021.

36 Legal Agenda, "Marj Bisri at the Heart of the Intifada," *Legal Agenda* 62, Special issue (January 2020): 5.

37 Legal Agenda, "Marj Bisri at the Heart of the Intifada," *Legal Agenda* 62, Special issue (January 2020): 11.

38 Legal Agenda, "Marj Bisri at the Heart of the Intifada," *Legal Agenda* 62, Special issue (January 2020): 8.

39 Legal Agenda, "Marj Bisri at the Heart of the Intifada," *Legal Agenda* 62, Special issue (January 2020): 14.

40 Legal Agenda, "Marj Bisri at the Heart of the Intifada," *Legal Agenda* 62, Special issue (January 2020): 12.

41 Legal Agenda, "Marj Bisri at the Heart of the Intifada," *Legal Agenda* 62, Special issue (January 2020): 12.

42 Legal Agenda, "Marj Bisri at the Heart of the Intifada," *Legal Agenda* 62, Special issue (January 2020): 13.

43 Legal Agenda, "Marj Bisri at the Heart of the Intifada," *Legal Agenda* 62, Special issue (January 2020): 22.

44 Legal Agenda, "Marj Bisri at the Heart of the Intifada," *Legal Agenda* 62, Special issue (January 2020): 11.

45 Legal Agenda, "Marj Bisri at the Heart of the Intifada," *Legal Agenda* 62, Special issue (January 2020): 14.

46 Legal Agenda, "Marj Bisri at the Heart of the Intifada," *Legal Agenda* 62, Special issue (January 2020): 9.

47 Legal Agenda, "Marj Bisri at the Heart of the Intifada," *Legal Agenda* 62, Special issue (January 2020):13.

48 Legal Agenda, "Marj Bisri at the Heart of the Intifada," *Legal Agenda* 62, Special issue (January 2020): 9.

49 Legal Agenda, "Marj Bisri at the Heart of the Intifada," *Legal Agenda* 62, Special issue (January 2020): 12.

50 Legal Agenda, "Marj Bisri at the Heart of the Intifada," *Legal Agenda* 62, Special issue (January 2020).

51 Interview with Roland Nassour by the author April 26, 2021.

52 Academics, experts, and journalists tried to oppose the Jenna Dam project in the 1990s, but they were silenced. The German Federal Institute for Geosciences and Natural Resources (BGR) declared the Janna Dam infeasible, but its construction began anyway, as per Marc Ghazali, "Roland Riachi: Lebanon applies a 70-year-old water dam policy," in his interview translated to English, in *Legal Agenda # 62*, 2020.

53 Roland Nassour, "An Open Letter to the World Bank's Board of Directors: Stop the Bisri Dam in Lebanon!" *Jadaliyya*, 2018.

54 See https://www.facebook.com/savebisri/, https://twitter.com/savebisri?lang=en and https://www.instagram.com/savebisri/?hl=en (accessed 2021).

55 The National Campaign to Protect the Bisri Valley, "Community · World Bank: Save the Bisri Valley أنقذوا مرج بسري," 2018.

56 The first one was incomplete and had been expired having been carried out more than two years before the start date of the implementation.

57 Nabih Bulos and Marcus Yam, "Climate Change and Corruption Endanger an Ancient Valley in Lebanon," *Los Angeles Times*, February 21, 2021.

58 Legal Agenda, "Marj Bisri at the Heart of the Intifada," *Legal Agenda* 62, Special issue (January 2020): 34.

59 Testimonials included a letter of support by the president of the Lebanese Landscape Association, the author.

60 Nabih Bulos and Marcus Yam, "Climate Change and Corruption Endanger an Ancient Valley in Lebanon," *Los Angeles Times*, February 21, 2021.

61 Barbara Bender, "Introduction: Landscape—Meaning and Action," in *Landscape: Politics and Perspectives*, ed. Barbra Bender (Oxford: Berg, 1995), 1–2.

62 Jala Makhzoumi, "Beirut's Public Realm and the Discourse of Landscape Citizenships," in *Landscape Citizenships*, eds. Tim Waterman, Jane Wolf and Ed Wall (London: Routledge, 2021), 184.

63 Kenneth Olwig, "The Right Rights to the Right landscape?," in *The Right to Landscape: Contesting Landscape and Human Rights*, eds. Shelley Egoz, Jala Makhzoumi and Gloria Pungetti (London: Ashgate, 2011), 40.

64 Jala Makhzoumi, "Unfolding Landscape in a Lebanese Village: Rural Heritage in a Globalizing World," *International Journal of Heritage Studies* 15, no. 4 (June 17, 2009): 317–337.

65 Barbara Bender, "Introduction: Landscape—Meaning and Action," in *Landscape: Politics and Perspectives*, ed. Barbra Bender (Oxford: Berg, 1995), 16.

66 Jala Makhzoumi, "Beirut's Public Realm and the Discourse of Landscape Citizenships," in *Landscape Citizenships*, eds. Tim Waterman, Jane Wolf and Ed Wall (London: Routledge, 2021), 182–204.

67 Stefanie Rixcker, "Re-conceptualizing Human Rights in the Context of Climtate Change: Utilizing the Universal Declaration of Human Rights as a Platform for Future Rights," in *The Right to Landscape. Contesting Landscape and Human Right*, eds. Shelley Egoz, Jala Makhzoumi and Gloria Pungetti (London: Ashgate, 2011), 27.

68 Shelley Egoz, Jala Makhzoumi and Gloria Pungetti, "The Right to Landscape: An Introduction," in *The Right to Landscape. Contesting Landscape and Human Right*, eds. Shelley Egoz, Jala Makhzoumi, and Gloria Pungetti (London: Ashgate, 2011), 5.

Chapter 5: Is Landscape Gendered?

Sonja Dümpelmann

Since the beginning of its widespread use in the English language in the early seventeenth century, *landscape*, like the nonhuman nature it depicted in Western culture, has been gendered. The term visualizes and embodies a system of power relations that are produced and shaped by gender, as well as class and ethnic identities. When the English word *landscape* was first used around 1600 to denote a scenic landscape painting, it described a predominantly male gaze upon the land that objectified female-gendered nature represented in a painting.[1] However, as scholars have shown since the 1990s, this gaze has also been appropriated by other genders. Besides presenting the gendered vision that has been fundamental for the meaning of landscape in the English-speaking world since early modern times, this chapter presents various ways landscape has contributed to forging gendered identities. It exposes how Western ideas of gender have shaped the physical landscape and vice versa.

Landscapes and their formations have been sexualized in many cultures for millennia. Yet, at the time of the early landscape gardens in Britain, which marked the materialization of pictorial landscape representations, these same landscapes began to be designed to express character traits attributed to male and female genders and for the use by these two recognized genders, respectively. This chapter lays out how landscapes have been shaped explicitly to support the social constructions of gender and how the land has been used as a medium and material of gender stereotyping. It also shows how designed landscapes have been both the instigator and result of gendered labor, with women both accepting their relegation to the private sphere and, since the nineteenth century, increasingly realizing the opportunities that landscape labor provided for transgressing the separation between private and public spheres. The question of whether and how female and male design work differ was first approached by feminists of the second wave, leading some to assume a gender-essentialist stance and/or use traditional female values and gender roles to foreground women's culture and spaces. However, while (eco)feminist voices shaped parts of female architectural culture, they were quiet in landscape architecture. The age-old association of nature with the female sex possibly underlies this relative lack of feminist arguments in this profession. The chapter concludes by showing how landscapes have literally become spaces of refuge, retreat, and resistance

DOI: 10.4324/9781003148142-6

for queer persons and how a wilderness aesthetic, which has often been associated not only with marginality and neglect but with empowerment, agency, and ecological health, has been appropriated to signify queer landscapes.

GENDERED VISIONS

In the 1600s, when *landscape* first gained currency in the English language, there were few women among the early landscape painters and even fewer who became famous. Nevertheless, by the time the writer Jane Austin—herself an avid gardener—penned her novels in the early nineteenth century, well-heeled women with easels, brushes, and colors had not only populated the lush British countryside and gardens in Austin's fiction. Landscape and flower painting had become an accepted female pastime in landscaped gardens on landed estates. Perhaps paradoxically, the estates themselves were an outcome of a patriarchal and colonial system that not only disregarded female rights but the basic human rights of many colonized and enslaved peoples.[2] Women adopted patriarchal viewing practices while often using alternative artistic media, for example, textiles.

However, despite the female appropriation of landscape in its long-standing visual sense as something that is "viewed" and represented on canvas, it was the male-coded gaze in Georgian and Victorian Britain (and beyond) that turned female-coded nature into landscape in the first place. Viewing devices such as the Claude glass; representational techniques like Humphry Repton's before and after watercolor paintings that revealed what land could become if his services as landscape gardener were employed; and tools such as the dendroscope[3] invented by French estate owner Amedée de Pérusse des Cars to prune trees into desired shapes, in the eighteenth and nineteenth centuries helped to paint, imagine, and transform landscape visions into three-dimensional space.

Nature, landscape, plants, and flowers have been associated with women, femininity, and female sexuality in many cultures. An explicit reminder is the many allusions found in Italian and French gardens of the seventeenth and eighteenth centuries by way of statues of Venus and Diana of Ephesus, which adorn grottos and fountains. In eighteenth-century English landscape gardens, Venus appeared not only as decorative statuary. Sexualized topography, such as the Venus Mound and Venus Temple at West Wycombe, was used as well. The mound and temple imitated in earth, vegetation, and stone a woman's belly and vagina, eliciting humorous contemporary comments.[4] Landscape gardening had a sexual dimension, given expression not only in the gardens' materiality and physicality but also in the many poems and literary texts written by male authors who conceived of and described the landscape gardens built throughout the century. They were, as literary scholar Carole Fabricant has argued, "statements about how power was conceived and wielded during this period."[5] Even though the landscape and its nature were often personified as a sensuous and active female, it was also seen as a passive material matter subjected to and acted upon by a god-like creator.[6]

The association of nature with women in Western culture is central to historian Carolyn Merchant's influential work showing how a paradigm shift occurred during the seventeenth-century Scientific Revolution, from conceptions and metaphors of the Earth as a nurturing mother to one as a natural resource that can be dominated and exploited.[7] An extension of the scopic regime further supported this early modern shift. In the late eighteenth century, the first manned balloon flights (only a few included women) offered views of the land from above and instigated a scenic shift from horizontal to vertical landscape. Of course, forms of cartography and mapping had, since antiquity, offered distanced overviews. However, direct observation, the perfection of aerial photography, and the development of powered flight in the early twentieth century further strengthened the vertical point of view that many associated with mastery, domination, authority, power, knowledge, and virility. This vertical male view of landscape supported yet another shift in its perception from scenic to systemic. Seeing the landscape from above facilitated its perception as a mechanistic system. If the late eighteenth-century descriptions and representations of landscape seen from above still adhered to the picturesque conventions and aesthetic philosophy of their time, by the early twentieth century, landscape had lost its picture frame and became a science. Photographic technology now offered the ability to simultaneously see landscapes at various scales, in both context and detail, facilitating the evolution of large-scale territorial planning.[8]

Despite the historical association of landscape with the male gaze and "its historical connections with heterosexual male power,"[9] some feminist geographers and art historians have in these last decades sought to deconstruct the often-presupposed gendered differences based upon a polarized distanced male and intimate female way of seeing by uncovering the straight female, lesbian, gay (as well as queer and transgender) "visual pleasure [in landscape]."[10] Given their socialization, women's gaze "c[ould] be transvestite." With their attention to detail and place, including sense perceptions like smelling, hearing, and touching, female artists have often assumed ways of seeing that others have ignored.[11] After all, as feminist cultural geographer Catherine Nash has noted, "Looking is never only or just masculine. . . . Its politics are contextual; there are different kinds of looking."[12]

THE SHAPE OF LANDSCAPE AND SOCIAL CONSTRUCTIONS OF GENDER

Throughout history, landscape has been designed and shaped in ways to support and promote the social constructions of gender, particularly of femininity and masculinity. Landscapes and outdoor spaces have been produced, constructed, and represented to foster gendered identities, and they have been used as medium and material of gender stereotyping. In Western society and beyond, they have also been employed to enforce heteropatriarchal power relations based upon the binary of domineering masculinity and subjected femininity. How patriarchy has shaped the living environment even in the last 100 years despite women's suffrage and recent initiatives to liberalize and decriminalize homosexual and transsexual lives and same-sex partnerships, becomes clear when we consider Ernst Neufert's *Bauentwurfslehre* (1936) and Le Corbusier's *Modular Man* (1940s), and the twentieth design norms that shape the spaces we inhabit. The *Modulor Man*, like its Renaissance and ancient forerunner, the Vitruvian Man, was based upon the proportions and sizes of an ideal (white) male figure that, in Le Corbusier's case, were grounded on subjective preferences. Man was the measure of all buildings, or vice versa; he had to fit all buildings.[13] It was never woman, although Le Corbusier's colleagues demonstrated that female rather than male proportions fit the golden section more accurately.[14] Neufert, who in his *Bauentwurfslehre* included women's abstract outlines among the human figures, animating the line drawings to illustrate his standard measurements, relegated them to the spaces and objects of domestic labor (Figure 5.1).[15]

Space allocation and distribution, as well as the form and materiality of design, have played important roles in the construction of gendered landscapes, which in the past were limited to accommodating male and female bodies. Bodies and gender identities beyond this binary had and still often have no official place, suffering injustice, disenfranchisement, and incrimination.

An example of the conception and enforcement of gendered spheres on the land are the beaches of early European coastal resorts. As soon as the first resorts were established at the end of the eighteenth century and bathing became popular among members of the growing middle class, beaches were separated into zones outfitted for use by men, women, and children. On the German North Sea island of Borkum, at high tide, the Atlantic took over the entire beach in equal measure, whereas at low tide, piers established the dividing lines (Figure 5.2). A series of steps formed the main beach access from the nearby village, leading onto the "so-called neutral beach" with a pier for sailing boats. A large "gentlemen's bathing beach" and a smaller "ladies' bathing beach" were located at opposite ends of the bathing area. Located between them but nearer to the ladies' facilities was the "children's bathing beach." In addition to the spatial separation, mobile bathing machines and separate washing and restrooms in each zone ensured that nineteenth-century gender norms, propriety, and sexual morality were maintained and enforced.

Figure 5.1 The female figure outlines are used to indicate measurements of kitchen space in Neufert's *Bauentwurfslehre*.
Source: Ernst Neufert, *Bauentwurfslehre* (Berlin: Bauwelt, 1936), 102.

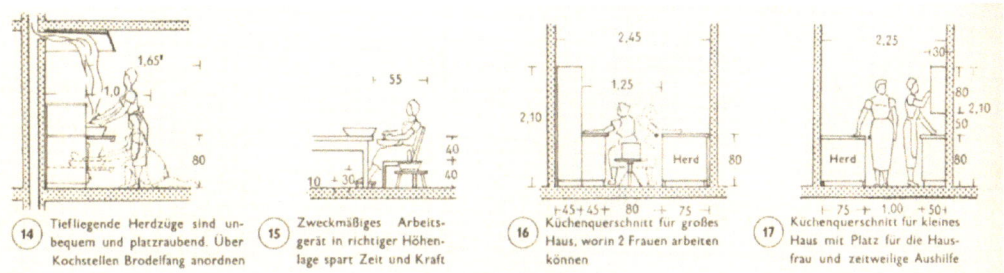

14 Tiefliegende Herdzüge sind unbequem und platzraubend. Über Kochstellen Brodelfang anordnen

15 Zweckmäßiges Arbeitsgerät in richtiger Höhenlage spart Zeit und Kraft

16 Küchenquerschnitt für großes Haus, worin 2 Frauen arbeiten können

17 Küchenquerschnitt für kleines Haus mit Platz für die Hausfrau und zeitweilige Aushilfe

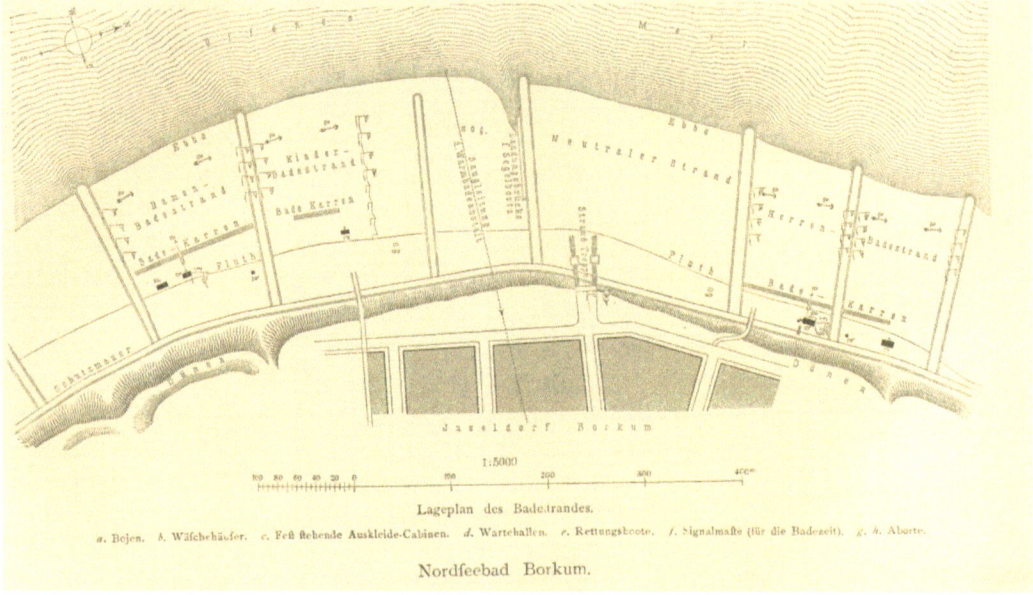

Figure 5.2 Gendered division of a beach on the North Sea island, Borkum, Germany.
Source: Felix Genzmer, *Bade- und Schwimmanstalten*, 5. Halbband, Heft 3 of *Handbuch der Architektur* edited by Josef Durm (Stuttgart: Bergsträsser, 1899), Fig. 124.

Besides the inscription of gender into existing landscapes, new landscapes were designed and constructed for gendered uses. As elsewhere in the Western world, in the nineteenth-century United States, aesthetic and gender ideology were interconnected. Landscape gardener Andrew Jackson Downing associated the aesthetic concept of the beautiful with character traits attributed to women and the picturesque with those attributed to men. He illustrated these select concepts of British eighteenth-century philosophy and presented two landscape scenes that included female and male figures as visual cues, respectively (Figure 5.3). In the beautiful scene, lawns and carefully manicured yet naturalistic-looking plantings of deciduous trees and shrubs provide the setting for two female figures (a mother or nanny and a young girl) walking along a curved pathway toward an Italianate mansion with a curved portico. The suggestive air of innocence emanating from this light, balanced, and blissful landscape scene with its gracefully flowing lines following artist William Hogarth's "line of beauty," contrasts with the darker picturesque scene that appears rugged, rough, rustic, even wild, suggesting adventure and strength. The picturesque is characterized by irregular shapes, sharp angles, and abruptly broken contours formed by evergreen trees in the distance and an old, declining canopy in the foreground. A lone male hunter carrying a gun and accompanied by a jumping dog walks through dense shrubbery toward a gothic-style mansion.

Literate in reading the eighteenth-century British landscape, designers in Europe and the United States soon produced beautiful and picturesque scenes in the early public urban parks. On occasion, the illustration of specific park views continued the association of the beautiful with femininity, as can be seen in the scene of the Mare d'Auteuil in the Parisian Bois de Boulogne produced for

Figure 5.3 Landscapes illustrating a beautiful or graceful scene (top) and a picturesque scene (bottom).
Source: Andrew Jackson Downing, Treatise on the Theory and Practice of Landscape Gardening Adapted to North America (London: Wiley and Putnam, 1844), 55.

LES PROMENADES DE PARIS

BOIS DE BOULOGNE — MARE D'AUTEUIL

Figure 5.4 A beautiful or graceful scene of the Mare d'Auteuil in the Parisian Bois de Boulogne.
Source: A. Alphand, *Les Promenades de Paris*, vol. II (Paris: J. Rothschild, 1867–1873).

Adolphe Alphand's *Les Promenades de Paris* (1867–1873). In the middle ground, a woman and a girl are seen walking along a curving lakeside path (Figure 5.4).[16] Together with other figures in the distance, they reveal something else: as Galen Cranz has shown for North American parks in this time period, women's presence, often together with their husbands and children, signified social stability and civilized space.[17] The Victorian gender ideology of separate spheres that was based upon gender essentialism—the belief in the differences between male and female character and in distinct fixed intrinsic qualities—was assigned to women for moral guidance. For the middle class, public urban parks became an extension of their private living rooms and gardens.

Nevertheless, women and girls quickly began to use parks independently, not only for passive recreation but also for active play and diversion. In the second half of the nineteenth century, a paradigm shift occurred, away from Frederick Law Olmsted's nineteenth-century conception of parks as scenery for passive or "receptive" recreation toward parks for active recreation. Although the intentions behind this shift focused particularly on boys and men, seeking to reduce delinquency and alcoholism, girls and women also profited. Finally, they were treated as "users in their own right" and not so much "as instruments through which to influence men's behavior."[18] This change had consequences. Had the parks' use for passive recreation by entire families still been relatively integrated (albeit mostly limited to the conventional middle class), their recreational use now enforced sex and age segregation.[19] Gendered landscapes with different forms and materials were the result.

A particularly clear example is the first US public open-air gymnasium, the Charlesbank playground, built along Boston's Charles River in the 1880s (Figure 5.5). Instigated by the women's Massachusetts Emergency and Hygiene Association and designed by Olmsted's firm, the playground was built to provide active recreation in a poor part of the city.[20] A men's and a women's open-air

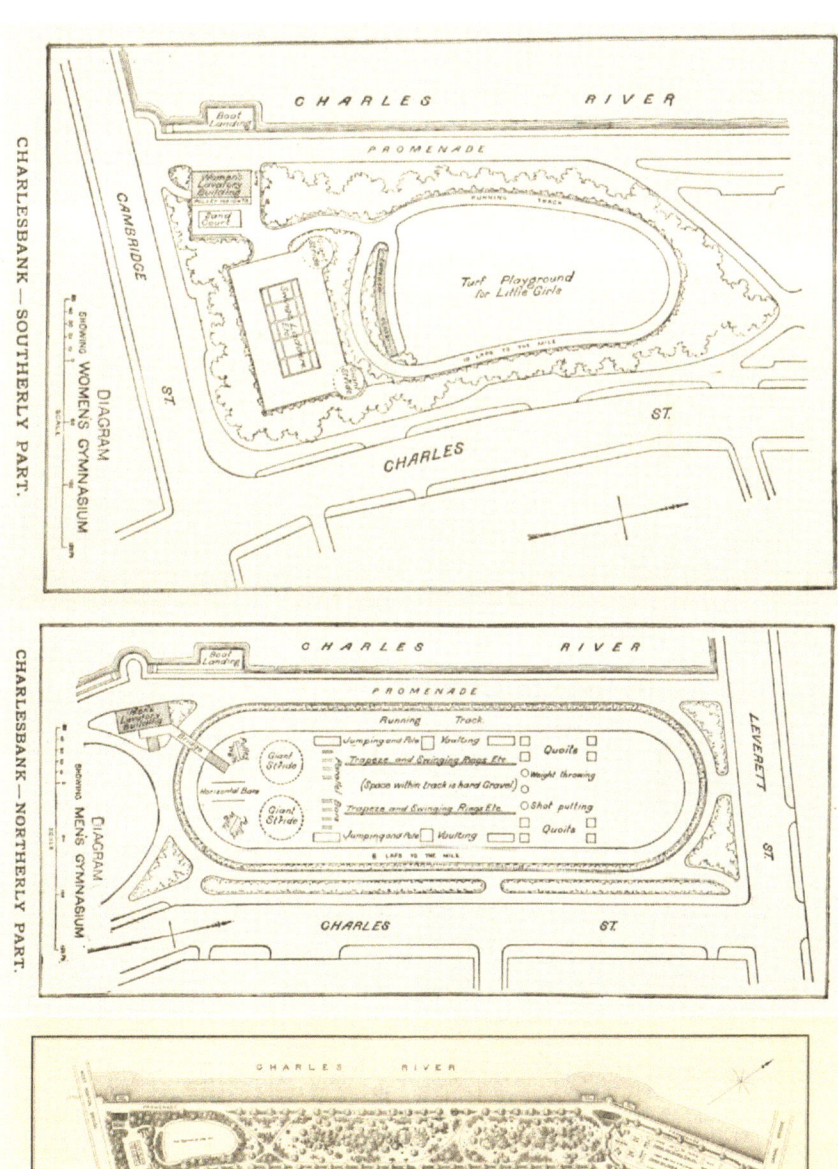

Figure 5.5 The Charlesbank playground was designed by F. L. Olmsted Landscape Architects and built along Boston's Charles River in the 1880s. Source: Top: The women's playground (Sylvester Baxter, *The Boston Park Guide* [Boston: Small, Maynard & Co., 1898], 36). Middle: The men's playground (Sylvester Baxter, *The Boston Park Guide* [Boston: Small, Maynard & Co., 1898], 34). Bottom: Comprehensive design (Olmsted Plans and Drawings Collection: Olmsted Job #907, Charlesbank, Boston, MA (plans); Olmsted Plan #907-z22, Plan of Charlesbank 1892; December 1892. [Courtesy: United States Department of the Interior, National Park Service, Frederick Law Olmsted National Historic Site]).

gymnasium were situated at opposite ends of a long plot of parkland. The men's gymnasium consisted of a cinder racetrack for competitive running and cycling (six laps to the mile) circumscribing a hippodrome-shaped area of rolled gravel, resistant to heavy and intense use by vaulting, jumping, weight throwing, shot putting, playing the game of quoits, as well as exercising on pulley weights, horizontal bars, trapezes, poles, and giant strides. Enclosed by turf, shrubs, and an iron fence, it allowed "a good view of the interesting and animated spectacle from outside."[21] In contrast, the smaller women's gymnasium with a children's crèche and sand court was "surrounded by a dense growth of shrubbery to screen them from public gaze, and provide the seclusion desirable for the sex that uses them."[22] It was divided into two parts: an area enclosed by wooden fencing and equipped with swings, giant strides, ropes, and other climbing apparatuses (pulley weights were discretely allocated along the side of the lavatory building next to the sand court so they could be operated while observing children); and an irregularly shaped, soft lawn surrounded by a sinuous dirt track, ten laps to the mile.

Form, material, size, and spatial character were enforced and determined by contemporary gender conceptions, norms, and expectations. Women and girls were generally considered to be less disciplined and more prone to playful movement, dancing, and singing than to systematic, rigorous exercise and competition.[23] The Charlesbank playground's gendered design was the product of these ideas, which also informed the designs the Olmsted Brothers offered for small neighborhood park playgrounds as part of Chicago's widely admired South Park System in the early 1900s.[24]

GENDERED LABOR IN THE LANDSCAPE

Although it was, for example, the women's Massachusetts Emergency and Hygiene Association that instigated the establishment of the Charlesbank playground in the first place, almost 100 years later, in 1980, historian Galen Cranz described female passivity, the fact that women have seldom advocated on their own behalf and that they have not constituted "a social problem" as being among the root causes for women's repeated instrumentalization by (male) city leaders to help solve urban problems.[25] However, despite this valid proposition, women, mostly under unequal conditions to their male colleagues, have been active and often very effective forces in shaping our built environments. As the Charlesbank playground and the following cases make clear, women have been significant agents in the gendered landscapes of labor for millennia as activists, workers, and designers. While they have often accepted the relegation of their roles to the private sphere, they have also actively transgressed the separation of private and public spheres.

As reported by Pliny the Elder in his Natural History, the cultivation and maintenance of vegetable and kitchen gardens during the Roman Republic and Empire belonged to women's duty. If such a garden was negligently cultivated, it was considered "a sign of a woman being a bad and careless manager of her family."[26] Closely related to the private domestic sphere, kitchen, herb, and flower gardens—in Britain, often comprehensively called cottage gardens—have been assigned to women throughout history. In one of the earliest English gardening handbooks for "country housewives," published in 1618, William Lawson advised them to establish a flower or summer garden and an herb or kitchen garden.[27] Market gardening was an occupation of both genders, but especially for country women.[28] Vegetable and kitchen gardens gained particular importance during times of crisis. Victory and war gardens established in the United States and many European countries during the First and Second World Wars are more recent examples in which women on the home front (together with children) "enlisted" in their respective country's efforts to secure self-sufficient food provision.

In contrast to domestic kitchen and vegetable gardens, public-facing plant nurseries and seed businesses were mostly run by men, even though their wives often supported or continued to run the business when their husbands fell ill or died, and it was often women gathering the seeds.[29] Weeding occupied a central position among the gardening tasks considered lighter or of a lesser kind, as it was deemed appropriate for the delicate female physiognomy, no matter how vernacular or noble the context. The housewife should weed her garden "as oft as needed," Anthony Fitzherbert cautioned in 1534, and Lawson noted that "The skill and pains of weeding the Garden with weeding knives or fingers" should be undertaken by the mistress herself, or by her well-instructed maids after a rain shower.[30] Estates and the court employed so-called "weeding women," and in the cherry garden at the estate of the Duke of Bedford at Woburn, a statue was dedicated to "an old weeder woman."[31] They were paid modest sums to weed and "clean up" gardens and provided with free food and drink.[32] These female occupations implying care and cleanliness continued in new forms as part of the capitalist economy in nineteenth and early twentieth-century industrial cities, where women were employed in street cleaning, for paper trash collection and the abatement of insect pests in public urban parks (Figure 5.6).

Flowers not only played a special role in female garden preoccupations because of their signification of beauty and fertility. The culture of flowers and their gardens were central to "female horticulture" and garden design because they were most likely to provide diversion and pleasure, as Charles Evelyn advised in his 1717 *Lady's Recreation*.[33] A century later, Jane Loudon explained that they implied "the lightest possible kind of garden labour."[34] In Victorian England, female pride and the protestant work ethic, however, forbade that any labor should be too light. The culture of flowers could be construed so as not to run this risk. As Loudon pointed out, "The culture of flowers is exactly in the happy medium between what is too hard and what is too easy."[35] Additionally, flower gardening was simply rendered as an occupation that provided pleasure to women and others,[36] and it, therefore, fit the image of benevolent and acquiescent female service. Yet, at times, flower symbolism, which also implied immoral

Figure 5.6 Two women collecting caterpillars of the pine-tree lappet moth in Berlin's Tiergarten, 1905. Photo by Otto Haeckel. © akg-images

sexuality and sin, cast female labor revolving around flowers in an ambiguous light. Furthermore, as historian Elizabeth Hyde has shown for seventeenth and eighteenth-century France,[37] male gardeners realized the distinction that could be gained through floriculture (Figure 5.7). Thus, for a time, they intruded upon flower gardening, which was previously and thereafter again considered "pre-eminently a woman's department."[38] Nevertheless, flower gardens to embellish homes and produce floral decorations have been a mainstay in women's activities shaping the environment. Early modern female flower sellers in France, the so-called *bouquetières*, in 1677 even formed their own guild,[39] and at the seventeenth-century French court, women worked on the floral orna-mentation of costumes and indoor decorations (Figure 5.8).[40] These decorative occupations were not the only horticultural realm women could enter at the French court. Although the estates of the European nobility were the domain of male court gardeners and the occupation was hereditary—its craft typically being passed on from father, or uncle, to son, or nephew—some French women replaced their deceased husbands and fathers. For example, Françoise Bouchard, André Le Nôtre's sister-in-law inherited her husband's position serving as orang-ery gardener in the Tuileries Garden after his death, and the daughters of court gardener François Le Juge, Elizabeth and Claude, directly inherited their father's position at the Tuileries Garden.[41] To be prepared for these situations, the eminent Jean-Baptiste de La Quintinie, kitchen gardener to Louis XIV at Ver-sailles, advised his colleagues to choose women as wives who would be capable and willing to assist their husbands and substitute for them if they fell sick or had to travel.[42] However, these French cases were exceptions.

Figure 5.7 Representation of a male "art and flower gardener" (Kunst und Blumengärtner), or "gardener of the parterre" (jardinier de parterre), as he is titled in French, with tools, plants, and a propped-up design for a parterre.
Source: Martin Engelbrecht, *L'Assemblage nouveau des manouvries habilles* (Augsburg, ca. 1730), plate 49.

Female empowerment only began to reach middle-class gardens in the nineteenth century, for example, through Jane Loudon's 1840 *Instructions in Gardening for Ladies*. Loudon was well aware "that knowledge is power,"[43] and her work exemplified a tactic female garden writers would use to reach a female audience well into the twentieth century.[44] She prefaced her book by attributing all her knowledge to the teachings of her husband, the well-known landscape

Figure 5.8 The title of this engraving—"flower gardener" (Blumengärtnerin), or "gardener for the parterre" (jardinière pour parterre)—suggests a female counterpart to the male art and flower gardener. However, the illustration shows a bouquetière, thus relegating women to an occupation in the shadows of their male colleagues. While the male art and flower gardener was the parterre gardener, she was the gardener *for* (pour) the parterre, supplying it with plants besides providing floral decorations to noble interiors and costumes.
Source: Martin Engelbrecht, *L'Assemblage nouveau des manouvries habilles* (Augsburg, ca. 1730), plate 50.

gardener and author John Claudius Loudon. She also made sure to stress her initial ignorance on the subject using a modest, subservient, even self-deprecating tone.[45] This could simultaneously appease male critics and appeal to female readers. Loudon did not spare detailed advice regarding both more and less physically demanding gardening operations, but accounting for the differences between the male and female physique, she introduced "female" tools to facilitate tasks like digging and pushing the wheelbarrow. Her "lady's spade"[46] was lighter than common models because its blade was "made of not more than half the usual breadth. . . of smooth polished iron," and its handle was "smooth and. . . slender," made of light willow rather than of the heavier ash usually used for gardeners' spades.[47] She advised that the "lady's wheelbarrow"[48] should have long and curved handles so that less strength was required to use it.[49] Acute instructions also include thematized postures and movements during gardening operations, as well as the necessary clothing. Clogs or strong shoes with iron plates under the sole belonged to a female gardener's outfit, as did "stiff leathern gloves. . . to protect her hands, not only from the handle of the spade but from the stones, weeds &c" (Figure 5.9).[50] Protection would also have been necessary when pruning. Described by Loudon as appearing "at first sight, a most laborious and unfeminine occupation,"[51] she was quick to divulge pruning as one of the easiest gardening operations overall. Nevertheless, a century later, the profession of pruner and climber was still out of reach for women.[52] Loudon

Figure 5.9 Gardening tools for women. Sources: Top: Jane Loudon's "Lady's Wheelbarrow" illustrated in German landscape gardener Ferdinand Jühlke's *Gartenbuch für Damen* (Berlin: Gustav Bosselmann, 1857), Fig. 1. Bottom: Lady's protective leather gloves (Mrs. Loudon, *Instructions in Gardening for* Ladies [London: John Murray, 1840], 10).

A LADY'S GAUNTLET.

was probably not yet thinking of scaling trees for this purpose. However, it has taken until recently for special tree-climbing courses for women to be offered by arboricultural institutions with the intent to dismiss misconceptions and diversify the field. Women—especially African American women—have long been discriminated against in urban arboriculture, contributing to the unjust distribution of urban trees, as Nik Heynen has demonstrated for Milwaukee.[53] Yet, in nineteenth-century Britain, the only garden work Loudon considered too strenuous for women was the mowing of lawns, regardless of whether this was done with the recently invented hand-driven lawnmower or a scythe.[54] Years before, her husband had described scything as an operation "requiring great force, and also twisting motion of the body" that brought "almost every muscle into action, and is. . . one of the most severe [operations] in vegetable culture."[55]

Despite Jane Loudon's pioneering emancipating ideas, in 1936, Neufert offered dimensions for a uniform spade and shovel (for male garden workers),[56] and the lack of agricultural tools and machinery appropriately designed for use by women has only recently led to the production of a series of "hergonomic" tools. Based on a brief online survey in 2008, the series began with the development of a hybrid spade/shovel.[57] While the need for differently designed tools and machinery is ubiquitous, it is especially pertinent in developing countries where predominantly women tend to fields and crops with few basic tools designed by and for men. Sexism oftentimes also denies women access to any kind of motorized machinery so that once mechanization does occur, female occupations are frequently taken over by men, pushing women back into the menial and often more laborious work.[58]

Back in the industrializing European world of the nineteenth century, middle-class women hardly got their hands dirty despite Loudon's encouragement, and only a few male contemporaries accepted any female heavy lifting in the garden. Saxon gardener Hermann Jäger prefaced his 1871 book *Frauengarten* (Women's Garden) by stating that garden books written for men were rarely suitable for women as they included too much specialized knowledge. His book, therefore, avoided anything that women "do not want to know, do not need to know, because it does not befit them."[59] Belittling ladies' garden tools and clothing as playful English fashion fads, Jäger maintained that there were enough graceful garden tasks befitting women that did not require heavy lifting and exertive activity. Although digging, hoeing, weeding, and carrying loads in the fresh air were increasingly attributed to health benefits and ultimately would become an accepted way for middle-class women to exercise in the turn-of-the-century German air baths,[60] Jäger considered them harmful for women, especially due to the required body postures.[61] While he certainly had a point regarding bodily attrition in the agricultural and horticultural occupations (in general); similarly patronizing views were expressed in the horticultural and landscape gardening literature throughout the late nineteenth and early twentieth centuries. As women became increasingly vocal and independent, they also came to be seen as a threat to what had previously been exclusively male labor markets. First, deriding the women training and working in the early horticultural schools,[62] many men watched with disdain as women increasingly took on design commissions. Even if male mentoring facilitated many a woman's entry into the young profession in the first place, discrimination prevailed.

In 1892, Boston's Arnold Arboretum director Charles Sprague Sargent questioned women's abilities regarding any activities beyond the small flower garden. For him, landscape gardening on a large scale was a "masculine art" requiring "a certain manly vigor of treatment, an unhesitating despotism, that the gentler sex deprecate as cruel and unnecessary."[63] Similarly, in 1894, the young Charles Eliot considered the new landscape profession to be an "essentially virile and practical. . . art" that had to "be founded in rationality, purpose, fitness."[64] In Germany, the First World War created contingencies of crisis that led to the (temporary) increase of female labor in landscape gardening and horticulture and its discussion by male professionals. In 1917, one year before German women received suffrage, the "women question" (Frauenfrage) was a topic at the 29th general assembly of the German Society for Garden Art (Deutsche Gesellschaft für Gartenkunst e.V.). Even if female occupation in the field was welcomed, at least temporarily and out of necessity, most men voicing their opinions wanted to keep women in their places. One of the arguments was the poor and inferior quality of female training at the women's horticultural schools. Some men deemed it sufficient that women would learn about gardening at the housekeeping schools or that they merely gardened for amusement. Others only wanted women in the laboring class to receive practical horticultural training. Female drawing and design education were considered secondary or even questionable. In short, to these men, women were of no importance in the profession and had to be kept out. In contrast, the Society's director, Carl Heicke, considered female professional occupation and participation necessary to gain broader societal support for garden art and culture. After all, he argued, horticulture was a natural fit for women.[65]

Meanwhile, many women remained unfazed by the landscape patriarchy, its discussions, and old-boy networks. They taught themselves the necessary skills, founded horticultural schools, wrote books and articles, helped each other, and collaborated, literally carving out a landscape for themselves.[66] They also became leaders in furthering landscape and environmental matters within the political sphere. In many cases, the purported "natural fit" worked for the women's own benefit. Landscape was not only a challenge but also a chance. In the mid-nineteenth-century United States, women founded the first village improvement societies, which engaged in what was then called "municipal housekeeping," i.e., the extension of female domestic care duties to the public realm. By 1891, what was recognized as the first women's gardening society, the Ladies Garden Club of Athens was founded; the first African American garden clubs were established in the 1930s. Already around the turn of the twentieth century, women had increasingly become active in landscape conservation efforts outside and inside cities. They used various initiatives focused on landscape, including the creation and protection of parks and playgrounds as well as urban tree planting campaigns, to transgress the separation of private and public spheres. Landscapes, in the form of plants, gardens, and parks, became a means and material for this transgression and a symbol of empowerment, emancipation, and even resistance.[67]

FEMALE AND ECOFEMINIST LANDSCAPES

Throughout the twentieth century, feminists have entertained the idea that female and male design work differ and that they are expressed in different design approaches and forms. The work of the pioneering American landscape architect Marjorie Sewell Cautley has been referenced as a case in point. Trained at Cornell in the 1920s and 1930s, Cautley collaborated with architect Clarence Stein and landscape architect Henry Wright on housing projects in New York City and New Jersey. Based on her own experiences as a housewife and mother, she paid special attention to design details that could have a significant impact on female lives. In Radburn, New Jersey, where she designed the transitional spaces between the suburban homes and the communal park spaces forming the interior of the neighborhood pods, Cautley determined the height of the hedges delimiting gardens so that they would provide spatial seclusion near the house without obstructing views of the communal children's play spaces, thus facilitating mothers' supervision. In her designs, productive gardens and orchards often crossed private plot boundaries to encourage communal lifestyles and mutual support between families.[68]

Concerns like Sewell Cautley's and the creation of "female landscapes" became the subject of discussion and critical scrutiny during the second wave of feminism. In 1984, some of Cautley's principles reappeared in urban historian Dolores Hayden's suggestions for the reorganization of suburban neighborhoods.[69] Hayden was hoping to inspire "more satisfactory patterns of housing, work, and family life in the United States. . . where the employment of women. . . has created. . . strains concerning outworn patterns of private and public life."[70] Although her cursory yet directed attention to Radburn neglected Cautley and the intricacies of her landscape design, Hayden recognized that future suburbs and housing projects needed "a sophisticated landscape plan for the whole block."[71]

As part of the second wave of feminism, the feminist movement in architecture intersected with the environmental movement and its concerns about environmental destruction and pollution. Assuming an ecofeminist stance, the purported female way of design and building, which was declared as more holistic, organic, flexible, and slowly growing, also received serious attention from men. Curvaceous earth and eco-architecture that blended with or even grew out of the landscape were among the architectural shapes promoted not only by feminist architects at the time. However, a century after the design of the first separate gendered spaces for male and female recreation, feminist activists and architects expressed their belief that men and women tended to relate to architecture with "different values." The German architect Margrit Kennedy was among the women who saw "a growing discrepancy between real social and psychological needs and the planned and built environment" as a result of building and architecture in the Western world lying firmly in male hands.[72] While Kennedy admitted that the shape of a purportedly "female architecture" was uncertain, there were, she argued, examples of the difference between environments shaped by men and male values and those shaped by women and female values. Rather than speaking in categories of male vs. female architecture, she preferred to differentiate between the "male and female principles in architecture"[73] that anybody could use, thereby softening binary distinctions. As the Italian feminist architect Paola Coppola Pignatelli[74] Kennedy maintained, women often design "from the inside out, beginning with the functions of a building and ending with its form."[75] Men frequently proceeded in the opposite way. Women were more likely to build with the user in mind and produce "ergonomic" as well as "anonymous architecture," whereas men were more fixated on representation and considered buildings monuments to the respective designer. Although seeking to steer clear of biologistic causation, Kennedy gave examples of oval, "rounded," incrementally developed "organic" architectural forms designed by women and of "male architecture" characterized by straight lines, right angles, and systematization. Translated into urban design, she found male principles dominant in Ludwig Hilbersheimer's streamlined proposal to redesign Chicago's Northside, which she described as a monofunctional faceless suburb characterized by order and boredom.[76] In contrast, she pointed to the "aesthetically and socially satisfying" labyrinthine town plan of the ancient Greek town of Hydra, where "male and female principles, even if weighted differently, had not yet been isolated from one another through specialization."[77]

Kennedy attempted to find a place between gender essentialism, the view that male and female behavior were biologically determined, and cultural feminism, which is the revaluing of traditional female values and gender roles to foreground women's culture and spaces. Her ideas reflected the messages published by some of her US-American colleagues, among them the feminist lesbian architects Noel Phyllis Birkby and Leslie Kanes Weisman.[78] During workshops Birkby and Weisman conducted across the United States beginning in 1974, they collected women's drawings of their "environmental fantasies," which were largely characterized by rounded and adobe shapes, sometimes explicitly womblike.[79] Birkby and Weisman encouraged women to fantasize and imagine the environments they would like to inhabit, to free themselves from being conditioned to adapt flexibly to the spaces created by and for men.[80] As Weisman firmly stated, "We will not create a new and integrated environment until our society values those aspects of human experience that have been devalued through the oppression of women."[81] Weisman and her colleagues fought for "women's environmental rights," which they saw suppressed by "the man-made environments which surround us reinforc[ing] conventional patriarchal definitions of women's role in society and spatially imprint[ing] those sexist messages on our daughters and sons."[82]

Together with architects Katrin Adam, Ellen Perry Berkeley, Bobby Sue Hood, Marie I. Kennedy, and Joan Forrester Sprague, Birkby and Weisman also ran the Women's School of Planning and Architecture (WSPA), an annual two-week program of seminars for students and professionals to empower women and "seek a clearer understanding of what women can contribute to making the profession more socially responsive."[83] Founded in 1974 by women for women in response to the first conference on women in architecture held in the United States, its workshops highlighted individual personal experiences and took place in different locations across the country from 1975 until 1981.[84] Despite these initiatives to overcome the dualisms of public/private, city/suburb, work/home, and production/reproduction, in 1989, environmental psychologist Karen A. Franck, in a brief survey of previous feminist thinking in architecture and related fields, could only speak of "a fledgling effort to outline a feminist approach to architecture."[85]

Although they were concerned with the built environment at large, it was mostly female *architects* and urban and architectural historians who voiced their feminist concerns and visions. At first sight, landscape architects and historians appear almost absent from these discussions. For example, when the journal *Heresies* published a 1981 special issue on women and architecture, Diana Balmori was the only trained landscape architect among the 43 contributors. Presenting research on the pioneering American landscape architect Beatrix Jones Farrand,[86] she argued that because Farrand had scarcely published, her work had remained little known. Balmori concluded: "Trapped in the private sphere, women and their work disappear and remain unacknowledged, ineffective."[87] The only other contribution dealing explicitly with landscape design was made by architect-activist Sharon Sutton, who would go on to become the first African American woman to be promoted to full professor in an accredited architecture program at an American university. Sutton's design for the Street Museum, a park in the Shippen-Locust neighborhood of Lancaster, PA, was a vacant-lot rehabilitation project that exhibited many of the character traits thought to be representative of a female practice. They included citizen and children's participation, the use of recycled local materials, and a focus on the participatory design process rather than its end-product.[88]

A first selective survey of the second generation of women in landscape architecture confirms the relative lack of feminist voices in the field during the second wave of feminism.[89] Although this preliminary observation warrants further research, we can speculate on some of the possible reasons. First, landscape architects' professional concerns dealt with the environment and nonhuman nature in the first place—they already employed some of the principles that feminist architects labeled as "female." Many women landscape architects would have considered their practice as a cultural activity that worked *with* nature rather than *against* it. In an ethically sound landscape practice, some of the binaries preoccupying architects were, therefore, already dissolved. Second, from the beginning, women had perceived landscape architecture quite literally as a space of opportunity, not least given the long tradition of female domestic gardening and the initial opportunity to enter a young profession without formal training or degree. Third, the youth and small size of the profession, as well as its obscurity paired with a relative lack of self-reflexivity in its early years, may not have been conducive to the development of an independent feminist practice or stance. A fourth reason could be the deceptiveness of the landscape medium itself. Landscape architecture deals with *open* space, or so it seems. At first sight, the gendered delineation of space within buildings (kitchen, "master bedroom," "powder room") can appear clearer than any spatial divisions outdoors, which often seem more porous and deceptively open to all.

Nevertheless, there would have been (and still are) good reasons for a feminist position in landscape architecture, not only given women's discrimination in the public urban realm and their historical relegation to suburbia. As in architecture, the low number of women in landscape architectural practice in the postwar United States was related to a masculinist culture. During this time period and until 1973, less than five percent of ASLA members were women, a number that had not increased since the 1930s.[90] As Thaïsa Way has suggested, several factors played a role in female invisibility and the decreased female presence in the profession in the years after the Second World War. There was an increasing interest in modernist formalism that negated the preoccupation with the profession's horticultural roots, its gradual affiliation with engineering and architecture, and other changes in professional culture, like the introduction of the corporate office.[91]

QUEER LANDSCAPES

Synchronous and partly intersecting with the second wave of feminism, the Gay Liberation Movement and later the LGBTQ rights movement have advocated for LGBTQ rights and have, among other things, finally legalized same-sex marriage in various countries beginning in 2001. Although opposition against LGBTQIA+ movements and the (often violent) discrimination of non-cis-heterosexual people remains widespread, in many parts of the world, public tolerance of different sexual orientations and (public) practices has increased in the last decades. For example, like numerous other queer internet guides for specific places, the website Queer Europe lists cruising areas in various European cities where both the public and police tolerate cruising activities. It notes that due to competing uses and less public tolerance, safety and privacy for cruising gays can be compromised.[92] After all, these spaces are sites of resistance to homophobic repression, which still exists in large parts of society. As many other spaces such as beaches, dance halls, saloons, and public restrooms (often verbally disguised as "tearooms"), which historically have played an important role in gay culture, parks have been sites for queer persons to meet and socialize, regardless of whether this included sexual activity.[93] However, unlike many other spaces, e.g., restaurants and cafeterias frequented by gay middle-class men, or saloons which in the late nineteenth and early twentieth centuries catered predominantly to the working class, parks were more democratic or porous in that they allowed queer persons of different classes if not ethnicities to meet. They were also more difficult for the police to surveil and control than indoor spaces.[94]

Landscapes used for gay cruising are often characterized by their marginal urban locations (or their marginality within public parks),[95] abandoned buildings, and by an overall seemingly "wild," uncanny, neoromantic, or even picturesque appearance. Stretches of more open, low vegetation alternate with woody areas, often with heavy undergrowth. In contrast to the pastoral scenery and clean-cut geometries of curated urban parks designed and maintained for more or less-programmed heteronormative uses, many queer landscapes exhibit a wilderness aesthetic resulting from neglect, spontaneous plant growth, high biodiversity, and pathways running along desire lines. They allow for spontaneous and unprogrammed use, which can be shaped by a "queer ecology" and by what cultural geographer Matthew Gandy has called "heterotopic alliances."[96] These describe the convergence of different interests in the same unprogrammed, un-curated, and "run-wild" public urban spaces, or heterotopias. For example, the spaces sought out by cruising gay men can be the same as those that urban ecologists seek out to study spontaneous urban wildlife and flora.[97] "Queer landscapes," in this sense, provide as much a safe haven for gays cruising as they offer a safe space for endangered animal and plant species inhabiting urban areas. The ailanthus, a pioneer tree species that can grow profusely on wastelands and neglected urban areas but is often displaced once urban development takes over, has, for example, been employed as a metaphorical stand-in for queer culture in urban space.[98] On occasion, highly curated sites like the rose garden in London's Hyde Park can become queer landscapes as well. According to Queer Europe, its "benches, quiet spots, and hidden spots. . . facilitate intimate encounters between men"[99] after dark when the daytime crowds have left. Landscape form and character, as well as its context, therefore, are not negligible when it comes to accommodating some gay men's interests. However, as artist Jean-Ulrick Désert has noted, "the definition of queer space by erotic program would be as limiting as the word *homosexual*."[100]

Gordon Brent Brochu-Ingram has noted that the privatization of public open space and its site design and management have been employed to "discourage contact and queer placemaking."[101] In fact, "landscape architecture. . . is too often used to exacerbate inequities [regarding sexual minorities] under the guise of balanced management and fiscal restraint."[102] New York City's High Line is an example of gay and green gentrification implicated in a mainstream homosexual culture that excludes other forms of queer culture previously existing on the site.[103]

Nevertheless, the vegetation and design of open space have played, and play, an important role in the production of queer landscapes in the first place and for their imagination and use by gay men and lesbian women.[104] Secluded spaces for private retreats and intimate encounters often stand at the center of these landscapes, and their creation and defense can be an act of resistance against heteronormativity. To this effect, in 1969, the first gay liberationist environmental group, Trees for Queens, formed in New York City to restore a park in Kew Gardens. Only a week before the Stonewall riots, trees and shrubs had been cut down by a citizen "vigilante committee" to discourage cruising in the park.[105]

There are forerunners of queer landscapes as well. Literary scholar Lisa L. Moore has suggested that the eighteenth-century English artist Mary Delaney employed landscape design, including grotto shellwork—partly developed in collaboration with her close, longtime friend Margaret Harley, the Duchess of Portsmouth—to express same-sex erotic desire and to create actual "spaces for female eroticism and intimacy."[106] For their male contemporaries, the various, more secluded features in the eighteenth-century landscape gardens also provided "privacy for voyeuristic delights and sexual fantasies"[107] of various kinds, which perhaps first and foremost found expression in their poetry and writing that in many cases followed the classical homoerotic pastoral literary genre exemplified by Catullus, Sappho, and Virgil and others.

In the nineteenth-century urban environment, curated "wild urban woodlands" like the Ramble in Central Park and special garden types such as the late nineteenth-century wild garden with their picturesque aesthetic may have offered similar attraction to the gay and bisexual urban subcultures of their time. At least as early as the turn of the twentieth century, several locations in Central Park, including the Ramble, became meeting areas for gay men (and in 1955, Robert Moses therefore suggested turning the Ramble into a recreational center for seniors).[108] The Ramble's designer, Frederick Law Olmsted, the American nineteenth-century master of picturesque and pastoral scenes within urban environments, had an eye for the beauty of spontaneous plant growth and, in 1882, declared that "neglect, if it continues not too long, may even have its advantages."[109] However, Olmsted himself can hardly be described in terms other than a white cis heterosexual male whose interest lay in conservative social reform. In contrast, his Irish colleague William Robinson, who popularized the idea of the wild garden in Britain, fashioned himself as a dandy. As a private person with one romantic engagement at the beginning of his career "to a young woman. . . who jilted him for a prosperous tradesman," Robinson "did not encourage intimacy," one of his first biographers wrote in 1951.[110] Later in life, Robinson, on occasion, posed lounging in dandy attire among tufted pansies—his favorites—on the lawn in front of his home, Gravetye Manor (Figure 5.10).[111] "Perhaps he preferred plants to people," the same biographer suggested.[112] In the early twentieth century, plants and their flowers provided Vita Sackville-West with verbal metaphors to express feelings for her lesbian lovers, and the gardens she built together with her husband Harold Nicolson at Sissinghurst more or less inadvertently disguised their unconventional marriage.[113] Throughout human

Figure 5.10 William Robinson lounging in dandy attire among tufted pansies in his garden at Gravetye Manor, West Sussex, England.
Source: Mea Allan, William Robinson 1838–1935: Father of the English Flower Garden (London: Faber and Faber, 1982), 156.

18 Among his Tufted Pansies

history, gardens and landscapes have not only provided metaphorical spaces for retreat, but they have also been heterotopias and actual physical spaces created and sought out by those marginalized because of their gender identification.

Although queer persons in some countries in the last 20 years have gained recognition and rights, and women since the second half of the twentieth century have begun to assume offices and leadership positions in various national and international organizations dealing with landscape, gender discrimination is a common phenomenon still today. It affects all persons who do not identify as hetero-cis males. The design of the physical space itself contributes to this discrimination. In the Western world, much of the environment has been and is shaped by norms developed by and for cis white men.[114] Even the conception of the seventeenth-century English term landscape itself bears witness to the heteropatriarchy. However, as much as the English term *landscape* here describes a gendered concept in the first place, both as the result of the male gaze and as the product of the imposition of human order upon nonhuman nature, that same landscape also holds an inherent power that is less easily controlled. Attention to nonhuman nature's inherent life forces, which co-produce our human landscapes, can lead to a heightened awareness toward those persons who have been marginalized in the histories of our built environment, be it through their gender, ethnicity, class, or a combination thereof. Vice versa, paying attention to marginalized people throughout human history, for example, women and queer persons, can also uncover new perspectives and raise awareness of concerns related to our landscapes and the environment more generally. The complexity underlying the relationship between landscape and gender identity becomes clear when we consider, for example, that the wilderness aesthetic mentioned above might have come to signal to some white gay men a welcome space of retreat (because other spaces are lacking) while signifying stress and potential danger to many straight and lesbian women as well as to many Black Americans and People of Color of all genders. Uncovering and understanding these nuances within the social, political, and cultural contexts of their time and place contributes to revealing landscape as a system of power relations that produce and are shaped by gender besides class and ethnic identities. Revising our landscape and environmental histories accordingly, as well as writing new histories exposing the role gender has played in the conception and construction of landscapes, is fundamental for raising gender awareness in the design professions so they may envision, design, and build more socially just landscapes and environments enabling multiple positions and views.

NOTES

1   For a feminist critique of the male gaze and its subversion through female spectatorship and the female gaze, see Susan Ford, "Landscape Revisited: A Feminist Reappraisal," in *New Words, New Worlds: Reconceptualising Social and Cultural Geography: Proceedings of a Conference* (Lampeter: St. David's University College, 1991), 151–155.

2   See, e.g., Tamar Garb, "The Forbidden Gaze: Women Artists and the Male Nude in Late Nineteenth-Century France," in *The Body Imaged: The Human Form and Visual Culture since the Renaissance*, eds. Kathleen Adler and Marcia Pointon (Cambridge: Cambridge University Press, 1993), 33–42; Deborah Cherry, *Painting Women: Victorian Women Artists* (London and New York: Routledge, 1993).

3   For the dendroscope and a feminist reading, see Sonja Dümpelmann, "Designing the 'ShapelyS City': Women, Trees, and the City," *Journal of Landscape Architecture*, no. 2 (2015): 6–17.

4   For Wycombe, see Stephanie Ross, *What Gardens Mean* (Chicago, IL and London: The University of Chicago Press, 1998), 66–70. For the role of Venus in eighteenth-century landscape gardens, see James G. Turner, "The Sexual Politics of Landscape: Images of Venus in Eighteenth-Century English Poetry and Landscape Gardening," *Studies in Eighteenth-Century Culture* 11 (1982): 343–366.

5   Carole Fabricant, "Binding and Dressing Nature's Loose Tresses: The Ideology of Augustan Landscape Design," *Studies in Eighteenth-Century Culture* 8 (1979): 109–135 (113).

6   Carole Fabricant, "Binding and Dressing Nature's Loose Tresses: The Ideology of Augustan Landscape Design," *Studies in Eighteenth-Century Culture* 8 (1979): 116.

7   Carolyn Merchant, *The Death of Nature: Women, Ecology and the Scientific Revolution* (San Francisco, CA: Harper & Row, 1980); Carolyn Merchant, *Earthcare: Women and the Environment* (New York: Routledge, 1995); Carolyn Merchant, *Reinventing Eden: The Fate of Nature in Western Culture* (New York: Routledge, 2003).

8   Sonja Dümpelmann, "Landscape: A Vignette," *AIJ Journal of Architecture and Building Science* 130, no. 1671 (2015): 20–23 (English and Japanese).

9   Catherine Nash, "Reclaiming Vision: Looking at Landscape and the Body," *Gender, Place and Culture* 3, no. 2 (1996): 156.

10 Catherine Nash, "Reclaiming Vision: Looking at Landscape and the Body," *Gender, Place and Culture* 3, no. 2 (1996): 156.

11 See Susan Ford, "Landscape Revisited: A Feminist Reappraisal," in *New Words, New Worlds: Reconceptualising Social and Cultural Geography: Proceedings of a Conference* (Lampeter: St. David's University College, 1991), 151–155; Catherine Nash, "Reclaiming Vision: Looking at Landscape and the Body," *Gender, Place and Culture* 3, no. 2 (1996): 156.

12 Catherine Nash, "Reclaiming Vision: Looking at Landscape and the Body," *Gender, Place and Culture* 3, no. 2 (1996): 167.

13 Nadar Vossoughian, "Standardization Reconsidered: 'Normierung' in and after Ernst Neufert's Bauentwurfslehre," *Grey Room* 54 (2014): 34–55 (44, 48); Anna-Maria Meister, "Ernst Neufert's 'Lebensgestaltungslehre': Formatting Life Beyond the Built," *BJHS Themes* 5 (2020): 167–185 (184); Kerstin Dörhöfer, "Der ,männliche' Blick in der Bauentwurfslehre," in *Ernst Neufert: Normierte Baukultur im 20. Jahrhundert*, ed. Walter Prigge (Frankfurt: Campus Verlag, 1999), 159–167.

14 On the invention of the *Modulor Man*, see Dave Tell, "Measurement and Modernity: Height, Gender, and Le Corbusier's Modulor," *Public Culture* 31, no. 1 (2019): 21–43.

15 See Ernst Neufert, *Bauentwurfslehre* (Berlin: Bauwelt, 1936), 97, 99, 101–103, 105.

16 A. Alphand, *Les Promenades de Paris, vol. II* (Paris: J. Rothschild, 1867–1873).

17 Galen Cranz, "Women in Urban Parks," *Signs* 5, no. 3, supplement (1980): S79–S95.

18 Galen Cranz, "Women in Urban Parks," *Signs* 5, no. 3, supplement (1980): S86.

19 Galen Cranz, "Women in Urban Parks," *Signs* 5, no. 3, supplement (1980): S79–S95.

20 On the role of women for the establishment of the Charlesbank playground and playgrounds in Boston generally, see Suzanne M. Spencer-Wood, "Gendering the Creation of Green Urban Landscapes in America at the Turn of the Century," in *Shared Spaces and Divided Places: Material Dimensions of Gender Relations and the American Historical Landscape*, eds. Deborah L. Rotman, Ellen-Rose Savulis (Knoxville: University of Tennessee Press, 2003), 24–61.

21 Sylvester Baxter, *Boston Park Guide* (Boston, MA: Small, Maynard & Co., 1898), 35.

22 Sylvester Baxter, *Boston Park Guide* (Boston, MA: Small, Maynard & Co., 1898), 35. On the Charlesbank playground also see Clarence Elmer Rainwater, *The Play Movement in the United States: A Study of Community and Recreation* (Washington, DC: McGrath Pub. Co. & National Recreation and Park Assoc., 1922), 28–29, 72–73. David Schuyler and Gregory Kaliss, eds., *The Papers of Frederick Law Olmsted: The Last Great Projects, 1890–1895* (Baltimore, MD: Johns Hopkins University Press, 2015), 34, 88–90; Steven Riess, *City Games: The Evolution of American Urban Society and the Rise of Sports* (Urbana and Chicago: University of Illinois Press, 1991), 136–137. Cynthia Zaitzevsky, *Frederick Law Olmsted and the Boston Park System* (Cambridge, MA: Belknap Press, 1982), 97–99.

23 See, e.g., Everett B. Mero, *American Playgrounds: Their Construction, Equipment, Maintenance, and Utility* (Boston, MA: Selling Agents American Gymnasia Co., 1908), 214.

24 Everett B. Mero, *American Playgrounds: Their Construction, Equipment, Maintenance, and Utility* (Boston, MA: Selling Agents American Gymnasia Co., 1908), 212–219; Galen Cranz, *The Politics of Park Design* (Cambridge and London: The MIT Press, 1982), 80–99.

25 Galen Cranz, "Women in Urban Parks," *Signs* 5, no. 3, supplement (1980): S94–S95.

26 *The Natural History of Pliny*, vol. IV, trans. John Bostock and H. T. Riley (London: Bohn, 1857), Book 19, Chapter 19, 153.

27 William Lawson, *The Country Houswifes Garden* (London: W. Wilson, 1653 [1618]), 85.

28 [Sir Anthony] Fitzherbert's *Booke of Husbandrie* (London: I.R. for Edward White, 1598 [1534]), 175–178.

29 Jane Brown, *The Pursuit of Paradise: A Social History of Gardens and Gardening* (London: Harper Collins, 1999), 107–108.

30 William Lawson, *The Country Houswifes Garden* (London: W. Wilson, 1653 [1618]), 85.

31 See Twigs Way, *Virgins Weeders and Queens: A History of Women in the Garden* (Stroud: Sutton Publishing, 2006), 8; Celia Fiennes, *Through England on a Side Saddle in the Time of William and Mary* (London: Field & Tuer, The Leadenhall Press, E.C.; New York: Scribner & Welford, 1888), 98.

32 Bea Howe, *The Life of Jane Loudon* (London: Country Life, 1961), 17.

33 Charles Evelyn, *The Lady's Recreation* (London: J. Roberts, 1717), 1–2.

34 Mrs. Jane Loudon, *Instructions in Gardening for Ladies* (London: John Murray, 1840), 244.

35 Mrs. Jane Loudon, *Instructions in Gardening for Ladies* (London: John Murray, 1840), 245.

36 Mrs. Jane Loudon, *Instructions in Gardening for Ladies* (London: John Murray, 1840), 244–245.

37 Elizabeth Hyde, *Cultivated Power: Flowers, Culture, and Politics in the Reign of Louis XIV* (Philadelphia: University of Pennsylvania Press, 2005), xix–xx.

38 Mrs. Jane Loudon, *Instructions in Gardening for Ladies* (London: John Murray, 1840), 244.

39 Elizabeth Hyde, *Cultivated Power: Flowers, Culture, and Politics in the Reign of Louis XIV* (Philadelphia: University of Pennsylvania Press, 2005), 25.

40 See Elizabeth Hyde, *Cultivated Power: Flowers, Culture, and Politics in the Reign of Louis XIV* (Philadelphia: University of Pennsylvania Press, 2005), 21.

41 Patricia Bouchenot-Dechin, "Hofgärtnerdynastien in Versailles und ihre Organisation," in *Preußisch Grün. Hofgärtner in Brandenburg-Preußen*, eds. Sonja Dümpelmann, Carsten Neumann, and Clemens Alexander Wimmer (Berlin: Henschel, 2004), 20–31 (25).

42 Jean-Baptiste de la Quintinie, *Instruction pour les jardins fruitiers et potagers* (Paris: Claude Barbin, 1690), 63.

43 Jane Loudon, *Botany for Ladies* (London: John Murray, 1842), vii–viii.

44 For other female garden writers using a similar tactic, see Dianne Harris, "Cultivating Power: The Language of Feminism in Women's Garden Literature, 1870–1920," *Landscape Journal* 13, no. 2 (1994): 113–123.

45 Mrs. Jane Loudon, *Instructions in Gardening for Ladies* (London: John Murray, 1840), dedication, v–vii.

46 Mrs. Jane Loudon, *Instructions in Gardening for Ladies* (London: John Murray, 1840), 9.

47 Mrs. Jane Loudon, *Instructions in Gardening for Ladies* (London: John Murray, 1840), 9.

48 Mrs. Jane Loudon, *Instructions in Gardening for Ladies* (London: John Murray, 1840), 399.

49 Mrs. Jane Loudon, *Instructions in Gardening for Ladies* (London: John Murray, 1840), 11–12.

50 Mrs. Jane Loudon, *Instructions in Gardening for Ladies* (London: John Murray, 1840), 11.

51 Mrs. Jane Loudon, *Instructions in Gardening for Ladies* (London: John Murray, 1840), 110.

52 Sonja Dümpelmann, *Seeing Trees: A History of Street Trees in New York City and Berlin* (New Haven, CT and London: Yale University Press, 2019), 82–83.

53 Nik Heynen, Harold A. Perkins, and Parama Roy, "Failing to Grow 'Their' Own Justice? The Co-Production of Racial/Gendered Labor and Milwaukess' Urban Forest," *Urban Geography* 28, no. 8 (2007): 732–754.

54 Mrs. Jane Loudon, *Instructions in Gardening for Ladies* (London: John Murray, 1840), 309–310.

55 John Claudius Loudon, *Encyclopaedia of Gardening* (London: Longman, Hurst, Rees, Orme, and Brown, 1822), 412.

56 See Ernst Neufert, *Bauentwurfslehre* (Berlin: Bauwelt, 1936), 84.

57 See the story of Ann Adams and Liz Brensinger who founded *Green Heron Tools. Garden & Farm Tools for Women*, https://www.greenherontools.com/ (accessed July 18, 2021).

58 See Maria Jones and Timothy Harrigan, "'Now We Can breathe'—Designing Mechanization to Benefit Women Smallholder Farms," https://agrilinks.org/post/now-we-can-breathe-designing-mechanization-benefit-women-smallholder-farmers (accessed July 18, 2021); Maria Jones, "Innovative Approaches to Including Gender with Agricultural Mechanization," https://agrilinks.org/post/innovative-approaches-including-gender-within-agricultural-mechanization (accessed July 18, 2021). Suman Agarwal, "Gender Involvement in Farm Mechanization–Issues for Extension and Research," in *Developments in Agricultural and Industrial Ergonomics: Women at Work,* vol. 2, eds. L. P. Gite, C. R. Mehta, Nachiket Kotwaliwale, Joydeep Majumder (2009), 50–57; Londa Vanderwal, Risto Rautiainen, Rex Kuye, Corinne Peek-Asa, Thomas Cook, Marizen Ramirez, Kennith Culp, Kelley Donham, "Evaluation of long- and short-handled hand hoes for land preparation, developed in a participatory manner among women vegetable farmers in The Gambia," *Applied Ergonomics* 42, no. 5 (2011): 749–756.

59 H. Jäger, *Frauengarten* (Stuttgart und Leipzig: Cohen und Risch, 1871), IX.

60 For female exercise in German air baths, see Sonja Dümpelmann, "Labyrinth, Hippodrome, Racetrack: Shaping Landscapes and Bodies in Nineteenth- and Early Twentieth-Century Berlin," in *Landscapes for Sport: Histories of Physical Exercise, Sport, and Health,* ed. Sonja Dümpelmann (Washington, DC: Dumbarton Oaks Research Library and Collection, 2022).

61 H. Jäger, *Frauengarten* (Stuttgart und Leipzig: Cohen und Risch, 1871), 8–9.

62 See Sonja Dümpelmann and John Beardsley, "Introduction," in *Women, Modernity, and Landscape Architecture*, eds. Sonja Dümpelmann and John Beardsley (London: Routledge, 2015)

63 Charles Sprague Sargent, "Taste Indoors and Out," *Garden and Forest* 5, no. 233 (1892): 373–374.

64 Charles W. Eliot, ed., *Charles Eliot Landscape Architect* (Boston, MA and New York: Houghton Mifflin, 1902), 546–547.

65 Carl Heicke, "Die XXIX. Hauptversammlung der Deutschen Gesellschaft für Gartenkunst e.V.," *Die Gartenkunst* 30, no. 11 (1917): 151–166 (159).

66 For the role of garden literature in creating female networks and empowering women in horticulture in late nineteenth-century and early twentieth-century England and the United States, see Dianne Harris, "Cultivating Power: The Language of Feminism in Women's Garden Literature, 1870–1920," *Landscape Journal* 13, no. 2 (1994): 113–123.

67 Sonja Dümpelmann, "Designing the 'Shapely City': Women, Trees, and the City," *Journal of Landscape Architecture*, no. 2 (2015): 6–17; Sonja Dümpelmann, *Seeing Trees: A History of Street Trees in New York City and Berlin* (New Haven, CT and London: Yale University Press, 2019), chapter 3.

68 Elizabeth K. Meyer, "The Expanded Field of Landscape Architecture," in *Ecological Design and Planning*, eds. George F. Thompson and Frederick Steiner (New York: John Wiley & Sons, 1997), 45–79; Thaïsa Way, "Designing Garden City Landscapes: Works by Marjorie L. Sewell Cautley, 1922–1937," *Studies in the History of Gardens & Designed Landscapes* 25, no. 4 (2005): 297–316.

69 Dolores Hayden, *Redesigning the American Dream: The Future of Housing, Work, and Family Life* (New York: Norton, 1984), 30; 186–191.

70 Dolores Hayden, *Redesigning the American Dream: The Future of Housing, Work, and Family Life* (New York: Norton, 1984), 15.

71 Dolores Hayden, *Redesigning the American Dream: The Future of Housing, Work, and Family Life* (New York: Norton, 1984), 191.

72 Margrit Kennedy, "Building 'Naturally' includes Women: Towards a Rediscovery of Female Principles in Architecture and Planning," *Natural Building, Symposium Report, Information of the Institute for Lightweight Structures (IL) University Stuttgart* 27 (1980): 78–85 (79).

73 Margrit Kennedy, "Building 'Naturally' includes Women: Towards a Rediscovery of Female Principles in Architecture and Planning," *Natural Building, Symposium Report, Information of the Institute for Lightweight Structures (IL) University Stuttgart* 27 (1980): 81.

74 See, for example, Paola Coppola Pignatelli, "Der Weg zu einer anderen räumlichen Logik," *Bauwelt* 70, no. 31–32 (1979): 1285–1287.

75 Margrit Kennedy, "Building 'Naturally' includes Women: Towards a Rediscovery of Female Principles in Architecture and Planning," *Natural Building, Symposium Report, Information of the Institute for Lightweight Structures (IL) University Stuttgart* 27 (1980): 79.

76 Margrit Kennedy, "Building 'Naturally' includes Women: Towards a Rediscovery of Female Principles in Architecture and Planning," *Natural Building, Symposium Report, Information of the Institute for Lightweight Structures (IL) University Stuttgart* 27 (1980): 83; Margrit Kennedy, "Seven Hypotheses on Female and Male Principles in Architecture," *Heresies 11*, 3, no. 3 (1981): 12–13. Also see the reception of feminism in architecture in Henning Eichberg, *Leistungsräume. Sport als Umweltproblem* (Münster: Lit Verlag, 1988), 45–46.

77 Margrit Kennedy, "Zur Wiederentdeckung weiblicher Prinzipien in der Architektur," *Bauwelt* 70, no. 31–32 (1979): 1279–1284 (1280–1281).

78 For lesbian feminism in architecture and Birkby's role in diversifying the profession of architecture, see Stephen Vider, *Queerness of Home: Gender, Sexuality, and the Politics of Domesticity after World War II* (Chicago, IL and London: The University of Chicago Press, 2021), 106–140. I thank M.C. Overholt for drawing my attention to Vider's pathbreaking work which was published after this chapter was written.

79 Phyllis Birkby, "Herspace," *Heresies 11*, 3, no. 3 (1981): 28–29; Noel Phyllis Birkby and Leslie Kanes Weisman, "Women's Fantasy Environments: Notes on a Project in Progress," *Heresies* 1, no. 2 (1977): 116–117. For women's search for "female form," also see Mimi Lobell, "The Buried Treasure: Women's Ancient Architectural Heritage," in *Architecture: A Place for Women*, eds. Ellen Perry Berkeley and Matilda McQuaid (Washington, DC: Smithsonian Institution Press, 1989), 139–157.

80 Noel Phyllis Birkby and Leslie Kanes Weisman, "A Woman Built Environment: Constrictive Fantasies," *Quest* 11, no. 1 (1975): 8–18. For the workshops and environmental fantasies, also see Stephen Vider, *Queerness of Home: Gender, Sexuality, and the Politics of Domesticity aft er World War II* (Chicago, IL and London: The University of Chicago Press, 2021), 123–129.

81 Leslie Kanes Weisman in Noel Phyllis Birkby and Leslie Kanes Weisman, "A Woman Built Environment: Constrictive Fantasies," *Quest* 11, no. 1 (1975): 117.

82 Leslie Kanes Weisman, "Women's Environmental Rights: A Manifesto," *Heresies 11*, 3, no. 3 (1981): 6–8.

83 Rita Reif, "Architecture: Feminist Ferment," *The New York Times*, August 9, 1975.

84 For the activities of the School, see Phyllis Birkby, "…I want it to help, not to hinder me … Neue Erfahrungen im Umgang mit Raum, Die 'Women's School of Planning and Architecture', USA," *Arch+* 56, (1981): 10–12; Leslie Kanes Weisman, "A Feminist Experiment: Learning from WSPA, Then and Now," in *Architecture: A Place for*

Women, eds. Ellen Perry Berkeley and Matilda McQuaid (Washington, DC: Smithsonian Institution Press, 1989), 125–135; Rita Reif, "Architecture: Feminist Ferment," *The New York Times*, August 9, 1975.

85 Karen A. Franck, "A Feminist Approach to Architecture: Acknowledging Women's Ways of Knowing," in *Architecture: A Place for Women*, eds. Ellen Perry Berkeley and Matilda McQuaid (Washington, DC: Smithsonian Institution Press, 1989), 201–216 (212).

86 See Diana Balmori, "Beatrix Farrand at Dumbarton Oaks," *Heresies 11*, 3, no. 3 (1981): 83–86. Balmori's research on Farrand would lead to a book publication: Diana Balmori, Diane Kostial McGuire, and Eleanor M. McPeck, *Beatrix Farrand's American Landscapes: Her Gardens and Campuses* (Sagaponack, NY: Sagapress, 1985).

87 Diana Balmori, "Beatrix Farrand at Dumbarton Oaks," *Heresies 11*, 3, no. 3 (1981): 86.

88 Sharon Sutton, "Street Museum," *Heresies 11*, 3, no. 3 (1981): 9.

89 See Sonja Dümpelmann and John Beardsley, "Introduction," in *Women, Modernity, and Landscape Architecture*, eds. Sonja Dümpelmann and John Beardsley (London: Routledge, 2015), 1–14.

90 See Helaine Kaplan Prentice, "A Century of Women: Evaluating Gender in Landscape Architecture, University of California, Berkeley, Department of Landscape Architecture and Environmental Planning," *Landscape Journal* 22, no. 1 (2003): 166–169 (166).

91 See Thaïsa Way, *Unbounded Practice: Women and Landscape Architecture in the Early Twentieth Century* (Charlottesville and London: University of Virginia Press, 2009), 261.

92 See https://www.queereurope.com/ (accessed August 14, 2021).

93 See George Chauncey, *Gay New York: Gender, Culture, and the Making of the Gay Male World 1890–1940* (New York: Basic Books, 1994), 183.

94 George Chauncey, *Gay New York: Gender, Culture, and the Making of the Gay Male World 1890–1940* (New York: Basic Books, 1994), 183.

95 Gordon Brent Ingram, "Marginality and the Landscape of Erotic Alien(n)ations," in *Queers in Space: Communities, Public Places, Sites of Resistance*, eds. Gordon Brent Ingram, Anne-Marie Bouthillette, and Yolanda Retter (Seattle, WA: Bay Press, 1997), 29, 31–32. Also see Aaron Betsky, *Queer Space: Architecture and Same-Sex Desire* (New York: William Morrow, 1997), 147–148.

96 Matthew Gandy, "Queer Ecology: Nature, Sexuality, and Heterotopic Alliances," *Environment and Planning D: Society and Space* 30 (2012): 727–747.

97 See Matthew Gandy, "Queer Ecology: Nature, Sexuality, and Heterotopic Alliances," *Environment and Planning D: Society and Space* 30 (2012): 727–747.

98 See Darren Patrick, "Queering the Urban Forest: Invasions, Mutualisms, and Eco-Political Creativity with the Tree of Heaven (*Ailanthus altissima*)," in *Urban Forests, Trees, and Greenspace: A Political Ecology Perspective*, eds. L. Anders Sandberg, Adrina Bardekjan, and Sadia Butt (New York and Abingdon: Routledge, 2015), 191–206.

99 See https://www.queereurope.com/gay-guide-to-london/ (accessed August 14, 2021).

100 Jean-Ulrick Désert, "Queer Space," in *Queers in Space: Communities, Public Places, Sites of Resistance*, eds. Gordon Brent Ingram, Anne-Marie Bouthillette, and Yolanda Retter (Seattle, WA: Bay Press, 1997), 17–26.

101 Gordon Brent Ingram, "Marginality and the Landscape of Erotic Alien(n)ations," in *Queers in Space: Communities, Public Places, Sites of Resistance*, eds. Gordon Brent Ingram, Anne-Marie Bouthillette, and Yolanda Retter (Seattle, WA: Bay Press, 1997), 45.

102 Gordon Brent Ingram, "Marginality and the Landscape of Erotic Alien(n)ations," in *Queers in Space: Communities, Public Places, Sites of Resistance*, eds. Gordon Brent Ingram, Anne-Marie Bouthillette, and Yolanda Retter (Seattle, WA: Bay Press, 1997), 51.

103 For the High Line, see Darren J. Patrick, "The Matter of Displacement: A Queer Urban Ecology of New York City's High Line," *Social & Cultural Geography* 15, no. 8 (2014): 920–941.

104 Also see Gordon Brent Ingram, "Marginality and the Landscape of Erotic Alien(n) ations," in *Queers in Space: Communities, Public Places, Sites of Resistance*, eds. Gordon Brent Ingram, Anne-Marie Bouthillette, and Yolanda Retter (Seattle, WA: Bay Press, 1997), 38; For the differentiation between marginalized and respectable homosexual subjects, see Matthew Gandy, "Queer Ecology: Nature, Sexuality, and Heterotopic Alliances," *Environment and Planning D: Society and Space* 30 (2012): 727–747, 741.

105 Gordon Brent Ingram, "Marginality and the Landscape of Erotic Alien(n)ations," in *Queers in Space: Communities, Public Places, Sites of Resistance*, eds. Gordon Brent Ingram, Anne-Marie Bouthillette, and Yolanda Retter (Seattle, WA: Bay Press, 1997), 46; David Bird, "Trees in a Queens Park Cut Down as Vigilantes Harass Homosexuals," *The New York Times*, July 1, 1969, 1; David Bird, "Queens Resident Says the Police Stood By as Park Trees Were Cut," *The New York Times*, July 2, 1969, 38; "Police did Answer Call on Cut Trees," *The New York Times*, July 3, 1969, 29; William J. Primavera, "Vigilante Vandalism," *The New York Times*, July 3, 1969, 30; David Bird, "Police Continuing Inquiry on Trees: Have Not Yet Queried Man Linked to Queens Mystery," *The New York Times*, July 16, 1969, 50; "Inquiries Still Open In Tree-Chopping At Park in Queens," *The New York Times*, September 18, 1969, 49; Donn Teal, *The Gay Militants* (New York: Stein and Day, 1971), 27, 31.

106 Lisa L. Moore, "Queer Gardens: Mary Delaney's Flowers and Friendships," *Eighteenth-Century Studies* 39, no. 1 (2005): 49–70; Lisa L. Moore, *Sister Arts: The Erotics of Lesbian Landscapes* (Minneapolis: The University of Minnesota Press, 2011), 29–38; 65–74.

107 Carole Fabricant, "Binding and Dressing Nature's Loose Tresses: The Ideology of Augustan Landscape Design," *Studies in Eighteenth-Century Culture* 8 (1979): 109–135 (121–122).

108 See George Chauncey, *Gay New York: Gender, Culture, and the Making of the Gay Male World 1890–1940* (New York: Basic Books, 1994), 182, 195–196; Roy Rosenzweig and Elizabeth Blackmar, *The Park and the People: A History of Central Park* (Ithaca, NY and London: Cornell University Press, 1998 [1992]), 479.

109 Frederick Law Olmsted, "The Spoils of the Park," cit. in *Frederick Law Olmsted Landscape Architect 1922–1903: Central Park*, eds. Frederick Law Olmsted, Jr. and Theodora Kimball (New York and London: Putnam's Sons, 1928), 117–155 (144).

110 Geoffrey Taylor, *Some Nineteenth Century Gardeners* (Skeffington: London, 1951), 71, 85.

111 See the photograph in Mea Allan, *William Robinson 1838–1935: Father of the English Flower Garden* (London: Faber and Faber, 1982), 156. The use of "pansy" to describe (often derogatively) homosexual, effeminate men only began in the 1920s, after the photograph was taken.

112 Geoffrey Taylor, *Some Nineteenth Century Gardeners* (Skeffington: London, 1951), 85. Also see Mea Allan, *William Robinson 1838–1935: Father of the English Flower Garden* (London: Faber and Faber, 1982), 162.

113 See Rebecca Nagel, "Naming Plants in The Garden by Vita Sackville-West," *Interdisciplinary Studies in Literature and Environment* 22, no. 2 (Spring 2015): 241–263.

114 Also see, e.g., Leslie Kanes Weisman, *Discrimination by Design: A Feminist Critique of the Man-Made Environment* (Urbana and Chicago: University of Illinois Press, 1992).

Chapter 6: Is Landscape Imperial?[1]

John E. Crowley

Empires required imaginative representation long before nations did. Imperial art asserted *imperium*, the rule by a charismatic figure over subject peoples. Aspirations of universal hegemony manifested in many forms, from triumphal columns and arches, such as Trajan's Column and the Arc de la Triomphe, to equestrian depictions, such as those of Marcus Aurelius and Napoleon. The Yongle Emperor's Forbidden City, Akbar's Fateh Sur Sikri, Ivan the Great's Kremlin, and Louis XIV's Versailles were imperial palaces with landscape gardens built to symbolize dynastic mystique.[2] Given its divine and hegemonic mandates, imperial art was over-determined, since it typically established the legitimacy of new dynasties (Figure 6.1).[3] Precisely because imperial expansion depended on conquests, others held countervailing fears of aspiring universal monarchs, with a near-endless list that included the Holy Roman Emperor Charles V, Pope Julius II, Sultan Suleiman I, the Spanish Hapsburg Philip II, the French Bourbon Louis XIV, and Emperor Napoleon I.[4] Awareness of latent anti-imperialism inclined imperial propagandists toward cosmopolitan promotions of universal rule with promises of peace and enlightenment.

In early modern European usage, "empire" was a loose term that could refer to a dynasty's domains generally.[5] For example, the first atlas of the British Isles was John Speed's *The Theatre of the Empire of Great Britaine, Presenting an Exact Geography of the Kingdom of England, Scotland and Ireland and the Isles Adjoyning* (1611–1612). In present-day historiography, "empire" refers primarily to European colonial regimes overseas. With the latter orientation in mind, this chapter privileges representations of early modern imperial spaces and landscapes overseas, focusing on a series of imperially scaled publications that featured landscape art, beginning with a German ecclesiastic's pilgrimage to Jerusalem and concluding with the Napoleonic *Description de l'Égypte*. In between are collections of hundreds of cityscapes, a 25-volume series of European narratives from the East and West Indies, a large-folio commemoration of the Dutch empire in Brazil, a print series celebrating Britain's conquest of France's North American empire, revelations of an unexpectedly picturesque Pacific, and a multivolume series of aquatints recording artistic pilgrimages throughout proto-British India. Each work had dozens of landscape prints in settings that were, as yet, only tenuously imperial. Most imperial landscape art was produced

DOI: 10.4324/9781003148142-7

Figure 6.1
The topmost caption reads: "What the Sun is in the heavens, Caesar is on Earth." Maximilian (Emperor 1508–1521) was the first, though not the last, Hapsburg to keep the title of Emperor after his father—in his case, Frederick III. Voltaire would observe dryly of Maximilian's great–great uncle, Charles IV, "This agglomeration which was called and still calls itself the Holy Roman Empire is neither holy, nor Roman, nor an empire"; *Essay sur l'histoire générale et sur les moeurs et l'esprit des Nations, depuis Charlemagne jusqu'à nos jours* (Paris, 1756), chapter 70.
Source: Albrecht Dürer, [The Great Triumphal Chariot of Maximilian I] (commissioned 1518; Nuremberg, 1522). Leftmost two panels of woodcut, eight sheets, 40 × 241 cm. Courtesy: Library of Congress, Prints and Photographs Division.

by artists who sought to acquaint metropolitan viewing publics with *potential* imperial expansions.

The beginnings of early modern European expansion overseas—from the 1480s to the 1580s—paralleled ambitious collections of *cityscapes* ranging across Europe and eventually to Africa, Asia, and the Americas. Bernard von Breydenbach's *Peregrinatio in Terram Sanctum*, the first illustrated travel book, applied northern European graphics to Venetian imperial locales. Breydenbach, Canon of the Cathedral of Mainz and an artist himself, commissioned an artist from Utrecht, Erhard Reuwich, to accompany him as a topographic artist along their 1483–1484 pilgrimage route from Venice to Jerusalem via the Venetian territories of Corfu, Modon, Heraklion, and Rhodes. The published results were

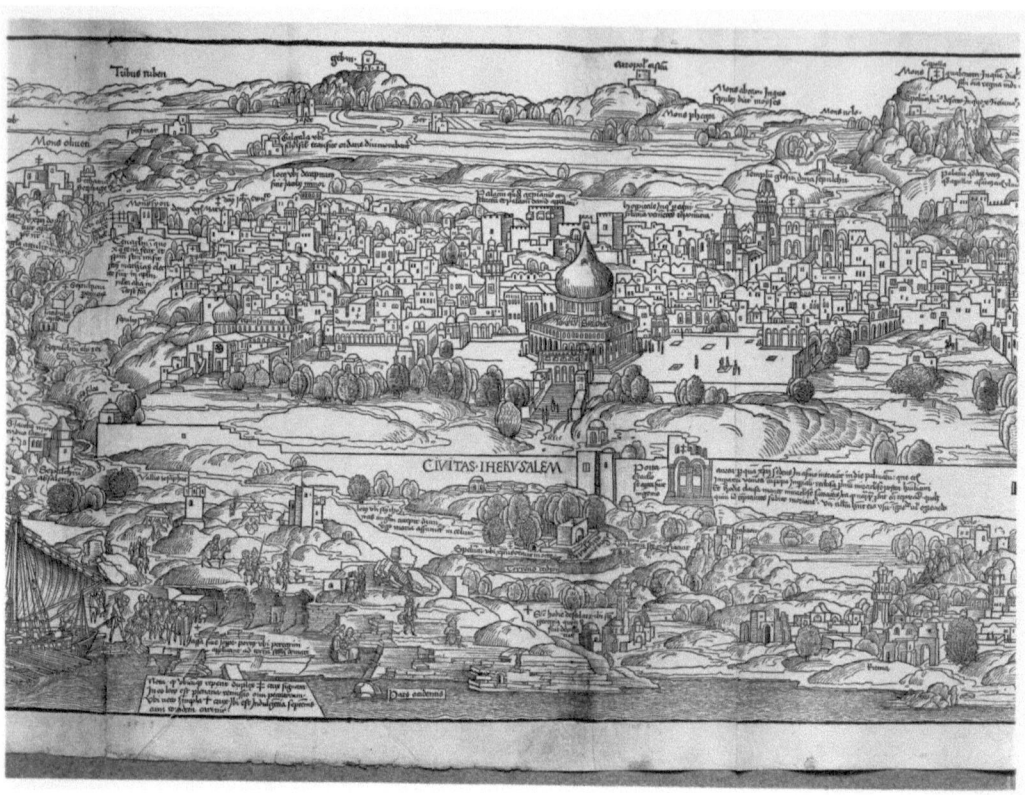

In the image: CIVITAS·IHERVSALEM

seven cityscapes, including the first foldout landscape prints in a book.[6] The book's first foldout centered on St. Mark's campanile and showed Venice from a high vantage point that allowed the identification of specific buildings. But the climactic landscape, Jerusalem, was beyond the Venetian empire (Figure 6.2). It also used an elevated viewpoint to give a topography of the Holy City and to identify viewpoints for Christians' indulgences. Breydenbach presented an exotic cosmopolitan adventure with images of Syrians, Jews, Greeks, Ethiopians, and Saracens. To the extent that the *Peregrinatio* was imperialistic, it was to encourage a crusade by Venice to recover the Holy Lands from the Ottoman Empire.

Breydenbach's book was a runaway success, with 13 reprintings in Dutch, French, German, Spanish, and the original Latin. (Its woodcuts traveled from printer to printer.) Reuwich's innovatively panoramic cityscapes immediately influenced the design of books with more comprehensive interests in landscape, notably Hartmann Schedel's *Libri cronicarum* (*Nuremberg Chronicle* to English readers), published in the year that Columbus completed his first American voyage.[7] It narrated the history of the world from Creation to the present and beyond. Its 1,809 illustrations juxtaposed 53 town views with portraits of historical figures. Many of the views of biblical towns, such as Jerusalem and Babylon, looked suspiciously like European walled cities, but more contemporaneous

Figure 6.2 This view of Jerusalem lay between maps of Palestine and Egypt in an enormous, three-woodblock, foldout print. It centers on the Dome of the Rock (al-Haram al-Sharif), accessible only to Muslims—here identified nostalgically as "Templus Solomonis."
Source: Erhard Reuwich, "Civitas Jherusalem," in Bernhard von Breydenbach, *Peregrinatio in Terram Sanctam*. Section of woodcut, 30 × 130 cm. Courtesy: Metropolitan Museum of Art, Rogers Fund, 1919.

Figure 6.3 Nuremberg, with identifiable cathedral, city walls, and riverside setting.
Source: Michael Wolgemut and Wilhelm Pleydenwurff, "Nuremberga," in Hartmann Schedel, *Registrum huius Operis libri cronicarum cum figuris et yagibus ab inicio mundi* (Nuremberg, 1493), folios 99v–100r. Woodcut, 34 × 50 cm. Courtesy: Metropolitan Museum of Art, Rogers Fund, 1906.

European ones were often recognizable from their landmark buildings and settings (Figure 6.3). Schedel dedicated the *Cronicarum* to Emperor Maximilian, but its landscapes transcended any specific empire since it embraced all of world history, when at the Last Judgment, Christ would rule.

The title of Sebastian Münster's *Cosmographiae universalis* promised a global geography. As the second most frequently published book in its era, after the Bible, with 35 editions from 1544 to 1628, in German, Latin, French, Italian, and Czech, it established cityscapes as a genre of landscape. The *Cosmographiae*, like the *Libri cronicarum,* was dedicated to a Holy Roman Emperor, with an elaborate title page schematizing Charles V's domains dynastically. However, it sidestepped using that association to shape its presentation of landscapes and ignored Charles' richest domains, Spain and the Netherlands. Instead, Münster concentrated explicitly on the *Deutscher Nation* (German nation), with 36 views of its cities, each of which had mobilized funds and artistic resources for its contribution (Figure 6.4).[8] If anything, the imperial ambitions of Maximilian and Charles V provoked the cityscape as a form of patriotic resistance rather than of imperial affirmation.

As a genre, the cityscape provided the visual means for Europeans to expand their topographic perspectives geographically.[9] Europeans' interest in

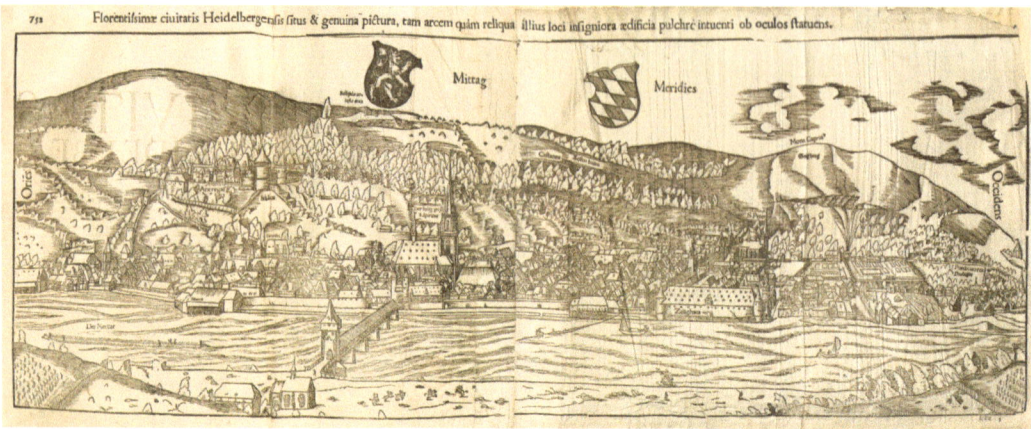

Figure 6.4 The Count Palatine of the Rhine, Ottheinrich of Palatinate–Neuberg provided Münster with this four-page foldout view of Heidelberg and its castle, along with textual materials to support his ancestral claim on the electorship of the Palatine, to challenge his uncle Friedrich II.
Source: Sebastian Münster, "Florentissimae ciuitatis Heidelbergensis," *Cosmographiae universalis lib. VI. In quibus, iuxta certioris fidei scriptorium traditionem describuntur, omnium habitabilis orbis partium situs, propriaeq[ue] dotes. Regionum topographicae effigies* (Basle, 1572), foldout plate between 752 and 753. Woodcut, 25 × 70 cm. Courtesy: David Rumsey Map Collection, Stanford University Libraries.
Source: "Plant & Portraict de l'Illstre Cité de Cuscho, ville Capitale du Royaume de Peru," in Antoine Du Pinet, Plantz, pourtraitz et descriptions de plusiers villes et
forteresses, tant de l'Europe, Asie, & Afrique, que des Indes, & terres neuues: leurs fondations, antiquitez, & manieres de vivre (Lyons, 1564), 292–293. Woodcut, 33 × 42 cm. Courtesy: David Rumsey Map Collection, Stanford University Libraries.

representations of unfamiliar places from first-hand observations coincided with new capabilities to publish these images in multiple copies, thanks initially to the woodcut print and then to copperplate engraving. Thus, the publisher Antoine de Pinet in Lyons sought to cash in on Münster's success by appropriating a few of his German views while adding ones from France to expand his market there. For exotic interest, he also included views of two American imperial capitals, the Aztecs' Tenoctitlán and the Incas' Cuzco (Figure 6.5).[10]

The woodcut illustrations in Münster's *Cosmographiae universalis* were soon surpassed by the copperplate engravings in the next topographic block-buster, Georg Braun and Franz Hogenberg's six-volume *Civitates Orbis Terrarum* (1572–1598, 1617), which had views of 546 towns.[11] Their work deliberately complemented Abraham Ortelius' contemporaneous cartographic atlas *Theatrum Orbis Terrarum.* The *Civitates Orbis Terrarum* was, by definition, a world of cities, overwhelmingly of cities known to Europeans before their commercial and imperial expansion overseas. It lacked a political center (Braun, the editor, was Catholic; Hogenberg, the engraver, was Protestant); views were ordered episodically, with just a few at a time from the same region. Dutch, German, Italian,

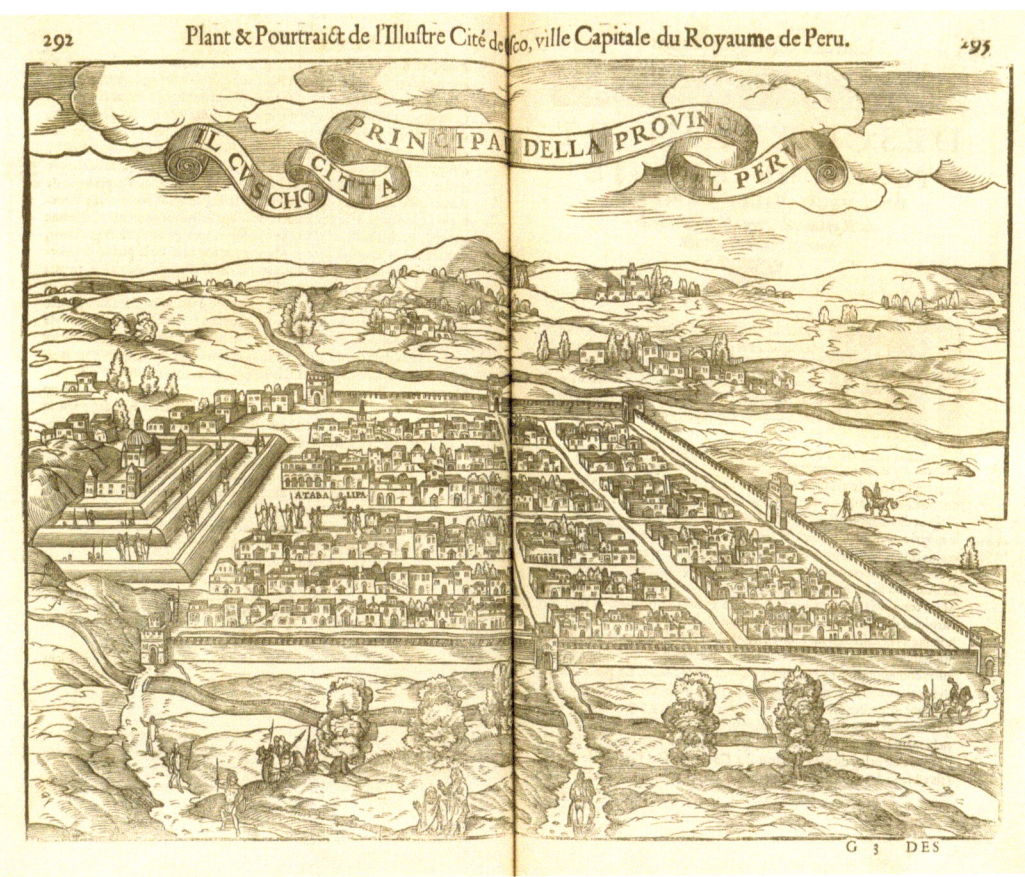

292   Plant & Pourtraict de l'Illustre Cité de Cco, ville Capitale du Royaume de Peru.   295

IL CVSCHO   CITTA   PRINCIPAL DELLA PROVIN   L PERV

G 3   DES

Figure 6.5 Bird's eye view of Cuzco, as the last Incan emperor, Atahualpa, departed.
Source: Georg Braun and Franz Hogenberg, *Civitates* Orbis Terrarum (Cologne, 1572), 57. Woodcut, 34 × 47 cm. Courtesy: David Rumsey Map Collection, Stanford University Libraries.

Andalusian, and eastern European cities predominated, in that order, with British and French ones slighted. North African, Mediterranean, Levantine, Indian Ocean, and American (still Cusco and Mexico) views followed in that order of frequency. The major expansion of geographic range from previous collections of cityscapes was into Mediterranean and Indian Ocean ports rather than to new colonial realms in the Americas (Figure 6.6).[12] To the extent that the *Civitates* privileged any empire, it was the Ottoman, with a large increase in views of North African towns in parallel with the expansions of Suleiman I and his successors.

Spain, with its "empire of towns," had the visual resources and political imperatives to carry the genre of cityscapes overseas to record Europeans' new Atlantic worlds. Anton van der Wyngaerde, who created many of the *Civitates*'s views, had previously worked for Philip II on a series of over 60 Spanish cityscapes based on drawing trips throughout the peninsula in the 1560s. The results were displayed prominently at Phillip's palaces, and there were plans to

have them engraved, though they went unrealized.[13] New Spain shared Phillip's topographic patronage. A royal questionnaire, drawn up in the late 1570s at Phillip's behest and then issued by the Office of the Cronista Mayor-Cosmógrafo, directed that New Spain and Peru be mapped architecturally in ways that amounted to cityscapes. Monastery-trained Mexican artists created hybrid Mexican–Spanish codices, the *Relaciones Geográficas* (Figure 6.7), but these great resources for imperial landscapes were ignored once they arrived in Spain.[14] Published landscapes of Spain's American empire only appeared a century later, under Dutch auspices (Figure 6.8).

The most fully realized publishing project dealing with Europe's overseas empires was Theodor de Bry and his sons' 25 volumes, the *India Occidentalis* and *India Orientalis* (1590–1634).[15] This lavishly illustrated series provided over 300 iconic images, based on previously published narratives of Europeans' encounters with peoples and places in the Americas, Africa, and Asia. De Bry apparently calculated that readers were more interested in narrative than landscape, and he had an eye for drama, so the settings were often theatrical rather than topographic. They conveyed strong impressions that nevertheless could only be *imagined* visually, since the De Brys' Europe-based artists produced images from textual descriptions rather than from first-hand drawings. Whether they served

Figure 6.6 This composite view showed Anfa (Casablanca) and nearby Azemour in conjunction with Diu and Goa in India—all of them Portuguese conquests. Source: Arnoldus Montanus, "Potosi," in Montanus, *De nieuwe en onbekende weereld, of, Beschryving van America en 't zuid-land* (Amsterdam, 1671), 352. Engraving, 27 × 35 cm. Courtesy: John Carter Brown Library, Brown University.

Figure 6.7 A map of Tehuantepec, a town in the Bishopric of Oaxaca, 1580, showing the church square, the surrounding villages, and the landscape of the Isthmus of Tehuantepec between the Gulf of Mexico and the Pacific Ocean. A mysteriously symbolic jaguar crowns the hill behind the church. Source: Addendum to Francisco Stroza Gall, Relación no. 102, Joaquín García Icazbalceta Collection of Relaciones Geográficas of Mexico and Guatemala, 1577–1585, Nettie Lee Benson Latin American Collection. Painting, 22 × 31 cm. Courtesy: Perry–Casteñeda Library, University of Texas at Austin.

any imperial interests is open to question (Figure 6.9). The De Brys published each volume in Latin and German—Latin for cosmopolitan humanists, German for more popular interests—with a view toward a wide cross-confessional, transnational readership of merchants, nobles, magistrates, and clerics. The accounts ranged from atrocities to conquests (to the extent that they could be distinguished), with a counterpoint of exoticism.

Thanks largely to the De Brys, by the 1630s, Europeans had rich visual familiarity with colonial settings. Just as the De Brys' project was concluding, the territorial ambitions of the Dutch East and West India companies began

P O T S I .

encouraging Dutch artists to represent landscapes overseas. In 1637, the Dutch West India Company sent Prince Johan Maurits van Nassau-Siegen to Dutch Brazil, as governor to deal with Spanish-sponsored insurgency. His self-image was both humanist and militaristic: along with troops, his expedition included dozens of natural philosophers and artists. He commissioned them to map the colony, construct an astronomical observatory, establish botanical and zoological gardens, and paint its landscapes.[16]

Prince Maurits had his successes commemorated with Caspar van Baerle's lavishly illustrated *Rerum per octennium in Brasilia et alibi nuper gestarum sub praefectura illustrissimi comitis I. Mavritii, Nassoviae* (1647)—translated as "the events that recently took place in Brazil and elsewhere over an eight-year period under the illustrious leadership of Johan Maurits of Nassau."[17] This oversized folio publication featured prints of the Company's establishments in Brazil and Africa, most of them designed by Frans Post, the first professional European landscape painter to work extensively in the Americas. Post paid exceptional

Figure 6.8 Dutch artists created the monumental landscape of Spain's mountain of silver at Potosi, while perhaps mockingly featuring a metonymically Dutch windmill.

Figure 6.9 Enslaved African miners deliver ore to Spaniards. De Bry's artists based this image on descriptions in Girolamo Benzoni, *Historia del mondo nuovo* (Venice, 1565), a denunciation of Spanish imperialism by a Milanese merchant who claimed to have traveled throughout Spain's American realms. Source: "Nigritae in scrutandis venis metallicis ab Hispanis in Insulas ablegantur," in Girollamo Bezoni, Urbain Chauveton, and Theodor de Bry, *Americae pars quinta nobilis & admiratione plena Hieronymi Bezoni Mediolanensis secundae sectionis Hispanorum* (Frankfurt am Main, 1595), pl. 1. Engraving with hand coloring, 16 × 20 cm. Courtesy: John Carter Brown Library, Brown University.

attention to colonial landscapes and vernacular architecture. A highly detailed map of Pernambuco, for example, featured a large illustration of a sugar mill and a manager's house (Figure 6.10). Post presented Brazil as a sparsely settled landscape of villages and isolated chapels, with roads winding through the partially cleared countryside, not entirely unlike the areas dear to Dutch landscape artists at home.[18] Similarly, the horizontality of coastal Brazil's vast riverine sweeps reflected the Netherlands' flatness in a tropical tone. Forts located strategically for their command of river bends—rather than high ground—would have been familiar from the home landscape as well. These images implicitly contrasted benign Dutch colonization with tyrannical Spanish imperialism.[19]

As the fortunes of the Dutch West and East India companies waned with the loss of Brazil and the rise of British dominance in the East, Dutch publishers began producing globally ranging collections of landscapes. While working with the preeminent atlas maker, Joan Bleau, the cartographer Johannes Vingboons produced the most impressive visualization of the Dutch global landscape. It included over 200 double-folio watercolor views of places Europeans, not just the Dutch, had encountered commercially and imperially around the world, from the Black Sea, along the coast of West Africa, around the Horn of East Africa, to Goa, Ceylon, Siam, China, Japan, Mexico, New Amsterdam, the Caribbean, and Brazil. Vingboons based his work on verbal, textual, and graphic information brought back to Amsterdam by agents of the East and West India companies and by more miscellaneous reporters, such as Jesuits. He channeled this information into exotic views, which he sold as landscapes (Figure 6.11).[20] The United

Figure 6.10 Frans Post naturalized the plantation landscape by showing the easy pace of work at the mill, dancing Afro–Brazilians, and the arrival of a party of Europeans, while masking its killing labor.
Source: Frans Post, "Praefecturae Paranambucae Pars Borealis, una cum Praefectura de Itâmaracâ" [copied from Caspar van Baerle, *Rerum per octennium in Brasilia*] in Joan Blaeu, *Atlas Major, sive Cosmographia Blaviana*, 11 vols. (Amsterdam, 1662), 11: 243/234 (detail). Colored engraving, 42 × 54 cm. Courtesy: John Carter Brown Library, Brown University.
Source: "Description du Saut, ou chuete d'eau de Niagara," in Louis Hennepin, *Nouvelle découverte d'un très grand pays situé dans l'Amérique entre le Nouveau Mexique, et la Mer Glaciale … & les avantages qu'on en peut tirer par l'établissem[ent] des colon: le tout dédié à sa Majesté britannique* Guillaume III (Utrecht, 1697), foldout plate after 44. Engraving, 12 × 17 cm. Courtesy: John Carter Brown Library, Brown University.

Provinces, exceptionally, had a market for colonial landscape art, both as prints and paintings, creating a market whose expansion coincided with the waning of Dutch imperial eminence.

Beginning in the 1660s and running through the middle of the eighteenth century, French publications dominated the literature on colonial spaces in North America.[21] French cartographers asserted claims to the interior of eastern North America between the Appalachians and the Mississippi. French military engineers applied their highly professional surveying and drafting skills to architectural plans, sections, elevations, and town plans throughout New France.[22] French illustrations privileged Indigenous inhabitants and natural history; landscapes had low priority.[23] The most recognizable among them, Louis Hennepin's often-reprinted view of Niagara Falls, was an exception proving the rule (Figure 6.12). He had viewed the Falls while accompanying Sieur La Motte de Lucière and Cavelier de La Salle's 1678–1681 exploration of the Mississippi. But he fell from grace with French authorities and eventually sought the patronage of William III, who sponsored the publication of the book with the famous illustration of Niagara.

Mainly under the impetus of British imperial agents, visual interest in European colonies shifted markedly toward the topographic. During the Seven Years' War (1756–1763), British artists began to claim visual authenticity for their

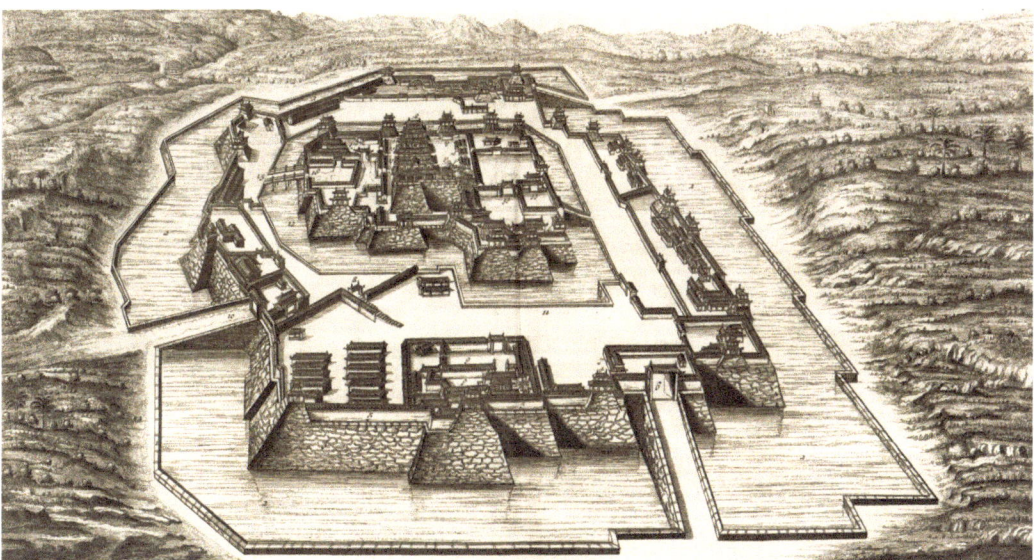

Figure 6.11 Montanus's published view of Osaka closely resembled that in Vingboon's manuscript atlas. Source: Jacob van Meurs, "Château d'Osacca situé dans Province de Quio," in Arnoldus Montanus, *Ambassades vers les Empereurs du Japon*, 2 parts (Amsterdam, 1680), between 1: 220 and 221. Colored engraving, 27 × 36 cm. Courtesy: John Carter Brown Library, Brown University.

Figure 6.12 Hennepin's published view of Niagara Falls gained verisimilitude from its close correspondence with his textual description: "Between Lake Ontario and Lake Erie there is a large and prodigious waterfall, whose falling water is completely overwhelming.... This unique waterfall is composed of two large flows of water around an outcropping of rocks in the middle. The waters which fall from this great height, foam and boil in the most terrible way in the world." There is no record of Hennipen's having provided a first-hand drawing for the print.

colonial landscapes more consistently and extensively than other Europeans had. Colonial landscapes claiming topographic accuracy based on first-hand observation became frequent at the same time that Britain extended its imperial interests globally by conquests, explorations, and accessions, in Canada, the Pacific, and India.

The diplomatic aftermath of the Seven Years' War had presented Britons with the prospect of a global territorial empire ruling non-British peoples. The simultaneous development of imperial landscape art linked three crucial developments in eighteenth-century British culture and politics: (1) the creation of a British identity available to peoples throughout the British Isles and in settler colonies, (2) the imperial assertion of this identity through commercial expansion and strategic success, and (3) the commodified representation of these identities and successes to a viewing public.[24] Viewing imperial spaces in the media of fine arts helped both colonists and metropolitans to maintain their identity and self-respect as civilized and civilizing Britons. Landscape art simultaneously created differences (topographically) and familiarity (aesthetically) among imperial spaces in ways that naturalized their appropriation as British rather than alien environments.

In conjunction with the empire's militarization, army and naval officers produced much of the topographic art representing Britain's new imperial landscapes. Yet their images sent home a curiously pacific, non-triumphalist message. Their aesthetically appealing representations of previously alien places helped naturalize imperial projects that sparked lively controversy in British politics. These disputes involved the eradication of the Scottish clans, the assimilation of a French-Catholic population in Quebec, the maintenance of peace between First Nations and European settlers in eastern Trans-Appalachia, the acquisition of additional Caribbean colonies dependent on slave labor, the intrusion of disease-ridden naval expeditions into unspoiled Polynesia, the suppression of militant patriotism in British North American colonies, the regulation of an empire-within-an-empire as the East India Company expanded territorially into Mughal realms, and the forcible transportation of free-born Britons to unimaginably distant New South Wales.[25]

The British global landscape first took shape in Canada.[26] Over the course of the British invasion, conquest, and occupation of New France, British officers deployed a demilitarized art for a benignly civilian, picturesque representation of a defeated enemy's landscape. In emphasizing the beauty of newly acquired domains and the picturesqueness of their inhabitants, British artists encouraged the metropolitan viewing public to consider their expanded North American empire as adding realms of amenable subjects.

The *Scenographia Americana* (1768) presented a 28-print series of topographical landscapes in an arc across North America from Montreal to Guadeloupe.[27] Its prints traced the geography of Britain's imperial strategy during the Seven Years' War. It began in Canada, with prospects of Montréal, Québec, and Louisbourg, plus views from the invading fleet in 1759 (the Montmorency Falls,

Cape Rouge, the Gaspé, Rock Percé, and the Miramichi Valley). It then moved southward for cityscapes of Boston, New York, and Charleston—reminders that the war in North America had been about the security and growth of Britain's colonies there. Next were four views of dramatic riverine landscapes in New York and New Jersey—the falls on the Passaic River in New Jersey, the highlands of Tappan Zee at the mouth of the Hudson, the Catskill Mountains farther upriver, and the Great Cohoes Falls on the Mohawk River in northern New York. From continental North America, the *Scenographia Americana* then shifted theater to the Caribbean, just as the war had, and concluded with six views of Havana, one view of the British attack on Roseau on Dominica, and three scenes around Guadeloupe.

Scenic attractiveness, as well as the success of British arms, united these places in their connection to Britain's global landscape. The *Scenographia Americana* was implicitly an imperial travelogue for the British North American colonies, drawing eclectically on previously published visual narratives of events during the recently concluded war. It privileged a scenic itinerary over a historical narrative (Figure 6.13).

The most important visual survey of Britain's new continental empire in North America was Joseph Frederick Wallet Des Barres' *Atlantic Neptune* (1777–1784), which published the results of a monumental coastal charting initiated shortly after the 1763 Treaty of Paris. For the first time, the Atlantic

Figure 6.13 The visual representation of the invasion balanced military events with picturesque landscapes.
Source: William Elliott after Hervey Smyth, *View of the Fall of Montmorency and the Attack Made by General Wolfe on the French Intrenchments near Beaufort* (London, 1760). Engraving, 34 × 52 cm. Courtesy: John Carter Brown Library, Brown University.

coast of North America would have a continuous series of navigational charts based on direct hydrographic surveys: from the tidal reaches of the Saint Lawrence in the west and the Strait of Belle Isle in the north, around Cape Breton Island and Nova Scotia, along the Atlantic seaboard of the British-settled colonies, and reaching south and westward to include the new British colonies of East and West Florida. The former New France would be seamlessly integrated with long-established colonies of British settlement and with recent accessions from Spanish America. The strategic importance of this project was self-evident: it made the Royal Navy's patrolling of a lengthy imperial coastline safer and more thorough. But its aesthetic priorities were more surprising. The *Atlantic Neptune* added scenic aquatints (the first ones of North American landscapes) to the conventional charts and coastal profiles dictated by its commission (Figure 6.14).

Representations of picturesque, sublime, and georgic scenes in Canada mitigated the forthrightness of its conquest and military rule. The supposed authenticity of topographic landscape art encouraged Britons at home to identify with Britain's new possessions overseas, while the familiar, picturesque style of these imperial landscapes masked their controversial aspects by naturalizing them as British scenes. Artists' representation of imperial landscapes harmonized the Empire's mayhem with its scenery.

As the Seven Years' War concluded, governments and corporations in France and Britain began to support the systematic exploration of the Pacific as a

Figure 6.14 The *Atlantic Neptune*'s favorite subjects for landscapes were rocky promontories—features that navigators had little interest in seeing closely. Source: Joseph Frederick Wallet Des Barres, "A View of the Plaister Cliffs, on the West Shore of Georges Bay," in Des Barres, *Atlantic Neptune*, 4 vols. (London, 1777–1784). Aquatint and etching, 43 × 54 cm. Courtesy: Library of Congress, Geography and Map Division.

A View of the Plaister Cliffs, on the West Shore of Georges Bay

new theater of Anglo–French geopolitics. Captain James Cook led the most famous explorations of the eighteenth century, three voyages throughout the Pacific that gave global dimensions to the British landscape.[28] Publication of Cook's account of his first Pacific voyage encouraged readers to visualize British voyagers' encounters with Pacific environments and cultures. These accounts included dramatic physical features such as "perforated rocks" at Tolaga Bay in New Zealand, distinctive Polynesian structures such as an altar on Huahine, and seaward views such as Matavai Bay in Tahiti as seen from One Tree Hill (Figure 6.15). Joseph Banks had hired Sydney Parkinson as a natural history illustrator for Cook's first voyage, but Parkinson developed his own artistic interests in picturesque aesthetics as they applied to the colors of tropical waters and the optics of tropical light and atmosphere.[29]

In contrast to representations of European explorations monumentalized by the de Brys' *Voyages*, which had relied on artists' imagining visually what first-hand accounts presented textually, the landscape prints accompanying Cook's *Voyages* privileged scenery drawn first-hand. Pacific spaces—extending over one-third of Earth's surface—became visually vivid to Europeans thanks to three artists aboard Cook's voyages: Parkinson, William Hodges, and John Webber. These artists applied a topographic imperative to represent how a place would appear to viewers if they were to go there themselves—a vast expanse of the world where Europeans had known themselves to be woefully ignorant after nearly 300 years of global imperial and commercial maritime expansion.

Figure 6.15 Parkinson's landscapes claimed topographic authority from being recorded on the spot, an impression reinforced here by the Maori leading two Britons to take in the view. Source: William Wollett after Sydney Parkinson, "A View of a Perforated Rock in Tolaga Bay in New Zealand," in John Hawkesworth, *An Account of the Voyages undertaken… for Making Discoveries in the Southern Hemisphere … by Commodore Byron, Captain Wallis, Captain Carteret, and Captain Cook*, 3 vols. (London, 1773), 2: pl. 17, between 318–319. Engraving, 20 × 25 cm. Courtesy: David Rumsey Map Collection, Stanford University Libraries.

Information from Cook's three exploratory voyages, from 1768 to 1780, showed Europeans the Pacific with a new, long-sought cartographic precision. However, the topographic representations were new information in a second way: not only did they show places of which Europeans had previously been ignorant, but they also showed those places with a new authority. What our modern visual culture takes for granted—access to naturalistic representations of unfamiliar places—was first prioritized in European visual culture during the era of Cook's voyages.

A pattern of scenic emphasis and political silence corresponded with the simultaneous scientific and imperial motives for Cook's voyages: to advance science *and* to discover natural and human commercial resources. The scientific work of the voyage was intended to enhance (1) European commodities through natural history, (2) commerce through cultural geography, and (3) navigational capability through astronomy. However, imperial priorities were deeply conflicted regarding territory and rule. The Admiralty's secret "Additional Instructions" directed that, after measuring the transit of Venus, Cook should search for the southern continent whose discovery "will redound greatly to the Honour of this Nation as a Maritime Power . . . and may tend greatly to the advancement of the Trade and Navigation thereof." Moreover, "With the Consent of the Natives," he was "to take possession of Convenient situations . . . in the Name of the King of England." Yet, a memo to Cook from the Earl of Morton, President of the Royal Society, which actually sponsored the voyages, could not have made it clearer that he should avoid any assertions of *dominion* over Pacific peoples:

> They are the natural, and in the strictest sense of the word, the legal possessors of the several Regions they inhabit. . . They may naturally and justly attempt to repel intruders. . . the Natives when brought under [control by violence] should be treated with distinguished humanity, and made sensible that the Crew still considers them as Lords of the Country.[30]

For Cook's second voyage, William Hodges came aboard as the first professional landscape artist in the Pacific. Hodges' appointment marked a new priority for landscape art in Pacific explorations. The Admiralty directed him to "make drawings and Paintings of such *places* in the Countries you may touch at in the course of the said voyage as may be proper to give a more perfect idea thereof than can be formed from written descriptions only."[31] Hodges had studied with Richard Wilson, Britain's foremost academic landscape artist, from whom he learned the advantages of working out-of-doors with oils, a practice that Wilson had adopted under the influence of Claude-Joseph Vernet while in Rome. Once in Polynesia, however, Hodges began to paint peoples and their landscapes in ways that did not need Italianate comparisons to recommend them (Figure 6.16). Cook readily conceded that Hodges' drawings recorded some information, particularly about material culture, more efficiently than he could convey in his journal. Reciprocally, Cook's journal of the second voyage had a new expressiveness about the beautiful and the sublime.

MONUMENTS IN EASTER ISLAND
Published Feb.¹ 1777, by W.Strahan New Street Shoe Lane, & Thoˢ Cadell in the Strand, London .

Drawn from Nature by W.Hodges .

Engraved by W.Woollett .
Nº XLIX.

Figure 6.16 The attribution of this surreal landscape declared its having been "drawn from nature by W. Hodges." Source: William Wollett after William Hodges, "Monuments in Easter Island," in James Cook, *A Voyage towards the South Pole, and Round the World. Performed in His Majesty's Ships the* Resolution *and* Adventure, 2 vols. (London, 1777), 1: pl. 49, facing 295. Engraving, 24 × 38 cm. Courtesy: David Rumsey Map Collection, Stanford University Libraries.

Cook's third voyage dispensed with scientists but not artists. The Admiralty hired John Webber as a "Draughtsman and Landskip Painter."[32] Webber drawings ranged widely in their settings, from panoramic landscapes to middle perspectives of settlements to close-ups of groups and structures. Webber was the first early modern European artist to take a sustained interest in representing Indigenous people's architecture. At settlements from the Nootka on Vancouver Island northward, Webber added interior scenes to his ethnographic agenda. Just as Hodges learned how to paint the unfamiliar light of tropical waters, in the sub-arctic and arctic North Pacific Webber came to terms graphically with representing a wide variety of vernacular architectures in their landscape settings (Figure 6.17).

Cook's colonialist gestures in the Pacific amounted to little more than leaving a few pigs to improve food resources and naming places after British aristocrats. However, artists aboard his vessels created a Pacific landscape based on direct observation, which helped naturalize the potential of Pacific spaces for British exploitation. The preface to the account of Cook's third voyage forthrightly asserted, "Every nation that sends a ship to sea will partake of the benefit; but Great Britain herself, whose commerce is boundless, must take the lead in reaping the full advantage of her own discoveries."

Unlike the Pacific, India had long been familiar topically with the British public, but only with the East India Company's conquest of a territorial empire in Bengal during the Seven Years' War did British visual culture develop a topographic interest there. Warren Hastings, Governor-General of Bengal, provided crucial

A VIEW of the TOWN and HARBOUR of St PETER and St PAUL, in KAMTSCHATKA.

patronage for Hodges, the first professional landscape artist in India as in the Pacific, to paint scenes on the spot. Hodges would show British viewers a readily imaginable India, with dozens of prominently exhibited paintings, an authoritative essay on comparative architectural history, a bilingual series of aquatint prints, and an illustrated travel account (Figure 6.18).[33] He identified strongly with Hastings' project to bring European and Indian cultures into conjunction:

> Whether we consider their high antiquity, grandeur, magnificence, or different style of architecture. . . their claim to the attention of our enlightened age is at least similar to that which the monuments and antiquities of Egypt have still amongst us.[34]

Vicarious nostalgia for the faded grandeur of Mughal rule implied that India's heritage needed British cultural salvage.

Hodges' success with Indian landscapes inspired and challenged the uncle–nephew team of Thomas and William Daniell to a more widely ranging effort, which resulted in *Oriental Scenery* (1795–1808), the most enduring presentation of British India's landscapes, which comprised four series of scenic views and two of architectural and archaeological antiquities.[35] After successfully marketing aquatint *Views of Calcutta* (1786–1788)—the first series of prints published in India and the most ambitious produced in any colony—they undertook three artistic expeditions throughout India. They initially retraced Hodges' itinerary up the Ganges, but once they passed northward of Delhi, they entered regions few Britons, and certainly no British artists, had seen. Not until Srinagar in the Garhwal mountains of the Himalayan foothills did they turn back. After a detour to Lucknow and the court of Nawab of Oudh, they continued down the Ganges, stopping at Bhagalpur, where they spent a year painting about 150 oils for sale in Calcutta. Their auction enabled the Daniells to tour southern India, all the way

Figure 6.17 The prints illustrating Cook's third voyage made an already enormous Pacific world even more varied culturally, here with the Kamchatkans' *balagans*, conical summer houses raised on stilts.
Source: P. T. Pouncy after John Webber, "A View of the Town and Harbour of St. Peter and St. Paul in Kamtschatka," in James Cook and James King, *A Voyage to the Pacific Ocean … for Making Discoveries in the Northern Hemisphere, to Determine the Position and Extent of the West Side of North America; its Distance from Asia; and the Practicality of a Northern Passage to Europe,* 3 vols. plus folio of prints (London, 1785), 4: pl. 74. Engraving, 22 × 50 cm. Courtesy: David Rumsey Map Collection, Stanford University Libraries.

Figure 6.18 Compressed depths of field encouraged Hodges' viewers to imagine themselves in proximity to the places illustrated, with low viewpoints, many of them as though from a boat.
Source: William Hodges, "A View of the Fort of Agra, on the River Jumah [with the Taj Majal in the left background]," in *Choix de vues de l'Inde, dessinées sur les lieux... Select Views in India, Drawn on the Spot in the Years 1780, 1781,1782, and 1783* (London, 1785–1788), pl. 15. Aquatint with etching, 29 × 45 cm. Courtesy: Yale Center for British Art.

to Cape Comorin. In February 1793, after seven years in India, they headed homeward. By March 1795, they were advertising subscriptions for *Oriental Scenery: Twenty–Four Views in Hindoostan*.

Though dedicated to the directors of the East India Company, the Daniells' first series began at the center of the *Mughal* Empire, not Calcutta. Its first print showed Delhi's Jama Masjid, the largest mosque in India, with a princely entourage of horses, elephants, and palanquins passing by its enormous eastern gate. Other scenes of Mughal grandeur included a garden wall of the Red Fort, elegant pavilions of the fortified palace at Allahabad, the Taj Mahal, and the mausoleum of Akbar (Figure 6.19). Mughal buildings appeared in near-pristine condition, while Hindu ones were typically in picturesque compositions, with ruins amidst encroaching forest, banyan trees crowding active temples, mysterious solitary structures sited on hilltops and riverside outcroppings, and the eclectic array of palaces, bathing stairs, and funeral platforms along Varanasi's *ghats*. The series took India on its own terms while playing down the British governing presence. After all, it was in the Company's interest to have the Mughal Empire still seem viable, lest Parliament take over British India outright.

In 1810, the Daniells published *A Picturesque Voyage to India by the Way of China*, a book whose title marked the integration of India into a global British

GATE OF THE TOMB OF THE EMPEROR AKBAR, AT SECUNDRA, NEAR AGRA.

landscape. Its introduction expressed unsurpassable self-congratulation for the ways that landscape art had mitigated the anxieties of conquest:

> It is an honourable feature in the late century, that the passion for discovery, originally kindled by the thirst for gold, was exalted to higher and nobler aims than commercial speculations. Since this new era of civilization, a liberal spirit of curiosity has prompted undertakings to which avarice lent no incentive, and fortune annexed no reward: associations have been formed, not for piracy, but humanity: science has had her adventurers, and philanthropy her achievements: the shores of Asia have been invaded by a race of students with no rapacity but for lettered relics; by naturalists, whose cruelty extends not to one human inhabitant; by philosophers, ambitious for the extirpation of error, and the diffusion of truth. It remains for the *artist* to claim his part in these *guiltless spoliations*, and to transport to Europe the picturesque beauties of those favoured regions.[36]

The Daniells' self-congratulation suggested that the British landscape art of India functioned as self-propaganda of empire for polite society among Britain's political nation. British armies had invested the Indian countryside to rule and to

Figure 6.19 The only visual reference to a British presence in the first series of *Oriental Scenery* was this scene of the Daniells' party, with lounging British officers and gentlemen, palanquin bearers and porters, and British sepoys and Mughal soldiers at ease.
Source: Thomas Daniell and William Daniell, "Gate of the Tomb of the Emperor Akbar, at Secundra, near Agra," in Daniell and Daniell, *Oriental Scenery: Twenty–Four Views in* Hindoostan (London, 1795), pl. 9. Colored aquatint; 47 × 60 cm. Courtesy: Yale Center for British Art.

exploit it; British artists had toured that same countryside to produce illustrations expressing respect for India's great civilizations and appreciation of its natural beauties. They typically represented locales closely associated with events in the expansion of British India, but in ways that kept the contradictions, exactions, and failures of British rule in India from being shown in too sharp a light.

India figured crucially in Charles-Maurice de Talleyrand-Périgord and Napoleon Bonaparte's strategy for invading Egypt. It deflected the Directory's plan to attack England by proposing an indirect way to cut off its crucial fiscal resource: trade with India.[37] In presenting their plan to the Directory, Bonaparte could not resist comparing his expedition with that of Alexander the Great (Figure 6.20). The Directory authorized a force of over 300 ships and 40,000 troops. Napoleon's conquest fizzled after Nelson destroyed the French fleet at the Battle of Aboukir. After a failed effort to stop Ottoman forces in Palestine, Napoleon retreated to Egypt (where he confronted an insurrection), abandoned his stranded army, escaped back to France, and soon became First Consul.

Napoleon initiated a vast survey of Egypt's natural history, geology, topography, material culture, and archaeology as a transformative exercise in cultural largesse for both Europe and Egypt. He poured artistic and scholarly resources into Egypt. The invasion party included over 150 *savants*—natural philosophers, engineers, surveyors, and humanists—who formed a *Commission des Sciences et Arts d'Égypte*. As Jean-Lambert Talien—Jacobin publicist, Terrorist, Thermidorian, and Napoleonic favorite—explained in the Commission's journal, *La Décade Égyptienne* (anticipating the Daniells' "guiltless spoliations"):

> The conquest of Egypt should not be useful for France solely in political and commercial terms. Arts and sciences should also profit. We no longer live in times when conquerors only know how to destroy; where the thirst for gold directs all their actions; where devastation, persecution, and intolerance shape everything. Today, the French respect not only the laws and customs of the land but also the prejudices of the peoples whose territory they occupy. They let time, reason, and education bring about the changes, which philosophy and the *lumières* of the century have prepared, and which every day come nearer to realization.[38]

Napoleon chartered the *Institut de l'Égypt*, modeled on the *Institut de France*, to publicize the Commission's findings. Once returned to Paris, after French forces in Egypt had surrendered to the British, the commissioners initiated publication of the *Description de l'Égypt* (1809–1829), which mobilized 2,000 artists, engravers, writers, and artisans.[39] Ten volumes, some over a meter in height, reproduced over 3,000 drawings in 837 plates. None of the prints portrayed the invasion itself, though many showed French troops moving easily about the countryside. Most prints in the five volumes of *Antiquités* abstracted subjects from their spatial context with recordings of hieroglyphics and plans and elevations of monuments, but topographic prints showed each site in its landscape

Figure 6.20 Napoleon upstages the pyramids at Giza. He reputedly rallied his forces by declaring, "from the tops of these monuments forty centuries are watching us." Source: [Napoleon at the Battle of the Pyramids, July 21, 1798], Philippe Joseph Auguste Vallot after Antoine-Jean Gros (original painting 1810; Paris, 1838). Etching with engraving, 67 × 49 cm. Courtesy: Library of Congress, Prints and Photographs Division.

setting (Figure 6.21). Moreover, the two volumes on the *État Moderne* represented vernacular Egyptian architecture and genre scenes of daily life, with views of bazaars, workshops, and interiors of houses and baths. The *Description de l'Égypt* sought to appropriate Egypt in its entirety, but the British got the Rosetta Stone (Figure 6.22).

From the era of Columbus' voyages to Napoleon's invasion of Egypt, no European empire successfully put landscape art to imperial purposes, except the British, whose artists used it to deflect criticisms of imperialism rather than to

Figure 6.21 Seven views in the *Description de l'Égypte* took the reader on a vicarious tour of Memphis' archaeological sites. The Great Pyramid and the Sphinx had only recently been excavated when this view was taken. Source: Friedrich Schroeder after Nicolas-Jacques Conté, "Pyramides de Memphis: Vue du sphinx et de la grande pyramide, prise du sud-est," *Description de l'Égypte, ou, Recueil des observations et des recherches qui ont été faites en Égypte pendant l'expédition de l'armée française*, Vol. 5: *Antiquités, Planches* (Paris, 1809–1828), pl. 11. Engraving, 46 × 69 cm. Courtesy: Beinecke Rare Book and Manuscript Library, Yale University.

assert *imperium*. For most of the early modern period, imperially scaled publications of landscape art told cosmopolitan rather than imperial stories. The development of woodcut printing in the late fifteenth century provided the visual resources for large-scale, highly successful publications of European, African, Asian, and American cityscapes. These collections were imperial in scale and sometimes dedicated to emperors, but they actually encouraged politically de-centered, cosmopolitan views of the world. As European empires expanded globally, landscapes of these new realms had a low priority, even in the most massive collection of extra-European images, the De Brys' *India Occidentalis* and *India Orientalis*. Short-lived Dutch Brazil was the exception, proving the rule. Dutch artists there, with their strong tradition of landscape art, readily represented colonial scenery in the context of imperial expansion, but by the time their work might have served imperial purposes back home, Dutch Brazil had been lost, and the United Provinces were giving up ambitions for territorial empire. Beginning with the Seven Years' War, however, British imperial authorities regularly commissioned artists to produce topographic landscapes of a newly expanded empire. These works typically prioritized places that were tenuously colonized and politically contentious, and they implicitly sought to mask imperial intents in aesthetically cosmopolitan terms. The *Description de l'Égypte* realized such masking on a monumental but profoundly ironic scale since the empire it was meant to glorify—Napoleon's—had disappeared before the project was completed. Imperial landscape art had not harmonized well with imperialism because it was usually over-determined to establish legitimacy in tenuous circumstances, and assertions of empire aroused anxieties about

Figure 6.22 The frontispiece of the *Description* framed an imaginary "perspective view of Egypt" with images of conquest: an Alexander-like figure in a chariot, with Arts and Sciences in his train, leads the Armée de l'Orient against cowering Mameluks; the side panels celebrate French victories; and in the lower panel Egyptian forces recognize Napoleon as Emperor. But the seeming landscape is actually a visual catalog of archaeological sites along the Nile and of looted antiquities, most of which ended up in the British Museum.
Source: Reville, Girardet, and Sellier after François-Charles Cecile, "Frontispice," *Description de l'Égypte, ou Recueil des observations et des recherches qui ont été faites en Égypte pendant l'expédition de l'armée française, publié par les ordres de sa Majesté l'Empereur Napoléon le Grand*, Vol. 1. *Antiquités, Planches* (Paris, 1809). Etching, 54 x 71 cm. Courtesy: Brown University Library.

self-defeating over-extensions. Cosmopolitan representations of imperial spaces mitigated such anxieties with depoliticized landscapes.

NOTES

1   This chapter draws on material in John E. Crowley, *Imperial Landscapes: Britain's Global Visual Culture, 1745–1820* (London and New Haven, CT: Yale University Press, 2011).

2   Benedict Richard O'Gorman Anderson, *Imagined Communities: Reflections on the Origin and Spread of Nationalism* (orig. publ. 1983; rev. edn., London: Verso, 1991);

Rites of Power: Symbolism, Ritual, and Politics since the Middle Ages, ed. Sean Wilentz (Philadelphia: University of Pennsylvania Press, 1985); Peter Burke, *The Fabrication of Louis XIV* (New Haven, CT and London: Yale University Press, 1992); Anthony Pagden, *Peoples and Empires: A Short History of European Migration, Exploration, and Conquest, from Greece to the Present* (New York: Modern Library, 2003); Jane Burbank and Frederick Cooper, *Empires in World History: Power and the Politics of Difference* (Princeton, NJ and Oxford: Princeton University Press, 2010); Holger Hoock, *Empires of the Imagination: Politics, War, and the Arts in the British World, 1750–1850* (London: Profile Books, 2010).

3 Larry Silver, *Marketing Maximilian: The Visual Ideology of a Holy Roman Emperor* (Princeton, NJ and Oxford: Princeton University Press, 2008).

4 Steven C. A. Pincus, "Popery, Trade and Universal Monarchy: The Ideological Context of the Second Anglo–Dutch War," *English Historical Review* 107, no. 422 (January 1992): 1–29; Anthony Pagden, "'Savage–Impulse–Civilized Calculation': Conquest, Commerce, and the Enlightenment Critique of Empire," in *The Burdens of Empire, 1539 to the Present*, ed. A. Pagden (New York: Cambridge University Press, 2015), 224–242.

5 Richard Koebner, *Empire* (1961; New York: Grosset & Dunlap, 1965), chapters 1–2; Anthony Pagden, *Lords of All the World: Ideologies of Empire in Spain, Britain and France c. 1500–c. 1800* (New Haven, CT and London: Yale University Press, 1995); David Armitage, *The Ideological Origins of the British Empire* (Cambridge: Cambridge University Press, 2000).

6 Bernard von Breydenbach, *Peregrinatio in Terram Sanctam* (Mainz, 1486); Elizabeth Ross, *Picturing Experience in the Early Printed Book: Breydenbach's* Peregrinatio *from Venice to Jerusalem* (University Park: Pennsylvania State University Press, 2014); Juergen Schulz, "Jacopo de' Barberi's View of Venice: Map Making, City Views, and Moralized Geography before the Year 1500," *Art Bulletin* 60, no. 3 (September 1978): 425–474.

7 Hartmann Schedel, *Registrum huius Operis libri cronicarum cum figuris et ymagibus ab inicio mundi* (Nuremberg, 1493).

8 The cityscapes first appeared in the fifth edition: Sebastian Münster, *Cosmographiae universalis lib. VI. In quibus iuxta certioris fidei scriptorum traditionem describuntur, omnium habitabilis orbis partium situs, propriaeq[ue] dotes. Regionum topographicae effigies* (Basle, 1550); Jasper van Putten, *Networked Nation: Mapping German Cities in Sebastian Münster's Cosmographia* (Leiden and Boston, MA: Brill, 2017).

9 Lucia Nuti, "The Perspective Plan in the Sixteenth Century: The Invention of a Representational Language," *Art Bulletin* 76, no. 1 (March 1994): 105–128.

10 Antoine Du Pinet, *Plantz, pourtraitz et descriptions de plusiers villes et forteresses, tant de l'Europe, Asie, & Afrique, que des Indes, & terres neuues: Leurs fondations, antiquitez, & manieres de viure* (Lyons, 1564).

11 *Civitates Orbis Terrarum* is a title of convenience; it appeared only for the first volume, with each of the subsequent volumes having its own title.

12 North Africa: Alexandria, Algiers, Arzilia, Azemmour, Cairo, Casablanca, Mahdia, Peñon de Velez, Safi, Tangier, Tunis. Mediterranean: Cyprus, Famagusta, Rhodes, Valetta. Levant: Aden, Damascus, Hormuz, Istanbul, Jerusalem. Indian Ocean: Aden, Kilwa, Calicut, Cannanore, Diu, Goa. Sub–Saharan Africa: Elmina, Mombasa, Sofala. Peter van der Krogt, "Mapping the Towns of Europe: The European Towns in Braun and Hogenberg's Town Atlas, 1572–1617," *Belgeo: Revue Belge de géographie* 3–4 (2008): 371–398.

13 Richard L. Kagan, *Urban Images of the Hispanic World, 1493–1793* (New Haven, CT and London: Yale University Press, 2000), 28 quoted; Egbert Haverkamp-Begemann, "The Spanish Views of Anton van den Wyngaerde," in *Spanish Cities of the Golden Age: The Views of Anton van den Wyngaerde*, ed. Richard L. Kagan (Berkeley: University of California Press, 1989).

14 Barbara E. Mundy, *The Mapping of New Spain: Indigenous Cartography and the Maps of the Relationes Geográficas* (Chicago, IL: University of Chicago Press, 1996).

15 Michiel van Groesen, *The Representation of the Overseas World in the De Bry Collection of Voyages* (1590–1634) (Leiden and Boston, MA: Brill, 2008). For the *India Occidentalis* and the *India Orientalis* online: https://bodmerlab.unige.ch/recits-et-images/debry/#/grands-voyages/Preface (accessed May 18, 2021).

16 *A Humanist Prince in Europe and Brazil, Johan Maurits Van Nassau–Siegl 1604–1679: Essays on the Occasion of the Tercentenary of His Death*, eds. E. van den Boogaart, H. R. Hoetink, and P. J. P. Whitehead (The Hague: Johan Maurits van Nassau Stichting, 1979); Whitehead and Boeseman, *A Portrait of Dutch 17th Century Brazil: Animals, Plants and People by the Artists of Johan Maurits of Nassau* (Amsterdam, Oxford, and New York: North-Holland Publishing Company, 1989).

17 For van Baerle's *Rerum per octennium in Brasilia* online: https://gallica.bnf.fr/ark:/12148/btv1b8556547p.r=octennium?rk=21459;2 (accessed December 16, 2022).

18 Catherine Levesque, *Journey Through Landscape in Seventeenth–Century Holland: The Haarlem Print Series and Dutch Identity* (University Park: Penn State University Press, 1994); Walter S. Gibson, *Pleasant Places: The Rustic Landscape from Bruegel to Ruisdael* (Berkeley, Los Angeles, and London: University of California Press, 2000).

19 Benjamin Schmidt, *Innocence Abroad: The Dutch Imagination and the New World, 1570–1670* (Cambridge and New York: Cambridge University Press, 2001).

20 For the view of Osaka in Vingboons' atlas: https://www.atlasofmutualheritage.nl/en/page/8519/vogelvlucht-van-het-kasteel-te-osaka#description-8519 (accessed December 16, 2022).

21 Philip P. Boucher, *Les Nouvelles Frances: France in America, 1500–1815—An Imperial Perspective* (Providence, RI: John Carter Brown Library, 1989).

22 Marc Grignon, *Loing du Soleil: Architectural Practice in Quebec City during the French Regime* (New York: Peter Lang, 1997).

23 Jean Baptiste du Tertre, *Histoire generale des Antilles habitées par les françois*, 2 vols. (Paris, 1667–1671); Jean Baptiste Labat, *Nouveaux voyage aux isles de l'Amerique contenant l'histoire naturelle de ces pays, l'origine, les moeurs, la religion & le gouvernment des habitants anciens & modernes*, 6 vols. (Paris, 1722); Claude–Charles Le Roy Bacqueville de La Potherie, *Histoire de l'Amérique septentrionale*, 4 vols. (Paris, 1722–1723); Joseph François Lafitau, *Moeurs des sauvages ameriquains comparés aux moeurs des premiers temps*, 2 vols. (Paris, 1724); Pierre-François-Xavier de Charlevoix, *Histoire et description générale de la Nouvelle France* (Paris, 1744).

24 P. J. Marshall and Glyn Williams, *The Great Map of Mankind: British Perceptions of the World in the Age of the Enlightenment* (Cambridge, MA: Harvard University Press, 1982); C. A. Bayly, *Imperial Meridian: The British Empire and the World 1780–1830* (London and New York: Longman, 1989); Linda Colley, *Britons: Forging the Nation 1707–1837* (New Haven, CT and London: Yale University Press, 1992); John Brewer, *The Pleasures of the Imagination: English Culture in the Eighteenth Century* (Chicago, IL: University of Chicago Press, 1997); Kathleen Wilson, *The Island Race: Englishness, Empire and Gender in the Eighteenth Century* (London: Routledge, 2003); P. J. Marshall, *"A Free Though Conquering People": Eighteenth–Century Britain and its Empire* (Aldershot, Hampshire: Routledge, 2003).

25 John E. Crowley, *Imperial Landscapes: Britain's Global Visual Culture, 1745–1820* (London and New Haven, CT: Yale University Press, 2011).

26 Hilda Neatby, *Quebec: The Revolutionary Age 1760–1791* (Toronto: McClelland & Stewart, 1966); Bruce Robertson, *"Venit, Vidit, Depinxit*: The Military Artist in America," in *Views and Visions: American Landscape before 1830*, eds. Edward J. Nygren and Bruce Robertson (Washington, DC: Corcoran Gallery of Art, 1986), 86–103; Colin M. Coates, "Like 'The Thames Towards Putney': The Appropriation of

Landscape in Lower Canada," *Canadian Historical Review* 74, no. 3 (September 1993): 317–343; Victoria Dickenson, *Drawn from Life: Science and Art in the Portrayal of the New World* (Toronto: University of Toronto Press, 1998).

27 *Scenographia Americana: Or, a Collection of Views of North America and the West Indies* (London, 1768); I. N. Phelps Stokes, *The Iconography of Manhattan Island 1498–1909*, 6 vols. (New York: Robert H. Dodd, 1915–1928), 1: 281–295.

28 J. C. Beaglehole, *The Life of Captain James Cook* (London: Adam and Charles Black, 1974); Bernard Smith, *European Vision and the South Pacific* (1960; 2nd ed., New Haven, CT: Yale University Press, 1985); *The Art of Captain Cook's Voyages*, eds. Smith and Rüdiger Joppien, 3 vols. (New Haven, CT and London: Yale University Press, 1985–1988); Smith, *Imagining the Pacific: In the Wake of the Cook Voyages* (Melbourne: Melbourne University Press, 1992); Nicholas Thomas, *Cook: The Extraordinary Voyages of Captain James Cook* (New York: Walker, 2003); *William Hodges 1744–1797: The Art of Exploration*, eds. Geoff Quilley and John Bonehill (New Haven, CT: Yale University Press, 2004).

29 *Sydney Parkinson: Artist of Cook's "Endeavour" Voyage*, ed. D. J. Carr (London and Canberra: British Museum Natural History and Australian National University, 1983); Sydney Parkinson, *A Journal of a Voyage to the South Seas, in His Majesty's Ship the Endeavour* (orig. publ. 1773; London, 1784).

30 For Cook's Admiralty instructions, see *The Journals of Captain James Cook on His Voyages of Discovery*, ed. J. C. Beaglehole, 4 vols., plus portfolio of charts and views edited by R. A. Skelton (Cambridge: Hakluyt Society, 1955–1974), 1: cclxxix–cclxxxiv (cclxxxii–iii quoted); James Douglas, Earl of Morton, "Hints offered to the Consideration of Captain Cooke, Mr Bankes, Doctor Solander, and the Other Gentlemen Who Go upon the Expedition on Board the *Endeavour*," in *Journals of Captain James Cook*, ed. Beaglehole,1: 514–515 quoted.

31 Bernhard Smith, "William Hodges and English *Plein–Air* Painting," *Art History* 6, no. 2 (1983): 143–152, 145 quoted (emphasis added).

32 William Hauptman et al., *John Webber 1751–1793: Landschaftsmaler und Südseefahrer / Pacific Voyager and Landscape Artist* (Bern: Kunstmuseum, 1996).

33 Isabel Combs Stuebe, *The Life and Works of William Hodges* (New York: Garland, 1979); Giles H. R. Tillotson, *The Artificial Empire: The Indian Landscapes of William Hodges* (Richmond, UK: Routledge, 2000); Geoff Quilley, "Hodges and India," in *William Hodges, 1744–1797: The Art of Exploration*, eds. Quilley and John Bonehill (New Haven, CT: Yale University Press, 2004); William Hodges, *Choix de vues de l'Inde, dessinées sur les lieux … Select Views in India, Drawn on the Spot in the Years 1780, 1781, 1782, and 1783, and Executed in Aqua Tinta* (London, [1785–]1788); William Hodges, *Travels in India, During the Years 1780, 1781, 1782, & 1783* (London, 1793).

34 William Hodges, *Dissertation on the Prototypes of Architecture, Hindoo, Moorish, and Gothic* (London, 1787), 7.

35 Mildred Archer, *Early Views of India: The Picturesque Journeys of Thomas and William Daniell 1786–1794. The Complete Aquatints* (London: Thames and Hudson, 1980); Hermione de Almeida and George H. Gilpin, *Indian Renaissance: British Romantic Art and the Prospect of India* (Aldershot: Ashgate, 2005).

36 Thomas Daniell and William Daniell, "Introduction," in *A Picturesque Voyage to India by the Way of China* (London, 1810), i–ii quoted (emphasis added).

37 Maja Jasanoff, *Edge of Empire: Lives, Culture, and Conquest in the East, 1750–1850* (New York: Vintage, 2006).

38 "Prospectus," *La Décade Egyptienne, Journal Littéraire et d'Économie Politique* (Cairo, 1798 [An VII de la République Française]), 6 quoted in translation; https://gallica.bnf.fr/ark:/12148/bpt6k106598h/f6.item (accessed May 30, 2021). Todd Porterfield, *The Allure of Empire: Art in the Service of French Imperialism, 1798–1836* (Princeton, NJ: Princeton University Press, 1998).

39 Fernand Beaucour, Yves Laissus, and Chantal Orgogozo, *The Discovery of Egypt: Artists, Travellers and Scientists* (Paris: Flammarion, 1990); Terence M. Russell, *The Discovery of Egypt: Vivant Denon's Travels with Napoleon's Army* (Stroud: Sutton Publishing, 2005); David Prochaska, "Art of Colonialism, Colonialism of Art: The *Description de l'Égypte* (1809–1828)," *L'Esprit Créateur* 34, no. 2 (Summer 1994): 69–91.

Chapter 7: Is Landscape Indigenous?

Ọlátúnjí Adéjùmọ̀

INTRODUCTION

The biosphere comprises diverse landscapes defined by geomorphology, climatic features, and different levels of anthropogenic influences. Environmental planners see landscape as beyond the spatial representation of overlapping ecosystems to accommodate functional values. That is, rather than describing landscape entirely in terms of its spatial characteristics alone, it is also defined in terms of its functional attributes. Landscapes are environmental systems that are species or species association specific and delimited. Each species thrives within supporting geomorphic characteristics that favor its existence. The fundamental components of a typical landscape are geographic and functional. These two components cannot be completely separated without minimizing important inherent information needed for sustainable planning. This portrays any landscape as a heterogeneous environment that influences the ecological diversity of a bioregion. If landscape is a spatial representation of ecosystems, then it is an agglomeration of ecological goods and services. Landscape is, therefore, a bank of natural assets on which a livelihood depends.

A livelihood comprises a person's assets, capabilities, and activities toward a satisfying means of living. Assets include stocks of natural resources in the landscape and environmental services provided by contextual ecological systems. A study of landscape to isolate ecological goods and services is an important strategy for meeting food security, reducing poverty, and addressing environmental injustice. Landscapes understood in this manner are considered social-ecological systems—a manifestation of the symbiotic relationship between humans and local ecosystems. When human imprints are accommodated, such natural landscapes transcend into cultural landscapes. Cultural landscapes are a symbiotic union of biophysical features of a geographically specific place and the overlay of human cultural activities. This union collectively defines a landscape's character and differentiates it from other landscapes. Human perceptions, belief systems, narrated myths, and collective experiences influence the form and meaning of cultural landscapes within a defined territory.

The word "territory" is particularly important when dealing with landscape as a spatial entity. Territory is a spatial area within landscapes with jurisdictional authority. In this way, the territory is a political and geographic framework that

DOI: 10.4324/9781003148142-8

legally accommodates a population. It covers the land within defined borders and "may extend to water bodies including the portions of sea, below the earth's surface, and above the earth's surface."[1] In modern parlance, territory is often discussed with the idea of territorial rights. Territorial rights cover other rights beyond jurisdiction. Moore, citing Simmons from "On the Territorial Rights of States," identified the rights to control resources within the geographical area, the rights to control borders and regulate the flow of people and goods across them, and the right to defend the territory against outside aggression.[2] This chapter asks the broad question: Is landscape indigenous? Or, more precisely, what makes landscape indigenous?

These questions will be examined within the context of Nigerian tropical landscapes. Emphasis will be placed on the southwestern geopolitical bioregion that supports Yoruba nationality. This chapter will investigate the relationship between sacred landscapes and the people as part of the process of land indigeneity. This inquiry is underpinned by the philosophy of territoriality, sacred landscapes as "spiritscapes," and the cartographic concept of ley lines. These understandings are explored through archival material and supported by interviews with local landscape custodians.

INDIGENEITY

The definition of the word "indigenous" is elusive. At surface value, it connotes a place of birth. However, indigenous is beyond a place of birth, cumulating in the place of residence. The literature often ties the word indigenous to another word, "people." In as much as this chapter is meant to address landscape and indigeneity, its discourse may be incomplete without accommodating the people who are the principal landscape modifying agents. Therefore, indigenous is about the experience garnered over a long period of time in inhabiting the place of birth. The experience is not about individuals but a group of people of the same ancestry bonded together by the same mythological narrative that influences their inhabited patch of the biosphere. Cunningham and Stanley referred to Te Ahukaramu Charles Royal for a proper understanding of the term indigenous.[3] His definition of indigenous from a global perspective is integral to this book and chapter. His proposal is that the word indigenous should connote people from cultural settings with a worldview that respects the union between humans and nature. These are places on Earth that emphasize biocentrism as a living philosophy. Royal suggests that this is why Western (Judeo-Christian) societies see God as an external supreme being that resides in heaven "above," while Asian societies situate a godhead internally, and the faithful rely on intense internal meditation. By contrast, an indigenous worldview considers the people an integral part of the ecosystem, with humans "having a seamless relationship with nature including seas, land, rivers, mountains, flora and fauna."[4]

Characteristics of Indigenous People

On a global platform, the International Labour Organization (ILO) Convention made the first attempt to define who indigenous people are.[5] This was followed by the United Nations Economic and Social Council's (ECOSOC's) Sub-Commission on Prevention of Discrimination and Protection of Minorities. The ILO convention defined indigenous people as those "whose ancestors have lived in a geo-referenced area before the settlement or the formation of the modern state borders."[6] In addition, the convention stated that indigenous peoples would have maintained either wholly or partly their own social, economic, cultural, and political institutions. In the same tune, Martinez Cobo's submission explains:

> Indigenous communities, peoples and nations are those which, having a historical continuity with pre-invasion and pre-colonial societies that developed on their territories, consider themselves distinct from the other sectors of societies now prevailing in those territories, or parts of them.[7]

Both the ILO and Martínez-Cobo provided the most accredited descriptions of indigenous peoples on a global political scale. The description characterizes indigenous people as having a strong attachment to a culturally distinct biophysical area with a preserved social, economic, and political way of life. Therefore, the term indigeneity revolves around a distinct historical narrative rooted in pre-colonialism; geo reference land area with adequate natural capital, distinct cultural system, especially language; an unwavering commitment to the relationship between ancestral systems relative to the supporting landscapes; and unique social, economic, and political systems. Indigenous people are thus bonded to their traditional landscapes beyond material goods and services. This is a pointer to understanding the importance of indigenous people in the conservation and management of regional landscapes. The starting block is to identify the values that indigenous people apply to landscapes on which they claim tenure.

Undercurrent in Indigeneity

Indigenous landscapes exhibit combined natural and cultural features on which the community's essences and means of livelihood depend. It is a deep-seated horizontal and vertical knowledge of ecological systems and plant associations. While horizontal knowledge respects symbiotic relationships in all biophysical activities, verticalism suggests a higher-dimensional web of life beyond this three-dimensional plane. Horizontal knowledge is traditional knowledge. The Indigenous knowledge system is the philosophy adopted by local people when living within a bioregion. Included are knowledge of subsistence-based traditional technologies, seasons and climate patterns, land management, ethnobotany, ecological knowledge, celestial navigation, ethnoastronomy, and eco-spirituality. This knowledge is rooted in a generational accumulation of empirical observations and interactions with the contextual environment. The knowledge oozes from a long-time human-land interaction fueled by a landscape-bound belief system. It manifests in immovable heritage resources, including shrines and seemingly insignificant waymarks to nonmembers of the community. These material signatures on the landscape are transmitted orally from one generation to the other.

Cloth Trees

This is true of totem species and their habitats. Tree species including, *Ficus exasperata*, *Spondias mombin*, *Chlorophora excelsa*, *Adansonia digitata*, *Ceiba pentandra*, *Chrysophyllum albidum*, *Irvingia grandifolia*, *Sterculia africana*, *Elaeis guineensis*, *Parkia biglobosa*, and *Dialium guineense* are socio-religiously classified as totem species set aside for common utilization. A totem animal or plant species is connected to supporting landscapes and represents a people. Typical examples of totem plant species are "Cloth Trees," which are community spiritual essences and way markers (Figure 7.1). Gurstelle, citing Agbaje-Williams, identified 24 sacred tree species in Nigerian Yorùbá human settlements.[8, 8b] They are plant species totems venerated for either environmental containment, territoriality, or historical consciousness. It is worth mentioning that not all totem species have cloth wrapped around the trunk (Figure 7.2). Clothed or sacred trees are not peculiar to the Yorùbá of Nigeria. Trees are common territorial and religious symbols in many African communities. They constitute a community biogeographic information system.

Often, totem animal species emerged in community zoomorphic ornaments and artworks, including tattoos. While most tattoos in the Global North are for fashion, zoomorphic tattoos in sub-Sahara Africa are eco-religious commitments to the supporting landscape. Totem species create community

Figure 7.1 Typical cloth trees. Courtesy: Yaagba Onilu.

Figure 7.2 *Adansonia digitata* sacred tree Yola. Courtesy: The author.

protection and conservation frameworks that restrain hunting for consumption and promote indigenous landscape management techniques. Values that indigenous people put on bioregional landscapes can be identified according to their importance on environmental, historical, spiritual, and livelihood dimensions.

Developmental Cosmogram

The values indigenous people put on their landscapes reflect on adopted spatial conceptualization and land resources management. This perception is a product of its worldview. A worldview is the window through which a group of people relate to their immediate environment. Then, the worldview addresses the perception of indigenous peoples to a geopolitical scenario and manifests in intuitive design and planning cosmograms. Smith (2007), while citing Clark (2004), defined a *cosmogram* as a "representation of the entire universe through symbolic shorthand or artistic metaphor."[9] Rapoport submitted that cosmograms in aboriginal settlements might be reflected on the city scale, land tenure system, land use, individual buildings, and even open space scales.[10] Cities and their components were built as cosmic images. The cardinal points were essential to Mesoamerica mythology, cosmology, and ritual practice. The cosmogram of the Yoruba people of Nigeria is a graphic representation of cosmic elements coded in the poetic Ifá myths of Earth's creation (Figure 7.3). It is indeed a mimesis of her cosmological worldview. The cosmogram is a 16-sided polygon with an inserted square that follows the cardinal orientation and is named after four primordial deities or "orisas," namely Ọbàtálá (Òrìsà-Ńlá) to the north, Ògún (Alákáayé) to the south, Ọ̀rúnmìlà (Ifá) to the west, and Ọ̀rámfẹ̀ (Ṣàngó or Jàkúta) to the east. The cosmogram is most apparent in pre-colonial Yoruba urban morphology, with

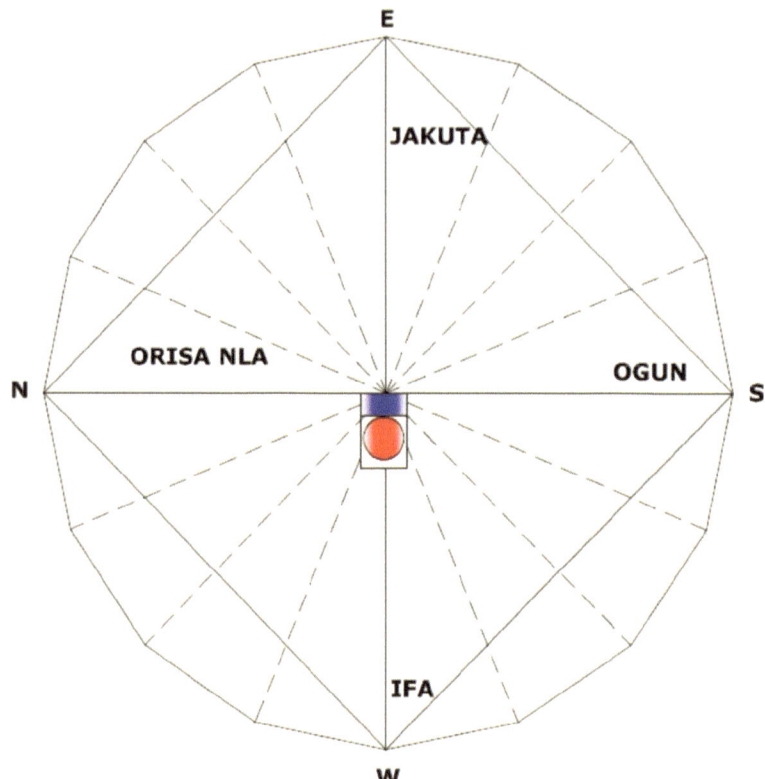

Figure 7.3 Yoruba philosophical cosmogram. Courtesy: The author.

the palace and King's market occupying the central dome and facing the rising sun.[11] Such areas in any twenty-first-century Yoruba human settlements are legible in core historical districts and rural settlements.

We should also acknowledge that most indigenous settlements in Africa across the Global South have their own form of cosmogram. Where the towns or villages are lost to war or natural disaster, semiotic studies can recreate their cosmograms, while archaeological works reveal how they were inscribed on the historic landscapes. That is, abandoned indigenous landscapes are a palimpsest. The vestiges of people's earlier use of the landscape remain. They are also either recorded in the surviving documented record or orally transmitted from one generation to the other in poems and folktales. It is about examining indigenous reality.

Popova-Gosart identified three approaches to examine the indigenous realities, namely legal, political, and social perspectives.[12] While legal and political realities focus on fighting for autonomy and are often traceable to human rights violations by a greater political order, the socio-ecological perspective emphasizes experience in analyzing indigenous realities. The latter approach highlights the position of "first settlers of a particular land" and the "dichotomy between modernity and tradition."[13] The first settlers' experience explores history and geosophy. This is within social perspectives of indigeneity made relevant by anthropology, heritage preservation, conservation biology, ecology, geology, and archaeology. These disciplines ascertain the territoriality of indigenous people established by totem species and spatiality informed by cosmogram. Though a political approach to defending indigeneity is very important, this chapter explores the windows of territoriality and geosophy as additional arguments to justify indigenous interpretations of landscapes.

TERRITORIALITY

Humans are social animals that are spatially conscious of their environment. Humans are territorial in their disposition. Lea, citing the works of Taylor, explained territory as the extent of geographical unit claimed by a community or individual to belong to the community.[14] Territory is, therefore, a geo-specific area on the landscape where a state, community, or individual can exercise the right of ownership. This supported Raffestin's thesis, cited by Klauser, that territoriality consists of an assemblage of relationships of individuals or social groups developed on different scales within the land. Such relationships may be social, spatial, and temporal.[15] As noted by Klauser, space on the landscape becomes territory when the components of the relationship become objects that trigger the attachment of the people to the ecosystems.[16] At this point, communities or states have territorial rights to biophysical and intangible resources that distinguish the land area. But territoriality is not static. Territories can be contested from the inside and outside boundaries. External aggressions manifest when "foreigners attempt to enter a country and when external bodies raise claims to the territory's natural resources."[17] At that point, the definition of the term indigenous as a place of birth and long-time occupancy is challenged. That is the scenario in the continent of America, Australia, New Zealand, and the Southern tip of Africa by colonial powers.

Colonial Doctrine and Territoriality

The Western law of conquest generated alternative legal doctrine to forcefully occupy these lands and establish new ownership of landscape resources. The British doctrine of "Terra Nullius" declares that these geographical entities belong to no one despite the presence of people living on the landscape. Contesting the right to territories is about free access to natural capital in and on the landscapes. The issue of natural capital eroding indigeneity occurs in different dimensions in all parts of Africa. An African scenario considers indigeneity from the perspective that nearly all inhabitants of African nations are indigenous to the continent. However, the definition that indigenous people have "historical continuity with pre-invasion and pre-colonial societies that developed on their territories" is tenable with respect to many tribes and nations in Africa.[18] Whereas the British used the Terra Nullius doctrine in Australia, New Zealand, and Southern Africa, a fragmentary developmental philosophy was engaged in Nigeria. Smith, Lopes, and Carrejo referred to this dualistic thinking as a

Cartesian paradigm separating human and natural systems.[19] The Cartesian paradigm is held accountable for human alienation, class differences, and social injustice.[20] It drove the implementation of the dualist British land management philosophy in Nigeria. The philosophy was to guarantee British interests in resource exploration and exploitation. The colonial process focused on reservation excision through dual land use principles—a rural version of the adopted Cartesian philosophy. A two-step approach was adopted: First, the creation of enclaves for indigenous communities who vehemently resisted resettlement schemes. Secondly, for those that were forcefully resettled, user rights were given to the families or the communities to explore and harvest wild biotic resources, use the reserves for controlled religious activities, and, in some cases, be permitted to bury their dead ones.

Customary Land Tenure System

British Dualism land management principles subdued the pre-colonial customary land tenure system. NIALS Dictionary of African Customary Laws defines "customary law" as,

> A mirror of accepted usage; rules handed down orally from generation to generation, which the persons living in a particular locality have come to recognize as governing them in their relationships between one another and between themselves and things. They must be in current usage.[21]

This was the law of the Nigerian indigenous people before colonial intervention. It defines their relationship with tangible and intangible things on the landscapes and regulates their lives and daily transactions. The Ọbà (King), Chief in Council and heritage Custodian of each community administers the unwritten law held in high esteem by the people. A customary land tenure system is a landholding system characterized by family and inheritance systems rooted in the communal concept of group ownership of absolute rights on the land, with individuals acquiring usufructuary rights.[22] If the land is held in trust for the family or community, then all members of the community are co-owners. Therefore, all co-owners have equal opportunity to satisfy basic needs without profit-oriented, colonial, destructive mechanisms that erode the collective ownership of resources. Natural resource utilization is at the heart of traditional land management. It is about livelihood sustainability to keep rural economies afloat. Natural resources form the bedrock of herbal medicine, recreation, food, subsistence income generation, and socio-religious activities. Therefore, the conservation of these natural resources through traditional frameworks based on norms, values, and religious essences occupy prime positions in the local governance system. Reservation of forests, rivers, and mountain ranges as sacred landscapes ensures the generational preservation of favored species as totems.

A major dictum of the customary land tenure system is that land is an ancestral trust that the living shares with the dead; hence, land is inalienable.

The British dualist land management philosophy eroded this singular principle. The result was the introduction of four overlapping tenure systems. These are "tenure under the received English Law, tenure under State Land Laws, tenure under Land Tenure Law, and tenure under Customary Law."[23] The customary land tenure system survived the plurality. The post-independent neo-colonial military and political elite's efforts to rectify the shortcomings of the colonial plural land tenure system introduced the Land Use Act of 1978.[24] It was a nationalization of all land in the country. The primary justification for the nationalization was that customary land tenure systems obstruct commercial agricultural estate development. Post-independent palliative measures to calm indigenous community agitation explored payment of royalties on some protected landscapes.[25] These interventions do not deny that the people own the land. Both Colonial and Post-independence developmental strategies in Nigeria were conscious that natural resources sustain rural livelihood and traditional medicinal system and contribute to their spiritual well-being. So, indigenous communities' interests were accommodated in a top-bottom decision-making process. Today, customary law is still one of the sources of the Nigerian legal system, especially in traditional land management of sacred landscapes in rural communities, urban fringes, and sometimes green enclaves within Nigerian cities, irrespective of tribal nomenclature.

SACRED LANDSCAPES AS INDIGENOUS LANDSCAPES

Experiencing the landscapes varies and differs from one culture to the other. Whereas the relationship between people and landscape in the Global North is often based on material assets, this relationship is more spiritual among Global South indigenous communities. The belief system sees the landscapes as sacred environments needed to fulfill both spiritual and livelihood obligations without over-exploitation. Landscapes are spiritual assets to indigenous people, and their worldview influences their deep meaning. The people's traditional knowledge of indwelling spirits relative to the ecosystems, places, and land use plays a major role in the meaning of landscapes. The material content of the biophysical environment is not totally jettisoned but is minimized to provide ecological goods and services for the people. The total effect of people's "perception, beliefs, stories, experiences and practices give shape, form, and meaning to the indigenous landscape."[26] The experiential relationship takes root in local mythology that relates both earthly and heavenly elements on a spiritual platform celebrated via rituals and festive activities. Within their origin story are spirit beings that were subsequently deified as human beings, transformed into animals or plant species, or now occupy hills and rivers. These stories are etched on the landscapes, providing windows for referencing and protecting diverse sacred elements in the ecosystem through community norms and values enforced by community-appointed guardianship.

Spiritscape—Worldview Approach to Land Management System

The International Union for Conservation of Nature's resurgence of interest in biocultural spaces pushed sacred groves to the global limelight.[27] Such landscapes are numinous places that indigenous people believe to be the abodes of spirit beings powerful enough to perform miracles, including the regeneration of the human body and soul, curing illnesses, childbearing, rainmaking, and land productivity. Byrne observed that indigenous people conceptualize the resident spirit(s) anthropomorphically.[28] That is, the spirit has rights, demands fairness in a relationship, and is often vengeful when harmed. The consciousness that a local landscape is a combination of spiritscape and physical topography is well established in their ethos. Intangibility is made material in daily activities by sacred grove custodians who are responsible for managing the spiritscape through religious undertakings, festivals, and biodiversity promotion. The definition of spiritual landscape in this chapter follows the basic ideas of a "sacred place" but accommodates both geo-diversity and biodiversity in combination with the human element of experiences and feelings. Therefore, sacred landscapes are "land or water" with special values for the people.[29] Values may be religious, provision of community cultural space, health, economic, educational, and psychological. Religious and cultural values are tied to the beneficial human well-being values experienced in such groves, including spiritual birth or forest birth. It is about the salutogenic effects of nature in the making of a total human. The sacred groves' psychological values shape the morals of local people to preserve age-old cultural ethos. These values are now threatened by developmental forces and misinterpreted by followers of the three Abrahamic religious doctrines that label anything indigenous as pagan undertakings. This is true of Nigerian communities and, indeed, entire African human settlements. However, spiritscapes are often connected to some historical, geo-heritage, and religious precincts. What drives the phenomenon in sacred places? Verschuuren, in a conversation with Mayan Spiritual Leader Felipe Gomez, deduced that "these sacred natural sites have great spiritual force, they are places where the heart of Fire, Earth, Water and Air manifests themselves and support all life in the universe."[30] Lane corroborates the view that specific points on Earth vibrate different types of subtle Earth energies.[31] Hagens' research depicts grid patterns all over the globe that infuse the earth with radiant energy.[32]

Ley Lines and Domes

This global gridiron phenomenon is referred to as planetary grids. Hagens labeled these grid lines as ley lines, while the points of intersections are called domes or ley centers.[33] Newman submitted that the planetary energy matrix directly relates to the human mind.[34] He explained that humans are "connected" and have "spiritual" experiences in earth energy centers. The recent findings are not new. Ancient humans knew this sacred, hidden body of earth energy and had settled on it in ways that took advantage of the visceral powers of the place.[35] Newman observed that major world megalithic structures, stone circles, pyramids, dolmens, volcanic spots, fault lines, ancient mounds, temples and shrines, ancient burial locations, purification sites, healing sites, ceremonial trails, meditational sites, specific plant associations, geo-heritage spots with unique stones and crystals, mythic and legendary historic spots were located on ley lines and domes to harness subtle earth energies that would dominate nature.[36] It attests to the deep interconnectedness of human processes and processes in nature by indigenous people. Recent studies of megalithic structures showed they were geomantically positioned at the domes, functioning as a receiving station for direct influences from heavenly constellations and earthly energies to achieve a harmonic relationship between heaven and earth.[37] Examples of such megalithic landscapes are Stonehenge and Avebury henge in England, the Pyramids of Egypt, and Fort Ancient in Ohio. Science is not comfortable with the "spiritual-cartographic concept of leylines."[38] However, this does not contest the fact that Neolithic man and people of old left a defined pattern of imprints on the landscape for this generation to unravel. Thurgill, referring to the 1921 submission by Alfred Watkins, admitted that the cartographic influence of ley lines is retracing history, uncovering site narratives, and reimagining the landscape.[39] It is about understanding a previous civilization's documented wisdom of the landscape. This is geosophy.

Geosophy

The concept of geomantic planning is a worldview approach to harmonious living with nature. Nature is both the material structure of the ecosystem and the intangible framework that drives it. Geomantic planning, therefore, recognizes humans as part of the system. It is a harmonious dependency on beneficial values entrenched in the biosphere that were assessed by Neolithic humans to create activity nodes. The geomantic concept is very relevant to human settlements in indigenous Nigerian communities. This is due to the belief in nature worship. The wisdom of living harmoniously with other structural members of the ecosystem revolves around geomantic planning that identifies the most auspicious location for every land use on a communal as well as individual basis. Chief Ẹlẹ̀bùrúìbọn is the Àràbà of Òṣogbo, one of the Yoruba nation Custodians. His submission is that Earth has a tremendous capability to hear, speak, and communicate with all living things.[40] It is the belief system that the locational choice of indigenous people is influenced by the presence of positive bio force that enhances peaceful coexistence, profitability in market settings, and communion between the living, the dead, and the deities. This is achieved through the Yorùbá Ifá divination system. Therefore, markets, houses, cloth trees, shrines, festival routes, and communal ceremonial spaces are aligned with ley lines and ley centers on the landscapes to indigenize the landscape.

Making of Yorùbá Indigenous Landscape—Ilé-Ọbà Narrative

This section examines a typical example of spatial engagements with the "landscape through [the] spiritual-cartographic concept of the ley line" as an indigenization process.[41] A closer look at Ilé-Ọbà village sheds light on how Yoruba landscapes are made indigenous. Ilé-Ọbà is about 20 kilometers northeast of Ọ̀yọ́ City in the Ọ̀yọ́ State of Nigeria. It lies between 12°13' – 13°35'N and 7°56' – 7°61' E. Ilé-Ọbà occupies a transitional position in the migratory narrative of Aláàfin of Ọ̀yọ́ (Paramount King of defunct Oyo Empire). Ọ̀yọ́-Ilé was the Ọ̀yọ́ Empire head-quarters founded during the golden era of the Yoruba city-building process. Ọ̀yọ́-Ilé was the administrative capital for over 1,000 years until internal political wrangling enhanced external forces in an attack in 1793. The study was carried out to investi-gate how Ilé-Ọbà landscapes were made indigenous as typical Yorùbá biocultural precincts. Chief Ọbadáísí, incumbent Ilé-Ọbà Custodian, historical narrative con-firmed Chief Olúfọ́n as one of the several defunct Oyo Empire Calvary leaders given the mandate to establish a military outpost around 1705.[42] This migration was accompanied by "Ikòkò Odù Òrìsà Olúfọ́n" mobile religious essence. On geo-mantic arrival at the pristine Guinean Savannah landscape, the Olúfọ́n deity was auspiciously rested at the present location. Ilé-Ọbà community settled around the shrine that is now housed in one of the 20 buildings in the hamlet. Chief Ọbadáísí submission was that the entire Ilé-Ọbà landscape is sacred, but there are three major areas with higher spiritual essences, namely Olúfọ́n Shrine, Yemo Shrine, and Òkè Aforo Shrine (Figure 7.4).[43]

Figure 7.4 Ile-Oba landscape essences. Courtesy: The author.

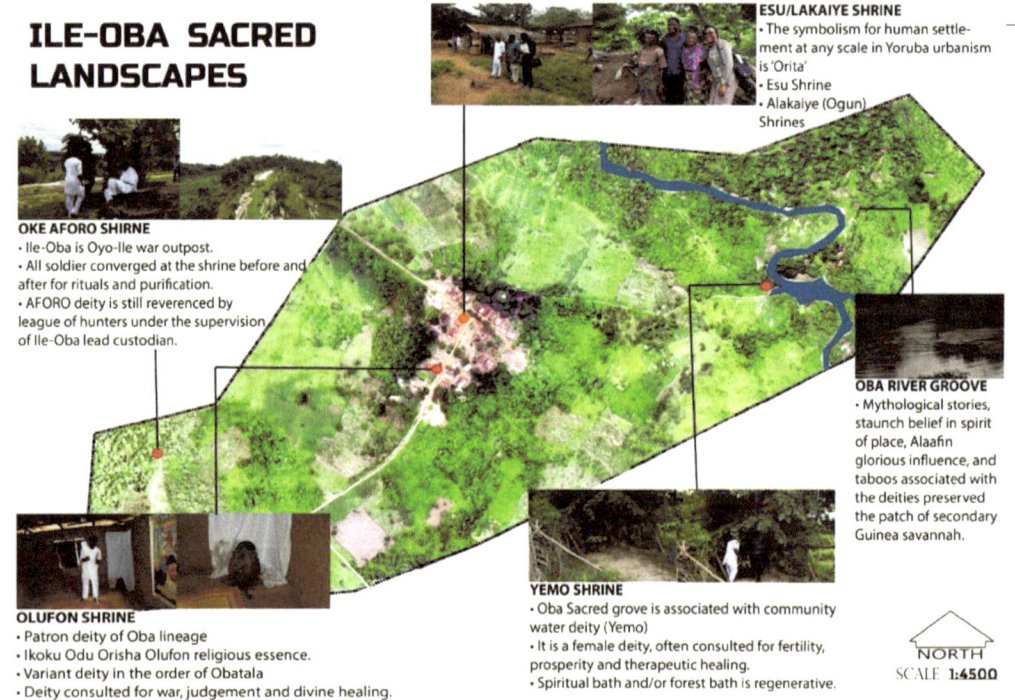

In addition, the Ile-Oba community Èṣù and Ògún shrines overlook the community market and adjoining ceremonial space according to the Yoruba human settlement cosmogram. The presence of *Ficus exasperata* as a totem tree around the village center further points to an active energy spot from Yoruba's city-planning perspective. Chief Ọbadáísí, Custodian of Ilé-Ọbà Sacred Landscape, confirmed Òrìsà Olúfọ́n as the village's principal deity. Òkè Aforo shrine is geomantically located at a panoramically auspicious hilly spot. Òkè Aforo deity is still reverenced by the community vigilante group under the supervision of the Custodian. Oba Sacred Grove is associated with the community water deity (Yemo) that played a dominant role in Aláàfin trajectory at the close of the Oyo Empire in the second decade of the nineteenth century. Yemo shrine is geomantically situated at the bank of the Ọbà River. It is a female deity, often consulted for fertility, prosperity, and therapeutic healing. A spiritual bath at the grove is considered regenerative and attracts faithful from far and near.

Aura-Meter Geomantic Analysis

This study aims to authenticate the shrines that were auspiciously located in the Yoruba geomantic system and to confirm the Custodian claim that the entire village is activated as a spiritscape during the annual Ilé-Ọbà February festive calendar. Simple dowsing using a handheld aura-meter put the energy center and ley line orientation in Ilé-Ọbà sacred landscapes in proper perspective. The four geo-referenced spiritual essence spots are:

1   Olúfọ́n Shrine 7°54′8.00″N 4° 5′1.013″E
2   Òkè Aforo Shrine 7°54′0.69 ″N 4° 4′31.46 ″E
3   Yemo Shrine-Oba Grove 7°54 ′4.30″N 4° 4′44.28 ″E
4   Èṣù & Alákáayé Shrines 7°54 ′6.02 ″N 4° 4′45.79″E

Aura-meter geomantic analysis shows that the Olúfọ́n, Oke Aforo, and Yemo shrines in Ilé-Ọbà sacred landscapes align in a straight line (Figure 7.5). Connecting ley lines seems to be a communication route synchronizing spiritual activities on the landscape. Though Èṣù and Ògún, energy spots are not directly on the line, the shrines double as boosters for the community's festive space. Mythological stories, staunch belief in the spirit of place, norms, values, and taboos associated with the deities preserved the patch of secondary Guinea savannah ecosystem typical of the northern fringe of Yorùbá nation since 1705.

Chief Obadaisi's closing remark was that "Orisha Olufon" must be propitiated before other shrines in the community are activated.[44] Rituals and ceremonies are major cosmic dimensions undertaken to acquire spiritual experience in sacred landscapes. Other sacred points are activated through necessary rituals, thereby setting the entire village up as a festive landscape with an inbuilt capacity to meet the needs of the faithful.

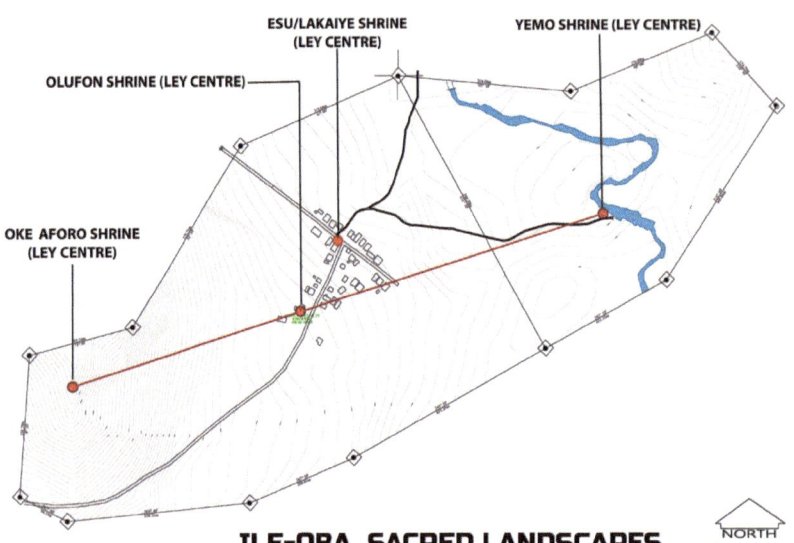

ESU/LAKAIYE SHRINE
(LEY CENTRE)

YEMO SHRINE (LEY CENTRE)

OLUFON SHRINE (LEY CENTRE)

OKE AFORO SHRINE
(LEY CENTRE)

ILE-OBA SACRED LANDSCAPES

NORTH
SCALE 1:4500

Figure 7.5 Ile-Oba sacred landscapes energy centers. Courtesy: The author.

INDIGENOUS LANDSCAPE DYNAMISM

The constituents of indigenous landscapes include vital community-oriented, bio-diverse ecosystems as a source of livelihoods, with heritage resources and extensive historical accounts etched on the landscapes narrating warfare, socio-cultural events, and languages. As noted by Bruno, Grabasch, and McIntosh, indigenous landscapes evolve through a process of human and land interaction underlain by religious activities that point to historical origin.[45] The normal land-scape is thus transformed into a traditional indigenous landscape. However, tra-ditional indigenous landscapes can be transformed into prevailing cultural landscapes. This is because the indigenous tenure system is not static. The transi-tional process of change from indigenous traditional landscapes to prevailing cul-tural landscapes may be triggered by changes in religious belief systems, population upsurges, political conquests, economic development, lifestyle, global financial systems, political and economic systems, and natural phenomena, espe-cially climate change.[46] In Nigeria, colonial influence and post-independent development policies in the last 175 years demanded the commercialization of agricultural production through cash crops for cocoa, rubber, oil palm, cashew, and coffee. The impact of monocultural plantation agriculture on indigenous landscapes is a major challenge for Custodians. This is because the colonial masters did not culturally acquire land areas for these plantations.

Totem trees were removed, rocks blasted, and the river course diverted. Most devastating is the exploration and exploitation of fossil fuel and mineral ore by highly profit-driven Western corporations without assessment of or respect for cultural impacts. Oil spillage degrades not only the land, creeks, rivers, and wet-lands but completely halts rural livelihoods and redefines culinary expectations. Yet the local population still believes they can restore their landscape after much-desired bioremediation. For now, their indigenous landscapes were recessed into mindscapes. Davidson-Hunt's insight into the "process of how an indigenous traditional landscape is replaced by a cultural landscape of managerial ecology" helps comprehend the conflict between indigenous landscapes and imposed profit-driven cultural resource exploitative landscapes.[47] This calls for a collaborative planning and design approach in contemporary projects that overlaps indigenous cultural landscapes. It is democratic planning that accommodates the community worldview, which was forcefully ignored by the colonial developmental apparatus across the Global South.

REFLECTIONS

Indigenous landscapes have cultural and historical meanings attributed to them by diverse peoples who settled, have traveled, used, and have interwoven these places into generations of practice. Beyond the biophysical components are the intangible, visual, and audio features that make the landscapes unique places. People make landscapes indigenous. The indigeneity is to meet their needs and wants in a way that is dictated by the prevailing worldview that appreciates nature-conscious geomantic planning. Indigenous people's environmental perception, values, belief systems, social institutions, available technologies, and governance apparatus dictate geosophic planning and bio-centric management goals for a contextual landscape. Such geomantic planning principles have a minimum respect for greedy managerial exploitative ecology. Rather, it is about harmony with both terrestrial and celestial environments on a worldview basis that generates attachment. Environmental professionals, especially landscape architects, need indigenous people's knowledge of localizing landscapes toward ecological restoration, productive landscape planning, and climate-sensitive sustainable living. Making manmade landscapes indigenous is re-etching progressive community values on landscapes with the singular goal of placemaking.

Placemaking versus Myth-Making

"Placemaking" is the way a "space" is transformed into a memorable landscape where humans have strong emotional, economic, social, and spiritual attachments without depreciating the natural capital, while enhancing their community's social capital. It is about people and activities that provide memorable feelings. Such activities highlight deeply held societal values displayed in arts, traditional festivals, religious undertakings, landmarks, and the biogeography of public open spaces. They constitute the ingredients that draw a variety of people to the same space from one generation to the other. Therefore, placemaking looks at heritage, culture, norms, values, and ideals that natives have developed in inhabiting a bioregion. The principle of placemaking seeks local design solutions over the current global developmental paradigm that solves human habitat problems through homogenous Western values-driven planning and design procedures. While sustainable physical planning strategy draws strength from scientific empirical data analysis in solution preferment, placemaking accommodates principles of natural resources, social capital, cultural belief, and spiritual inputs in transforming a "space" into a "place."

Therefore, place conceptualization is nothing but a simulation of mythological models dear to the people.[48] Symbolization of verbal myth of a people generates a design cosmogram that assists in etching societal values on the landscape. Myth is about a sacred or secular narrative that evokes attachment to a communal belief system. When rooted in events and festivities, myth gradually generates symbolic meaning. Wortham-Galvin, referring to the submission of Bronislaw Malinowski, reiterates that myths are stories with the inbuilt capability to justify the present toward social stability.[49] Activating such a myth in spatial configuration redefines the graphic vision and mission to which the people are attached. Placemaking on this platform dissolves into "myth-making" in the conscious reconstruction of local people's spatial identity.[50]

CONCLUSION

Is landscape indigenous? Yes, it is. Indigenous landscape is a fusion of the natural and cultural features a community manages through "traditional ecological knowledge."[51] The United Nations favors political and legal indigenous realities to protect the rights of marginalized people.[52] From that window, indigenous people are those who, having a historical continuity with pre-invasion and pre-colonial societies that developed on their territories, consider themselves distinct from other sectors of the societies now prevailing on those territories or parts of them. They form, at present, non-dominant sectors of society and are determined to preserve, develop, and transmit to future generations their ancestral territories and their ethnic identity as the basis of their continued existence as a people in accordance with their own cultural patterns, social institutions, and legal systems.[53]

While legal and political dimensions are critical in Nigeria, social and biocultural realities are more meaningful in understanding indigenous landscapes in Africa. This is traceable to a colonial perception of landscape as a bank of natural resources that must be completely exploited on a for-profit basis with minimum consideration for an ecosystem's structure. On the contrary, the indigenous landscape in Nigeria is a spiritscape managed bio-centrically through a customary land tenure system on a communal basis. Access to land is based on membership of a landholding community by birth. The land is bequeathed from one generation to another through customary rules of succession. Again, the right to natural resources is equally communal. Therefore, such communal tenure still explains security rights to biotic resources in rural areas. Indigeneity is secured through world-view-influenced spatial definitions that appreciate nature-conscious geomantic planning. It is rooting the people in the landscape to harness viscera energies through socio-religious undertakings in harmony with the ecosystem.

Typical examples are Stonehenge and Avebury henge megalithic landscapes that were geomantically positioned at the domes and ley lines.[54] They function as receiving stations for direct influences from the heavenly constellation and earthly energies that have been driving the harmonic relationship between heaven and earth since the time of Neolithic man.[55] These Neolithic structures still seasonally support social and religious enthusiasts. Stonehenge and Avebury henge are indigenous British landscapes. Cohen noted the prime position of Wailing Wall Plaza, a western extension of the Temple Mount in the urban morphology of Jerusalem.[56] Since the purchase of 24 acres of Mr. Ornan's threshing floor by King David to erect an altar that would stop an angel from destroying the people, this mountaintop in Jerusalem continued to attract men.[57] Solomon's temple with all the flanges was built there in 955 BC. Even though Nebuchadnezzar destroyed it in 587 BC, Alexander the Great in 333 BC, and Emperor Titus in 70 AD, the entire Temple Mount never ceased to attract Judaism, Christian, and Islamic faithful.[58] A combination of cultural, historical, spiritual, political, and landscape attributes give the Wailing Wall open space symbolic meanings to Israelis. Temple Mount is a Jewish indigenous urban landscape. Thousands of biocultural landscapes in Sub-Saharan Africa are indigenous landscapes, too. To invoke Terra Nullius, a Cartesian paradigm, or post-independent neo-colonial land-grabbing legislation in Africa is to erode the little morals left in double standard global political economy. Landscape is indeed indigenous.

NOTES

1   Margret Moore, "Territorial Rights and Territorial Justice," in *The Stanford Encyclopedia of Philosophy 2020*, eds. Edward N. Zalta, Center for the Study of Language and Information (CSLI) (Stanford, CA: Metaphysics Research Lab, 2020), 2.

2   John A. Simmons, "On the Territorial Rights of States," *Philosophical Issues* 11 (2001): 300–326, doi:10.1111/0029–4624.35.s1.12.

3   Chris Cunningham and Fiona Stanley, "Indigenous by Definition, Experience, or Worldview," *BMJ* 327 (2003): 403–405.

4   Chris Cunningham and Fiona Stanley, "Indigenous by Definition, Experience, or Worldview," *BMJ* 327 (2003): 403.

5   Erika Sarivaara, Maatta Kaarina, and Uusiautti Satu, "Who is Indigenous? Definitions of Indigenuity," *European Scientific Journal* 1 (December 2013): 369–378.

6   Erika Sarivaara, Maatta Kaarina, and Uusiautti Satu, "Who is Indigenous? Definitions of Indigenuity," *European Scientific Journal* 1 (December 2013): 370.

7   Erika Sarivaara, Maatta Kaarina, and Uusiautti Satu, "Who is Indigenous? Definitions of Indigenuity," *European Scientific Journal* 1 (December 2013): 370.

8   Andrew Gurstelle, "Sacred Trees of the Savè Hills Cultural Landscape," *University of Michigan Working Papers in Museum Studies*, no. 10 (2013): 1–18.

8b Babatunde Agbaje-Williams, "Clothed Ritual Trees: An Insight into Yoruba Religious Thought," in *Yoruba Religious Textiles: Essays in Honour of Cornelius Oyeleke Adepedgba*, eds. E. Renne and B. Agabje-Williams (Ibadan: Book Builders, 2005), 157–187.

9   Michael E. Smith, "Form and Meaning in the Earliest Cities: A New Approach to Ancient Urban Planning," *Journal of Planning History* 6, no.1 (2007): 3–47; John E. Clark, "Mesoamerica Goes Public: Early Ceremonial Centers, Leaders, and

Communities," in *Mesoamerican Archaeology: Theory and Practice*, eds. Julia A. Hendon and Rosemary Joyce (Oxford: Blackwell, 2004), 43–72, 61.

10 Amos Rapoport, "Levels of Meaning in the Built Environment," in *Cross-Cultural Perspectives in Non-Verbal Communication*, ed. F. Poyatos (Toronto: C. J. Hogrefe, 1988), 317–326.

11 Ọlátúnjí Adéjùmọ̀, "Geosophic Urbanism: A Localized Urban Developmental Philosophy," *Urban Crisis and Management in Africa—A Festschrift for Prof. Akin Mabogunje*, eds. Isaac Olawale Albert and Taibat Lawanson, Society for Peace Studies and Practice (Austin, TX: Pan-African University Press, 2019), 469–487.

12 Ulia Popova-Gosart, "Indigenous Peoples: Attempts to Define," in *Biomapping Indigenous Peoples*, eds. Susanne Berthier-Foglar, et al. (N.p.: Brill Rodopi, 2012), 87–116.

13 Ulia Popova-Gosart, "Indigenous Peoples: Attempts to Define," in *Biomapping Indigenous Peoples*, eds. Susanne Berthier-Foglar, et al. (N.p.: Brill Rodopi, 2012), 92.

14 Lea Ypi, "A Permissive Theory of Territorial Rights," *European Journal of Philosophy* 22, no. 2 (2014): 288–312; P. J. Taylor, *Political Geography: World-Economy, Nation-State, and Locality* (London: Longman, 1985) 95–140.

15 Francisco Klauser, "Thinking through Territoriality: Introducing Claude Raffestin to Anglophone Socio-spatial Theory," *Environment and Planning Digest: Society and Space* 30 (2012): 106–120.

16 Francisco Klauser, "Thinking through Territoriality: Introducing Claude Raffestin to Anglophone Socio-spatial Theory," *Environment and Planning Digest: Society and Space* 30 (2012): 110.

17 Lea Ypi, "A Permissive Theory of Territorial Rights," *European Journal of Philosophy* 22, no. 2 (2014): 1.

18 African Development Bank Group, "Development and Indigenous Peoples in Africa," *Safeguards and Sustainability Series* 2, no. 2 (August 2016): 7–12, 10.

19 Chad Smith, Vicente Lopes, and Frank Carrejo, "Recasting Paradigm Shift: True Sustainability and Complex Systems," *Human Ecology Review* 18, no. 1 (2011): 67–75.

20 Chad Smith, Vicente Lopes, and Frank Carrejo, "Recasting Paradigm Shift: True Sustainability and Complex Systems," *Human Ecology Review* 18, no. 1 (2011): 70.

21 NIALS Dictionary of African Customary Laws, *NIALS Dictionary of African Customary Laws*, eds. Azinge Epiphany and Oluchi Azoro (Abuja: Nigerian Institute of Advanced Legal Studies, 2013), 39.

22 Nkiruka Chidia Maduekwe, "The Land Tenure System under the Customary Law," *SSRN Electronic Journal* (May 2014): 1–19, https://ssrn.com/abstract=2813056.

23 Nkiruka Chidia Maduekwe, "The Land Tenure System under the Customary Law," *SSRN Electronic Journal* (May 2014): 5, https://ssrn.com/abstract=2813056.

24 Nkiruka Chidia Maduekwe, "The Land Tenure System under the Customary Law," *SSRN Electronic Journal* (May 2014): 12, https://ssrn.com/abstract=2813056.

25 Kayode Adeyoju, *Forestry and the Nigeria Economy* (Ibadan: Ibadan University Press, 1975), 35.

26 Australian Heritage Commission, "Protecting Local Heritage Places: A Guide for Communities," in *Australian Heritage Commission* (Canberra, ACT: Australian Heritage Commission, 1998), 115.

27 Robert Wild and Christopher McLeod, eds., *Sacred Natural Sites: Guidelines for Protected Area Managers*, Series No. 16 (Gland, Switzerland: IUCN, 2008), 4.

28 Dennis Bryne, "The Enchanted Earth: Numinous Sacred Sites," in *Sacred Natural Sites Conserving Nature and Culture*, eds. Bas Verschuuren, Robert Wild, Jeffrey McNeely and Gonzalo Oviedo (London: Earthscan Ltd, 2010), 53–61.

29 Robert Wild and Christopher McLeod, *Sacred Natural Sites: Guidelines for Protected Area Managers* (Gland, Switzerland: IUCN, 2008), 5.

30 Bas Verschuuren, "A Conversation with Mayan Spiritual Leader Felipe Gomez," *Langscape: Sacred Natural Sites; Sources of Biocultural Diversity*, ed. Ortixia Dilts, *Terralingua* 2, no. 11 (2012): 23–27, 26.

31 Melissa Lane, *Plato: The Republic*, trans. Desmond Lee (London: Penguin Books Limited, 2007), 45.

32 Bethe Hagens, "The Divine Feminine in Geometric Consciousness," *Journal of American Anthropological Association* 17, no. 1 (2006): 1–34.

33 Bethe Hagens, "The Divine Feminine in Geometric Consciousness," *Journal of American Anthropological Association* 17, no. 1 (2006): 17.

34 Hugh Newman, *Earth Grids—The Secret Patterns of Gaia's Sacred Sites* (Glastonbury, Somerset: Wooden Books Limited, 2008).

35 William Becker and Beth Hagens, "The Planetary Grid: A New Synthesis," in *Anti-Gravity and the World Grid*, ed. David Hatcher Childress (Kempto: Adventures Unlimited Press, 1999), 27–50.

36 Hugh Newman, *Earth Grids—The Secret Patterns of Gaia's Sacred Sites* (Glastonbury, Somerset: Wooden Books Limited, 2008).

37 Hugh Newman, *Earth Grids—The Secret Patterns of Gaia's Sacred Sites* (Glastonbury, Somerset: Wooden Books Limited, 2008).

38 James Thurgill, "A Strange Cartography: Leylines, Landscape and 'Deep Mapping' in the Works of Alfred Watkins," *Humanities* 4, no. 4 (2015): 637–652.

39 James Thurgill, "A Strange Cartography: Leylines, Landscape and 'Deep Mapping' in the Works of Alfred Watkins," *Humanities* 4, no. 4 (2015): 637.

40 Ifáyẹmí Ẹlẹ̀bùrúìbọn, "Geomancy and City Ground Breaking by the Yoruba People," interview by Tunji Adejumo, March 2, 2010.

41 James Thurgill, "A Strange Cartography: Leylines, Landscape and 'Deep Mapping' in the Works of Alfred Watkins," Humanities 4, no. 4 (2015): 637.

42 Musiliu Obadaisi, "*Spirituality of Ile Oba Landscapes,*" interview by Tunji Adejumo, August 5, 2018.

43 Musiliu Obadaisi, "*Spirituality of Ile Oba Landscapes,*" interview by Tunji Adejumo, August 5, 2018.

44 Musiliu Obadaisi, "*Spirituality of Ile Oba Landscapes,*" interview by Tunji Adejumo, August 5, 2018.

45 Marques Bruno, Greg Grabasch, and Jacqueline McIntosh, "Fostering Landscape Identity through Participatory Design with Indigenous Cultures of Australia and Aotearoa/New Zealand," *Space and Culture*, 24, no. 1 (June 27, 2018): 37–52, https://doi.org/10.1177/12063312187839.

46 Iain Davidson-Hunt, "Indigenous Lands Management, Cultural Landscapes and Anishinaabe People of Shoal Lake, Northwestern Ontario, Canada," Environments 31, no. 1 (January 2003): 21–41.

47 Iain Davidson-Hunt, "Indigenous Lands Management, Cultural Landscapes and Anishinaabe People of Shoal Lake, Northwestern Ontario, Canada," *Environments* 31, no. 1 (January 2003): 24.

48 Brook D. Wortham-Galvin, "Mythologies of Place Making," *Places Journal* 20, no. 1 (2008): 32–39.

49 Brook D. Wortham-Galvin, "Mythologies of Place Making," *Places Journal* 20, no. 1 (2008): 32.

50 Brook D. Wortham-Galvin, "Mythologies of Place Making," *Places Journal* 20, no. 1 (2008): 32.

51 Deanna Beacham, Suzanne Copping, John Reynolds, and Carolyn Black, "Indigenous Cultural Landscapes: A 21st-Century Landscape-scale Conservation and Stewardship Framework," *The George Wright Forum*, 34, no. 3 (2017): 343–353, https://www.jstor.org/stable/26452977.

52 Erika Sarivaara, Maatta Kaarina, and Uusiautti Satu, "Who is Indigenous? Definitions of Indigenuity," *European Scientific Journal* 1 (December 2013): 369–378.

53 Erika Sarivaara, Maatta Kaarina, and Uusiautti Satu, "Who is Indigenous? Definitions of Indigenuity," *European Scientific Journal* 1 (December 2013): 369.

54 James Thurgill, "A Strange Cartography: Leylines, Landscape and 'Deep Mapping' in the Works of Alfred Watkins," *Humanities* 4, no. 4 (2015): 637–652.

55 Hugh Newman, *Earth Grids—The Secret Patterns of Gaia's Sacred Sites* (Glastonbury, Somerset: Wooden Books Limited, 2008).

56 Daniel Cohen, *The Holy Land of Jesus* (Yehuda, Israel: Doko Media Limited, 2008).

57 Mathew Ashimolowo, *What Is Wrong in Being Black: Celebrating Our Heritage, Confronting Our Challenges* (Shippensburg: Destiny Image Publishers Inc., 2007).

58 Peter Vasko, *Pilgrimage Journey in the Holy Land* (Jerusalem: Mount of Olive Press, 2007); Daniel Cohen, *The Holy Land of Jesus* (Yehuda, Israel: Doko Media Limited, 2008).

Chapter 8: Is Landscape Elitist?

Alison B. Hirsch

Elitism is undoubtedly a loaded term, compounded by the recent US politiciza-
tion wielded to stoke fear and describe social stratification. As it pertains to landscape in this volume, it will be examined as indicative of power, privilege,
practices of exclusion, and even oppression. The following will attempt to trace
the term *landscape* as elitist, and the practice of landscape architecture and the
study of designed landscapes as historically for the privileged few to the exclusion of most.

In a book focused on the meanings of landscape and landscape architecture, it is important to understand the origins of landscape as a European invention and to question how the legacy of this invention has evolved and can
continue to evolve to serve those it has historically ignored, excluded, oppressed.[1]

There is much to be unpacked in this coupling of landscape with its
deep-rooted elitism, so the following starts briefly with the term's etymological
and ideological roots before addressing a few landscape histories that have
created the foundation for landscape architecture as a discipline and profession. This chapter will then assess what landscape is and who it is truly for, in
addition to what landscape architecture is, who landscape architects are, and
how expertise is defined through the lens of those who have challenged its
elitist canons. Finally, the chapter will conclude where we are *now*—in a
(not-so-post) pandemic world transformed and how we might arrive at a
more inclusive future.

Author Disclaimer: Landscape architecture is actively evolving. The following traces conventions and expectations of traditional forms of practice that are,
indeed, elite (a word that makes me uneasy because of its catchall lack of
nuance). It then considers how this elitism has been challenged historically and in
the present to demonstrate how we might move forward. It is this moving
forward toward more justice-oriented models of applied research and practice
that I have committed my professional life.

ORIGINS AND IDEOLOGIES

Landscape as a term and an "idea," or ideology, is deeply bound in European
worldviews of modernity and conquest. While we might trace arguments about

DOI: 10.4324/9781003148142-9

its etymological roots to practices of dwelling in a northern European form of medieval settlement (see *Landschaft*; Olwig et al.) or as a "way of seeing," a scopic regime emerging parallel to and in conjunction with the rise of capitalism and colonial conquest (Cosgrove et al),[2] it is popularly accepted today as a term that refers to representation rather than the material manifestation of practices of inhabitation and use.

The British geographer Denis Cosgrove explained that landscape is a "way of seeing" and is connected to European practices of projection or the development of linear perspective in the fifteenth century, a development that paralleled and emerged from the transition of feudalism to capitalism. Landscape as "a way of seeing" is also associated with mapping as a means of ordering, controlling, and claiming territory by objectifying land for profit and other forms of value extraction, rather than use.[3] As a form of representation expressed either in painting or in physically designing the land as a manifestation of particular values, landscape has abstracted, omitted and/or erased that which is outside (or Othered from) the dominant scenic or spatial order.

In his 1973 book, *The Country and the City*, cultural theorist Raymond Williams uses the term "working country" as the flip side of landscape. He argues the realities of the working country make these fictions (omissions, concealments, etc.) of landscape possible and manifest.[4] Those fictions are what inspire cultural geographer Don Mitchell's description that

> 'landscape' fully mystifies [the] contentiousness [of its production], creating instead a smooth surface, a mute representation, a clear view that is little clouded by considerations of inequality, power, coercion, or resistance—at least until the moment when those struggles over power become overt.[5]

This smooth, mute character of landscape is easily determinable today in neoliberal practices of "greenwashing" or using a sanitized version of "nature" in the city as a marketing strategy for urban real estate development.

As referenced often, a more obvious example of the fictionalizing power of landscape, or how landscape is wielded as an instrument of power and erasure, is the considerable brutality behind the construction of the US National Parks, which included the violent dispossession of indigenous peoples who had been integral in the shaping of those places. The invention of these parks as "untouched wilderness" enforced a fictionalized notion of Nature (as landscape or something obscured through our distancing from it).[6] The suppression of their production makes these landscapes seem naturally occurring or naturalized. This led to scholars Emily Eliza Scott and Kirsten Swenson's investigation of art practices that "denaturalize the American landscape" in their book *Critical Landscapes*, referencing examples such as those by the *New Topographics*, who photographed unromanticized views of the everyday United States as a counterpoint to distanced landscape photographs by artists such as Ansel Adams (Figures 8.1 and 8.2).[7]

Figure 8.1 Everyday landscape of the *New Topographics* in contrast to fictions of "pristine" wilderness. Robert Adams, *Mobile Homes,* Jefferson County, Colorado, 1973, George, Eastman House Collections.

Figure 8.2 Everyday landscape of the *New Topographics* in contrast to fictions of "pristine" wilderness. Ansel Adams, *The Tetons and Snake River, Grand* Teton National Park, Wyoming, 1942. Gelatin silver print. © The Ansel Adams Publishing Rights Trust.

These fictions are a continued extension of the Colonizing (and Capitalist) treatment of land as something to be possessed, conquered, pillaged, and exchanged or consumed, thus prompting feminist analyses of landscape as a construction of the male gaze. Landscape's often interchangeability with Nature—as feminine, passive, docile Other; as chaos to be ordered; or as a resource for exploitation—has been explored by landscape theorist Elizabeth Meyer, who looks to environmental historian Carolyn Merchant and other ecofeminists of the 1970s and 1980s that connected the rise of modern science with the patriarchal exploitation of nature, specifically objectified for purposes of extraction.[8]

LANDSCAPE HISTORY IS ELITIST!

The history of landscape, as traced to Northern European painting practices and perpetuated through the invention of perspective and coordinate geometry as it othered and neutralized space, is the historical foundation for the creation of landscape architecture. While rapidly evolving today, histories that have served as the foundation of landscape architectural education largely trace European garden-making practices for a political and economic elite emerging from these representational inventions. Those histories have typically been told through the lens of their patrons and designers, outside an examination of the structural forces behind their development or an interrogation of how systems of power and exclusion enabled these iconic spaces and places. Theories of the Picturesque developed in eighteenth century English gardens continue to inform contemporary standards of landscape taste culture. Most notably, "landscape improver" Lancelot "Capability" Brown sought to "correct" nature to conform to a pastoral aesthetic ideal of smooth contours and planting manipulated to look "natural" (Figure 8.3). Formulas for Brownian landscapes were later translated by Paris' Chief Engineer of Parks, Jean-Charles Adolphe Alphand, appointed by Baron Haussmann in 1854 to transform royal gardens into "public" parks fashioned to the tastes of the bourgeoisie, or newly wealthy Parisian industrialists. As landscape scholar Heath Massey Schenker argued, for Alphand, picturesque meant "like a picture"—this visual experience of nature was commodified as a standard of taste and a speculative real estate market for bourgeois investors (Figure 8.4).[9]

The translation of English garden-making ideals arrived in the United States under extremely different circumstances, most forcefully through the writings and teachings of Andrew Jackson Downing. With the dramatic growth of US cities in the mid-nineteenth century, Downing found it imperative that a nation founded on agrarian values contain open spaces to protect public health and provide suitable areas for recreation. Most essentially, Downing and Frederick Law Olmsted, Sr., whose writings on public parks Downing had published in *The Horticulturalist*, believed parks were an essential part of a Reformist Program that would provide the poor with "an education to refinement and taste and the mental and moral capital of gentlemen."[10] While parks would supposedly serve as democratic institutions offering social freedom, Downing equally argued for their "social influence"

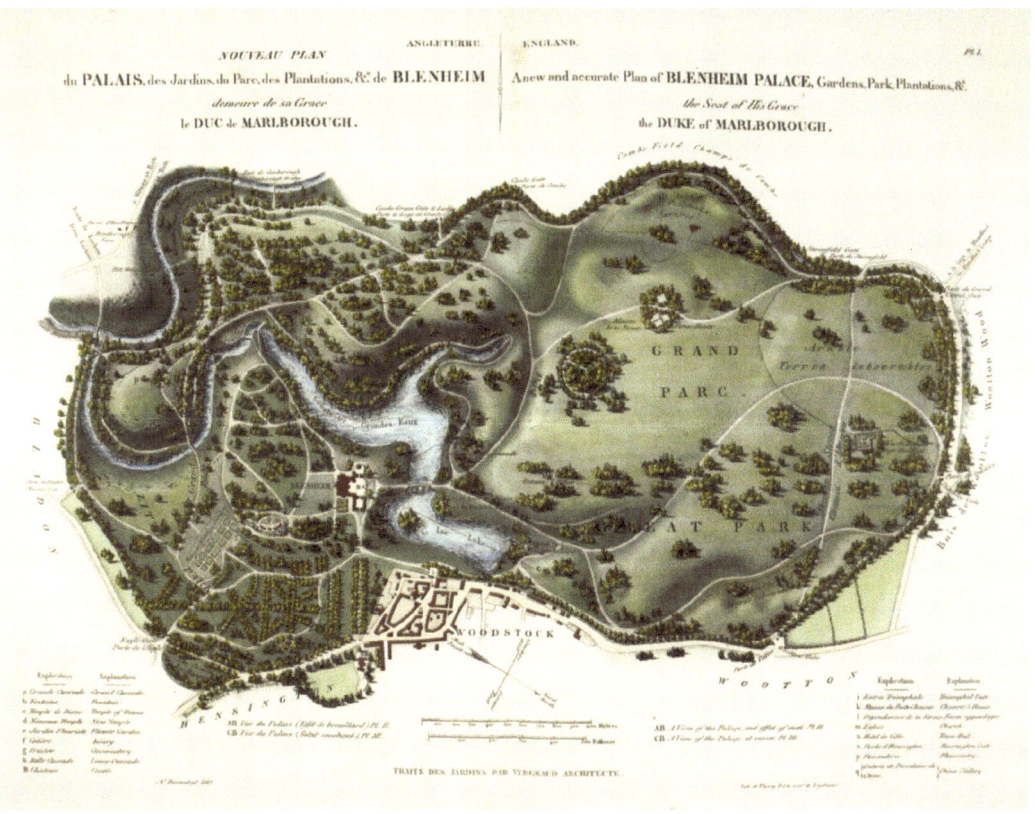

NOUVEAU PLAN
du PALAIS, des Jardins, du Parc, des Plantations, &ᶜ de BLENHEIM
demeure de sa Grace
le DUC de MARLBOROUGH.

ANGLETERRE.   ENGLAND.

A new and accurate Plan of BLENHEIM PALACE, Gardens, Park, Plantations, &ᶜ
the Seat of His Grace
the DUKE of MARLBOROUGH.

Figure 8.3 Lancelot "Capability" Brown redesigned the grounds for Blenheim Palace in Oxfordshire in the 1760s, creating standards of taste. "A new and accurate plan of Blenheim Palace, Gardens, Park, Plantations" for the Duke of Marlborough, 1835. Source: Nicolas Vergnaud, L'Art de Créer les Jardins (Paris, 1835). Courtesy: The British Library/Wikimedia Commons.

as agents of moral improvement, which could "soften and humanize the rude, educate and enlighten the ignorant, and give continual enjoyment to the educated."[11] It is nearly impossible to write anything about landscape architecture without bringing up Frederick Law Olmsted, and he continues to receive important critical and celebratory reflections. Yet in terms of the elitism of landscape, specifically in the United States, his use of pastoral scenery introduced both the commodification of nature (Central Park set off rampant real estate speculation) and, with Downing, its moralizing influence.[12] Wealth accumulation and establishing or imposing codes of morality and standards of behavior are clear means and expressions of the consolidation of power, a critical definition of elitism.

The ultimate devolution of the picturesque to "little more than 'a pleasing scene,'"[13] as it was translated and reduced to the formulaic or reproducible visual scenery, became the standard of taste applied to suburban developments throughout the United States and is still a sought-after ideal for developers (often the landscape architect's client), who are often eager to anesthetize publics and sanitize difference.

The voices left out of these dominant histories of designed landscapes are too numerous to mention. Revisionist histories have begun to incorporate the

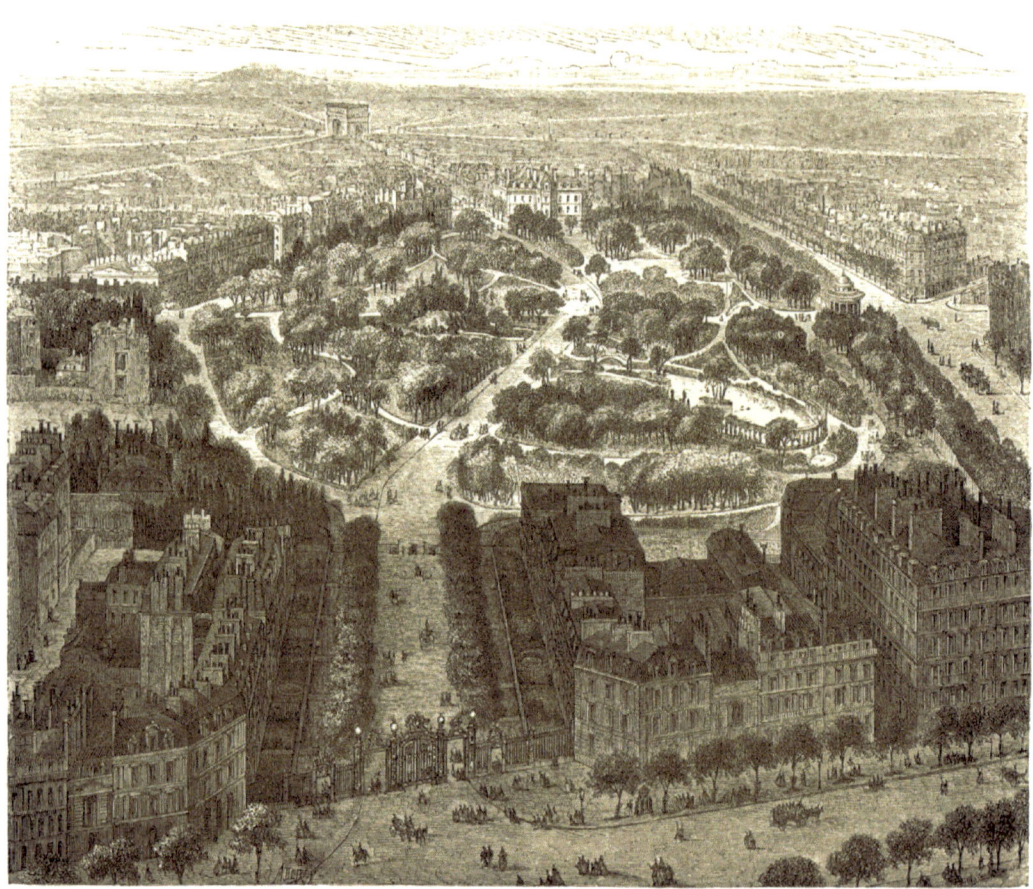

contributions of women, of cultural practices of landscape formation from outside the European canon, of indigenous land practices that shaped the places that colonizers settled and brutally transformed to conform to their own landscape conventions, as well as ongoing Native practices that have continued despite violent dispossession. These efforts have begun to open this elite history or single narrative on which the profession has been founded to the range of possible ways to look back as a means to look forward. This said, landscape architecture is, nonetheless, built out of the Colonial Project, since designers are tasked with architecting land that, at least in the Americas, is not rightfully "our" own (I use "our" to indicate those outside indigenous communities from which the land was stolen). Throughout what became known as the Americas, European settlers despoiled Native land while attacking indigenous ecological knowledge and practice, forcefully imposing Western modes of private property and land use that are perpetuated today. While a much larger project than the landscape architect's alone, this history of land needs to be told and reckoned with as critical to the field's development and a means to challenge the elitism upon which the profession and field were founded.

Figure 8.4 Bird's eye view of Parc de Monceaux with entrance from the Boulevard de Malesherbes in the foreground. From Adolphe Alphand, *Les Promenades de Paris*, vol. 2 (Paris: J. Rothschild, 1867–1873). Courtesy: University of Southern California, on behalf of the USC Libraries Special Collections.

In order to dismantle the elitism of the field's historical foundations, it is thus imperative to acknowledge the fundamental role that colonialism and enslavement have played in the development of modernity and, by extension, its landscape manifestations. As an educator who teaches landscape history courses, students and I attempt to localize ourselves by understanding whose land we are standing on, settler modes of land domination and ownership, and "the 'sorrow and suffering' of labor that was, and continues to be, extracted from [racialized and gendered] bodies for the purposes of mass profit and the propagation of empire."[14] While landscape architecture indisputably emerged out of European traditions and imperialist conceptions of land and space (and peoples), challenging, problematizing, and broadening the historical canon for landscape architecture is fundamental, including looking at sites and practices outside the "industrialized" world, as well as looking beyond Western worldviews and ways of seeing to understand the construction of space and time through other knowledge systems. This radical reshaping of the elitist foundations of the field is one fundamental way to imagine a more inclusive landscape and practice of landscape architecture.

WHAT IS LANDSCAPE? WHO IS IT FOR?

The conventions and expectations inherited from a common disciplinary reliance on Eurocentric models of landscape have perpetuated in how designers, trained in landscape architecture, shape urban space. Through the scenic and spatial ordering of our cityscapes according to these particular norms and conventions, struggle and resistance, as well as forces of oppression ever-present, are rendered invisible yet under a tenuous landscape veneer. Don Mitchell sees no other means of dispelling this mystification than the transformation of landscape into public space or the claiming of space "through the practices and political actions of people using it."[15] Landscape's elitism is often attributed to the prioritization of aesthetic values that determine what *and who* should be visible in urban space.[16] With a focus on anti-homeless laws, in his passage on "Landscape or Public Space?" Mitchell traces the symbolic language of exclusion and entitlement expressed in the urban landscape. He continues, as Cosgrove and others have argued,

> "Landscape" implies a particular way of seeing the world, one in which order and control over surroundings takes precedence over the messy realities of everyday life. A landscape is a "scene" in which the propertied classes express "possession" of the land and their control over the social relations within it. A landscape in this sense is a place of comfort, of relaxation perhaps, of leisurely consumption, unsullied by images of work, poverty, or social strife. Landscape, Cosgrove shows, developed from, and reinforces a "bourgeois, rationalist conception of the world."[17]

Living in Los Angeles, this tension could not be more apparent—between the city, with the private interests that sustain its Global City status, and the reality of a homelessness epidemic spiraling out of control. The exclusionary landscapes that criminalize homelessness as an extension of the laws and codes set forth to protect private persons and interests have inevitably become part of the curricular considerations for the landscape architecture program I directed at USC. These aspects of "public space" are undeniably connected to a profession entrenched in the market and working in partnership with institutions that exclude and often violently displace those who inhabit public space *through* the construction of "landscape." Aggressive homeless sweeps continue in parks throughout the city, including in Echo Park in 2021, where a long-established encampment that had created some communal forms of self-policing, food provisions, and sanitation was ultimately removed by police for "park repairs"— repairs that were undoubtedly designed and sanctioned by landscape architects (Figure 8.5).[18]

Working to question and challenge the field's role in creating "elite" spaces for only those deemed worthy, Jared Edgar McKnight, a 2021 graduate of the USC landscape architecture master's program, studied the hostility of these urban landscapes as institutionalized in LA's Municipal Code (Figure 8.6). LA's codes render almost every aspect of one's daily life and routine punishable while unhoused (sitting, sleeping, lying in public places, storing personal property. These codes become physically manifest in urban landscapes that are brutal for unsheltered persons to navigate. Streetscapes, civic spaces, and parks are the very spaces landscape architects design as spaces intended to reflect order and specific standards of behavior. Rather than isolate, exclude, and target the vulnerabilities of a community without other options, USC students have been asked how to design spaces that support a range of publics and not perpetuate suffering and its convenient erasure.[19]

Figure 8.5 A "Love Garden" outside one of the remaining tents at the Echo Park homeless encampment amid a homeless sweep that eventually displaced all encampment residents for the council district to make "park repairs." Echo Park Lake, Los Angeles, March 24, 2021 (AP Photo/ Damian Dovarganes).

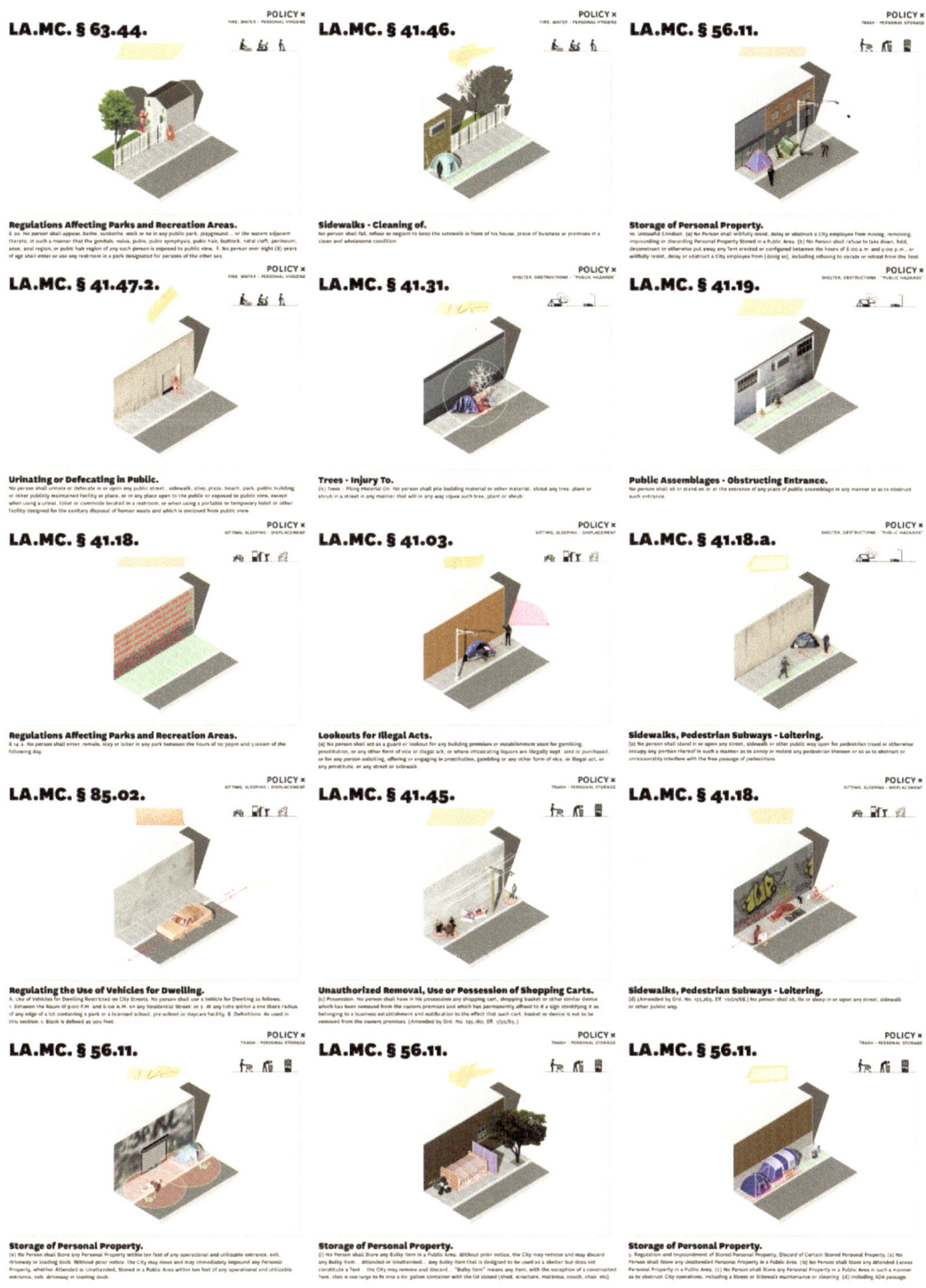

Figure 8.6 For his USC graduate thesis, Jared Edgar McKnight (MLA '21) traced the spatial manifestations of Los Angeles' Municipal Code to demonstrate how almost every aspect of one's daily life and routine becomes punishable while unhoused.

Railing against the political lethargy of the field, landscape architect and planner Billy Fleming accuses the field of being a "largely project-driven enterprise, dependent upon the elite, private interests that now shape urbanization, even in ostensibly public spaces."[20] Landscape architects work closely with urban developers and cities seeking capital investment to create artful urban spaces that might recall ecological histories or even perform important ecological functions, including contaminant remediation. Many of these spaces are some of the most memorable experiences one might take away from a visit to a new place (think Teardrop Park in New York City or Tanner Springs Fountain in Portland, etc.[21]), but the field has not reckoned with how these sites have been used as props to entice investment that leads to increased land values and the displacement, job loss, and exclusion of communities. In other words, landscape elitism is most tied to economic priorities rather than enriching experience, and stabilizing and caring for communities in place.

Jane Hutton, in her book *Reciprocal Landscapes*, critiques "post-industrial" site remediation and reinvention through landscape architecture, asking, "In addition to remediation and providing recreational access to new publics, how might this type of project confront the evacuation of industry and the economic devastation and locally experienced environmental risks that were left in its wake?" She asks how landscape architects might become increasingly involved in simultaneously addressing "local economic development, working-class employment, and ecological regeneration [to] foster a new kind of 'post-industrial' development, one that creates socially and ecologically supportive forms of work."[22] Her assessment of the High Line as "an archetypal neoliberal space" is equally compelling to understand how landscape's elitism is manifest in "post-industrial" projects, and in how public spaces are increasingly funded, maintained, and programmed through private organizations:

> [The High Line] is emblematic of new landscape projects built under neoliberal urban agendas, which, through different mechanisms, shift public services towards the private sector, all taking place within the aesthetics of obsolescence. Many of the most-acclaimed North American landscape architecture projects of the last two decades are sited in more affluent cities, in the remnants of twentieth-century industry. . . Like the High Line, these sites index the migration of industry out of urban centers (and often out of the country), their replacement with service and financial industries, and the displacement of the urban working class.[23]

She continues to address recent financial models for urban development where parks are no longer operated, maintained, and programmed by city agencies but must be made self-sufficient through private organizations (Conservancies and "Friends of" groups), who raise money from private, often corporate, sponsors.

> Corporate partners contribute as park stakeholders and tend to support higher end consumption and passive activity—preferences enforced by heightened levels of security and managerial control. On the one hand, parks like the High Line are of exceptional design quality. . . and offer unique park experiences. But on the other hand, many of the

park's most desirable qualities are made possible through measures of control that ultimately exclude. That visitors to the High Line are disproportionately white is just one significant example of exclusionary privilege in this type of post-industrial landscape. While a visitor to the High Line might enjoy a late spring nap on a wooden chaise longue, feeling secure in the complex of guards, park attendants, and vendors, in other parts of the city, someone attempting to sleep in a park might be fearful of their personal security and police aggression. The result is a very different type of park than exists in poorer or racialized neighborhoods, where less social capital exists for developing such management systems, policing is more aggressive, and park budgets dwindle.[24]

I include this lengthy reference to demonstrate just how implicated landscape architects can be in systems of exclusion and oppression, even unintentionally. While these assessments might seem unfair given (1) how remarkable a place like the High Line is and (2) landscape architects are not responsible for the structures that create these injustices, it is still important to recognize how entrenched landscape architecture is in perpetuating inequity and even violence.

Landscape architects historically benefited from the same policies and projects that have not only excluded but actively traumatized communities of color. While active in New Deal public works initiatives in the 1930s, landscape architects, particularly postwar, profited tremendously from the Federal Housing Administration and the Veterans Administration established under the GI Bill, which secured loans to white veterans returning from the war and led to the country's suburban building boom that segregated the nation. Even those few landscape architects like Lawrence Halprin and Paul Friedberg—both of whom had notable social consciences and contributed to the rise of participatory design in the 1960s and 1970s—and who were working in the vastly restructuring American city, benefited directly and indirectly from Title I of the Housing Act of 1949 (slum clearance and urban renewal) and the Highway Act of 1956. Meanwhile, these policies devastated Black and immigrant communities throughout the country through their systemic forced removal (Figure 8.7).[25] Clearly, this racialized form of elitism (i.e., racism) perpetuates through uneven investment in public environments across cities nationwide, whereby the redlining maps created by the Home Owners' Loan Corporation in the 1930s are often still legible by where shade trees, parks and other forms of "green infrastructure" exist and do not (addressed again below).[26] Questions of race, racial oppression, and landscape architecture are complex subjects that need their own standalone analysis.[27]

Finally, while not a disabilities scholar, it is not difficult to recognize how, despite the achievements of the Americans with Disabilities Act of 1990, the urban landscape has been designed to exclude a range of bodies. Disabilities range from physical to sensory, neurocognitive to mental. Yet, standards are written for a much more limited scope and with the lowest baselines for access, rather than focusing on how a range of bodies can thrive and enjoy the public landscape. In many ways, the landscape designer is in a position of power and control to determine if a landscape serves only elite bodies that conform to

Figure 8.7 Lawrence Halprin & Associates was invited by redevelopment agencies nationwide in the 1960s and 1970s to create the centerpiece of "renewed" areas that had been cleared of what were often the only communities where Black residents could live. In Rochester, New York, the firm designed Manhattan Square Park for the Rochester Department of Urban Renewal and Economic Development as the center of the Southeast Loop Urban Renewal Area. This project devastated Black and diverse working-class communities: Lawrence Halprin Collection, Architectural Archives, University of Pennsylvania.

normative ideals for sensing, perceiving, and moving through space. Only recently has scholarship and its application in the built landscape attempted to challenge the culturally constructed ideal body as the only considered inhabitant of the designed landscape.[28]

CULTURAL LANDSCAPE IS INCLUSIVE!

In the late 1990s, James Corner wrote of landscape's "recovery" from what he identified as a field that had lost itself to nature stewardship (primarily via followers of McHarg) and the "restoration of an essentially cultureless natural world."[29] He writes about landscapes' cultural recovery through the retrieval of memory and history specific to particular places. Yet, the specificity of *whose* memories and histories were to be considered remained mostly unexplored. However, one history Corner specifies as providing a new creative challenge is the reinvention of once industrial sites (he later becomes the landscape architect whose firm is responsible for the High Line, described above). An earlier quintessential example of this "new" landscape project is Landschaftspark Duisburg-Nord, the transformation of part of the massive steelworks in Germany's Ruhr Valley, by landscape architect Peter Latz. Resonant within Hutton's critique referenced above, in a symposium brief titled "Where is the Outrage?" geographer David Harvey asks, "Converting an old iron and steelworks into a magnificent park or play space is one thing, but then I thought to myself, have we stopped using iron and steel? Where is the iron and steel now being made?" To this, he explains the working conditions of Chinese manufacturing in the Yangtze and Pearl River deltas where much of this industry has moved:

> Having witnessed 16-year-old girls working, usually for the grand total sum of $15 a month… you realize our world is being constructed in a certain kind of way and that there is a certain luxury which is allowing certain areas of the advanced capitalist world to somehow or other engage in this act of atonement, (i.e., transforming decommissioned industrial sites into play space.)[30]

As landscape historian John Beardsley additionally notes, the site is sanitized of "the close connection between the steel industry and the Nazi party during World War II, [and]. . . the widespread use of prisoner of war and slave labor in German factories in those years."[31] The hiddenness of forced, enslaved, and low-wage labor is the primary impetus for my own research to uncover the back-of-house landscape realities in California. The industrial agriculture that characterizes California's Great Central Valley, depends on unfree and racialized labor, while the Coastal Range mountains separate this "working country" from the "landscapes" associated with the palm tree-lined coastal cities and continuous beaches of the California imaginary (Figures 8.8 and 8.9). Landscape scholar Kofi Boone has introduced how enslaved people on Southern plantations adapted techniques from Western Africa to design and engineer these landscapes of production. Using the example of Middleton Place, whose unique landscape design is typically included in the field's history textbooks without mentioning those responsible for building it, Boone brings to light the techniques adapted from Senegal by the Wolof people to optimize rice production.[32]

Thus, an extension or a primary component of landscape's elitism has been the limited representation of cultural experience and the muting of labor, conflict, oppression, and resistance that shapes the social realities around us. Challenging the deployment of design and public art to distract or mute social realities in the city, Dolores Hayden presented cultural landscape studies as an alternative. In her 1997 essay "Urban Landscape History: The Sense of Place and the Politics of Space," Hayden revisited the term cultural landscape and landscape's shifting convergences with the contested terms *place* and *space*. Hayden

Figure 8.8 Screenshots of select news coverage on the slow violence of toxic exposures in the San Joaquin Valley (lower third of the Central Valley), where farmworkers experience the country's worst air quality, lack of access to drinking water, failing septic systems, and nitrates, arsenic, and agricultural chemicals impacting the soil, water, and air. Clockwise from upper left: Rory Carroll, "Life in San Joaquin Valley, the Place with the Worst Air Pollution in America," *The Guardian*, May 13, 2016; Robin Meadows, "Living in California's San Joaquin Valley May Harm Your Health," *The New Humanitarian*, July 5, 2017; Jacques Leslie, "What's Killing the Babies of Kettleman City," Mother Jones, July/August 2010; Jose A. Del Real, "'Brown Water for Brown People': Making Sense of California's Drinking Water Crisis," *New York Times*, November 29, 2019; Jose A. Del Real, "They Grow the Nation's Food, But They Can't Drink the Water," *New York Times*, May 21, 2019; and "Twilight Greenaway, Climate Change-Fueled Valley Fever Is Hitting Farmworkers Hard," *Civil Eats*, June 17, 2019.

recognized the political dimensions of urban space as an arena of contestation and negotiation, and the aesthetic dimensions of place as emergent from material practices experienced via the body. She insisted on broadening the definitions of the cultural landscape to include the political dimensions of territory and space, as well as the dynamics of power and resistance ever legible in the urban landscape. She explains cultural landscapes tell "the story of how places are planned, designed, built, inhabited, appropriated, celebrated, despoiled, and discarded. Cultural identity, social history, and urban design are here intertwined."

Figure 8.9 The rural community of Allensworth in the lower Central Valley of California. While Allensworth has a unique history as a Black Town founded, financed, and governed by African Americans in the early twentieth century, like most unincorporated settlements in the lower Central Valley, residents today are mostly Latino farmworkers facing the array of stresses on rural communities (see caption for 8.8). The Landscape Justice Initiative at the University of Southern California is currently working with the Allensworth Progressive Association to use its unique past to leverage possibilities for a thriving future. Drone photograph by Yuliang Jiang, research assistant to author.

Converse to Williams' distinction between landscape and "working country," Dolores Hayden argues for the latter's interpretation—through public history, preservation, and art practices—of "working landscapes," including spaces of subsistence, as well as capitalist production and social reproduction. While the study of urban history might seem remote from the design of landscape, at the end of her article, Hayden calls on landscape architects, designers, and planners to draw more resonant connections to public memory in efforts to achieve a more inclusive city.[33] Revealing the invisible processes of landscape production through new approaches to design has the potential to overturn landscape's structural elitism.

Designing places through cultural landscape frameworks means understanding how to research the history of sites and their social meanings. While part of the landscape architect's creative project might be the curation of stories we want to bring forth about site and land, it is landscape architects' obligation to acknowledge and reckon with the dark and often violent histories of land that have shaped the United States (and many other "industrialized" countries). Landscape architecture as a form of revisiting and reckoning opens up new ways of thinking about practice as socially productive and inclusive of histories, memories, and voices that have been long silenced. As landscape historian Thaisa Way has put it,

> History is a means to embark on a rigorous reckoning in order to unravel the inherent complexity of places and their becoming as we know them today. Designers cannot be ignorant. Instead, designers might lead a more rigorous reckoning of past narratives of the sites and places, leading to more responsible and responsive designs for all.[34]

This is the story of people... working together to create a new environment in an old neighborhood

NEIGHBORHOOD COMMONS

LANDSCAPE ARCHITECTURE: WHO ARE THE LANDSCAPE ARCHITECTS?

The distinction of the landscape architect from the gardener—as an extension of Alberti's privileging of the architect from the craftsman—is an obvious component of landscape's elitism.[35] Landscape scholar Julian Raxworthy argues that this "class divide," whereby landscape architects have attempted to distance themselves from what they see as an amateur hobby, has resulted in a landscape that is "denatured,"[36] a complete disengagement with plants. The field has worked hard to distinguish itself from horticulture and gardening in its insistence that large public works and infrastructural projects cannot be concerned with the fussiness of planting arrangements. Clearly there is a gendered subtext here that undoubtably associates gardening as part of domestic work rather than important enough to structure our public spaces. Thankfully, this denatured approach to landscape seems to be changing as more landscape architects join those who knew all along that plants were not just decorative means to "green" or "soften" architectural spaces, but have specific qualities, characteristics, and behaviors that have significant public health, biodiversity, climate, and aesthetic benefits. Many of these benefits of what is often referred to as "green infrastructure" have yet to reach the communities who need it most, however, as can be interpreted by recent studies of the correlations between historically redlined communities and the impacts of extreme heat in these same areas due to the lack of shade.[37] Beyond the elitism of human-centric thinking, these "benefits" move beyond "ecosystem services" and have the capacity to support more-than-human life.

Like the gardener, the community designer, despite professional credentials, has also often been cast off as a service provider rather than a landscape Visionary. Community designers such as Karl Linn, whose work set the foundation for the Community Design Centers springing up throughout the nation

Figure 8.10 Outreach material for the "Neighborhood Commons" Karl Linn constructed with community members in Harlem, New York: Karl Linn Collection, Environmental Design Archives, University of California, Berkeley.

starting in 1963, have received almost no scholarly attention within landscape architecture, which I have, in previous writings, attributed to his predominant focus on community development and participatory design rather than leaving us with iconic physical spaces or a signature material inheritance. He introduced "Neighborhood Commons" in declining areas of North Philadelphia in 1960 and then subsequently in similar districts of Washington DC, New York, Baltimore, Chicago, and other US cities; yet, most of these places have long since disappeared (Figure 8.10). His attention to community capacity-building and empowerment through the establishment of organizational entities, such as the Neighborhood Renewal Corps, had wide-ranging impacts outside his field, including on government-sponsored antipoverty programs and the growing advocacy planning and Community Design movement in architecture, which itself was marginalized by the architectural establishment for its lack of guiding aesthetic principles.[38]

This marginalization of practitioners and forms of practice that do not fit within the traditional designer-client project model has much to do with insecurity over challenged expertise and the desire for control over how a project should look and be used. Landscape architects have to fight for their rightful place in urban and regional-scale projects that are too often led by architects and engineers who know less about social and ecological systems, so their desire to demonstrate and defend their expertise as designers is not unwarranted. Yet it is typically class and race privilege that enables one to become a designer, and while the hope is this is changing with improved access to educational opportunities, landscape design practice needs a significant ideological shift to shed itself of the elitism that has left so many out of decisions about the future of their own physical environments.

The community-based and activist work of landscape architects such as Anne Whiston Spirn (who has recently spoken up about her marginalization by academic colleagues for her community-based work in the Mill Creek neighborhood of Philadelphia[39]), as well as Randy Hester, Laura Lawson, Diane Jones Allen, Jeffrey Hou, David de la Peña, Kofi Boone, and Julie Stevens, among many others, has historically been situated outside the conventions of practice and capital-D design.[40] Yet the field has so much to learn from such efforts at democratizing design. Professional expertise based on education and experience still guides the contributions of the aforementioned practitioners to projects, yet the co-generating processes they employ allow communities to step into ownership of these projects as part of broader efforts at self-determination. It is such forms of practice that inspired my establishment of the Landscape Justice Initiative (LJI) at USC, which is largely focused on developing sustained pathways to learn from local knowledge and expertise derived from lived experience, as essential to a more inclusive form of practice, and quite different from the limited scope of community meetings and public presentations required of public projects today[41] (Figure 8.11).

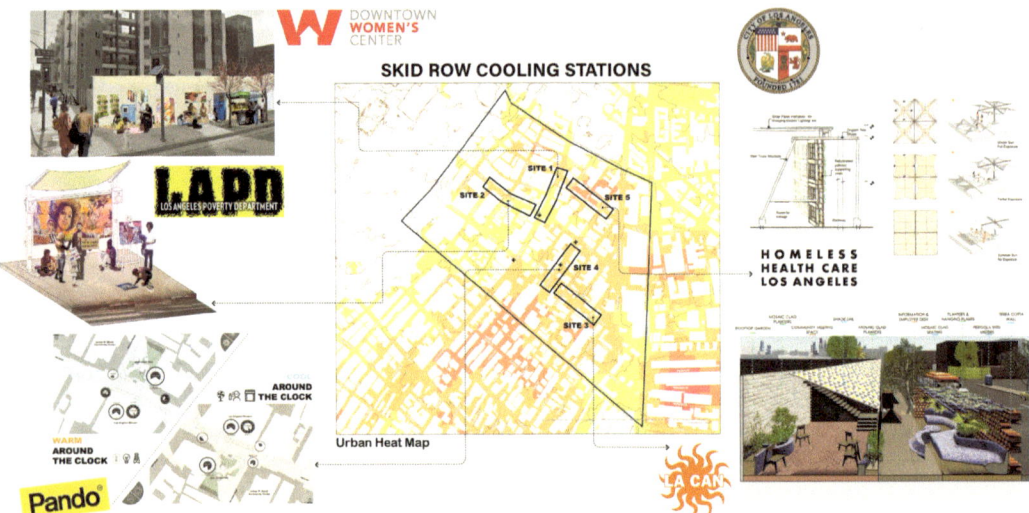

CONCLUSION

Project-based incremental change can have resounding impacts to provide models for future practice. Yet, landscape architecture must be reframed and radically restructured to truly transcend the field's deep-seated elitism. How we encourage youth to discover and understand landscape and the empowering capacities of landscape architecture is one place to start, as is transforming how we teach landscape architecture in accredited programs, offering emerging designers the opportunities to listen and collaborate, partner, and co-produce. Landscape architecture cannot "be both an instrument of neoliberalism *and* an activist force in the fights against climate change and for social justice."[42] Fleming suggests that landscape architects must become more politically engaged to truly challenge the status quo, and while we watch some of the world's most vulnerable people drown and burn on account of actions and decisions made by the global elite, it is undoubtedly true that to have real impact and make the field more relevant, inclusive, and resoundingly impactful, we cannot remain apolitical. It is potentially through federal infrastructure plans and public works that we can start to reach beyond the current exclusivity of the profession. In addition, we can continue to find ways of operating outside and challenging capitalist frameworks whose timescales of "progress" do not match the extended timelines required for meaningful work—especially with the landscape medium. Building trust and taking *care,* particularly in communities that have been harmed time and again by racist spatial policies, must be "uncoupled from the timeframe of a project," in Kofi Boone's words.[43]

Care is a word I would like to dwell on in the context of dismantling the elitism that has long-defined landscape and landscape architecture, particularly as we slowly *and unevenly* emerged out of the global COVID pandemic.

Figure 8.11 Plans and partners for the USC Landscape Justice Initiative Workshop on urban heat and cooling infrastructure in Los Angeles' Skid Row neighborhood. Partner Skid Row organizations included Downtown Women's Center (DWC), Center for Harm Reduction, Homeless HealthCare Los Angeles (HHCLA), Los Angeles Community Action Network (LACAN), Los Angeles Poverty Department (LAPD), The ReFresh Spot, and Los Angeles Office of City Homelessness Initiatives. All 2021 MLA students participated with support from Pando Populus.

No other time than the peak of the pandemic had the work of care felt so primary, as we attempted to care for young and old, sick and well, while being asked to continue without interruption the "work" fueling the grander economic machine. Feminist approaches to care show how the work of reproduction and maintenance of life has traditionally been considered marginal to "value-creating" work. It was revealed during the year plus of the pandemic's peak to see whose work was deemed "essential" and essential to what—to life or to profit. While profit was prioritized at the structural level, mutual aid and other community care infrastructure emerged locally and challenged the legacy of the elite few who made decisions for the less entitled many. How landscape architects might contribute to stabilizing and strengthening those infrastructures seems a productive way forward and still leaves room for Design and the craft of spaces and experiences that tap into cultural memory and meaning.

After the global economic collapse in 2008 and the Occupy Movement that challenged the vastly imbalanced concentration of wealth in a few elite hands, landscape scholar Ed Wall questioned if we might be reaching a "post-landscape condition," in which "new designs, representations, and physical forms provide for collective actions and alternative relations with where we live, work, and visit?"[44] Though at the time I concluded decidedly no, we are not yet there, today we seem closer. After a year and a half severed from public space (beyond cyberspace) and those outside our immediate networks, it was clear we emerged starved of contact and engagement with our physical world and each other. National and international reckoning over racial injustice has likewise cracked open debates and some tangible action around deeply entrenched structures of inequality, exclusion, and violence. Perhaps the forced localization of communities and individuals will render landscape practice more local or more invested in the inequalities latent and blatant in our own immediate contexts. At the same time, it undoubtedly forced us all to recognize the global interdependence and consequences of prioritizing economic agendas over questions of survival.

Our cities are different now. The pandemic revealed so much about what it means to be part of a local and global community, but it also exposed the vast inequalities palpable at all scales. Despite the supposed rebound of the global economy, the pandemic left behind a tangible wake of suffering. Landscape, which has been wielded as an instrument of control, power, authority, and erasure, could be transformed in this period of reckoning into meaningful places of compassion and care and the acceptance of the "messy realities of everyday life" as part of what it means to live together.

NOTES

1   James Corner, "Eidetic Operations and New Landscapes," in *Recovering Landscape: Essays in Contemporary Landscape Architecture*, ed. James Corner (New York: Princeton Architectural Press, 1999), 157.

2   See particularly Ken Olwig, "Recovering the Substantive Nature of Landscape," *Annals of the Association of American Geographers* 86, no. 4 (1996): 630–653; and Denis Cosgrove, *Social Formation and Symbolic Landscape* (Madison: University of Wisconsin, 1998). The latter was first published in 1984. Denis Cosgrove traced "links between landscape representation and capitalist alienation from the land," specifically looking at the relationships between the development of linear perspective in the early Italian Renaissance and the materiality of property relations in simultaneously emerging capitalist and mercantile economies. The detached, fixed and singular envisioning subject made possible by linear perspective, objectified land as commodity, converging with shifts in modes of production from feudal to capitalist systems.

3   Denis Cosgrove, *Social Formation and Symbolic Landscape* (Madison: University of Wisconsin, 1998).

4   Raymond Williams, *The Country and the City* (New York: Oxford University Press, 1973), 120.

5   Don Mitchell, *Cultural Geography: A Critical Introduction* (Oxford: Blackwell Publishers, 2000), 113.

6   Mark David Spence, *Dispossessing the Wilderness: Indian Removal and the Making of the National Parks* (Oxford: Oxford University, 1999); see also William Cronon, "The Trouble with Wilderness," *Uncommon Ground: Rethinking the Human Place in Nature*, ed. William Cronon (New York: W.W. Norton & Co., 1996), 69–90.

7   Emily Eliza Scott and Kirsten Swenson, "Introduction: Contemporary Art and the Politics of Land Use," *Critical Landscapes: Art, Space, Politics*, eds. Emily Eliza Scott and Kirsten Swenson (Oakland: University of California, 2015), 4; see "New Topographics: Photographs of a Man-Altered Landscape" exhibition at the International Museum of Photography of the George Eastman House in 1975–1976.

8   Elizabeth Meyer, "Landscape Architecture as Modern Other and Postmodern Ground," *The Culture of Landscape Architecture*, eds. Harriet Edquist and Vanessa Bird (Melbourne: Edge Publishing, 1994), 13–34; see especially Carolyn Merchant, *The Death of Nature: Women, Ecology and the Scientific Revolution* (New York: Harper Collins, 1980).

9   See Heath Massey Schenker, "Parks and Politics During the Second Empire in Paris," *Landscape Journal* 14, no. 2 (Fall 1995): 201–219, on the lack of political innocence in the cooption of English landscape ideals in Haussmann's Paris; see Ann Bermingham, *Landscape and Ideology* (Berkeley: University of California, 1989), on the ideology of the picturesque.

10 Olmsted to George Loring Brace, December 1, 1853, quoted in David Schuyler, *The New Urban Landscape: The Redefinition of City Form in Nineteenth-Century America* (Baltimore, MD: Johns Hopkins Press, 1986), 66.

11 Andrew Jackson Downing, "A Talk about Public Parks and Gardens," *The Horticulturalist* (October 1848); reprinted in *Rural Essays* (New York: G. P. Putnam, 1853), 192; See also Schuyler, *The New Urban Landscape*; see Matthew Gandy, *Concrete and Clay: Reworking Nature in New York City* (Cambridge: MIT Press, 2002) on symbolic representation of abstract normative ideals connected to the ideological legitimation of American society.

12 Matthew Gandy, *Concrete and Clay: Reworking Nature in New York City* (Cambridge: MIT Press, 2002), 77–114.

13 Elizabeth Meyer, "Situating Modern Landscape Architecture: Theory as Bridging, Mediating, and Reconciling Practice," *CELA Proceedings, Design + Values* 4, (1992): 167–178.

14 Leilani Nishime and Kim D. Hester Williams, "Afterword: Collective Struggle, Collective Ecologies" *Racial Ecologies*, eds. Leilani Nishime and Kim D. Hester Williams (Seattle: University of Washington, 2018), 251.

15 Don Mitchell, "The Annihilation of Space by Law: The Roots and Implications of Anti-Homeless Laws in the United States," *Antipode* 29, no. 3 (1997): 303–335. Quotation from Ed Wall, "Post-landscape or the Potential of Other Relations with The Land," *Landscape and Agency: Critical Essays*, eds. Ed Wall and Tim Waterman (New York and Abingdon: Routledge, 2018), 189.

16 Don Mitchell, "The Annihilation of Space by Law: The Roots and Implications of Anti-Homeless Laws in the United States," *Antipode* 29, no. 3 (1997): 322. Citing Sharon Zukin, *The Cultures of Cities* (Oxford: Blackwell, 1995).

17 Don Mitchell, "The Annihilation of Space by Law: The Roots and Implications of Anti-Homeless Laws in the United States," *Antipode* 29, no. 3 (1997): 323; Citing Denis Cosgrove, *Social Formation and Symbolic Landscape* (Madison: University of Wisconsin, 1985), 49.

18 There are a number of parks throughout the United States that have been removed more or less because they have become overrun with unhoused and "undesirable" populations. Skyline Park in Denver, designed by Lawrence Halprin in the 1970s during a period of downtown "renewal," was redesigned in the early 2000s and fulfilled all tenets of Oscar Newman's "defensible space," which promoted principles of "crime prevention through environmental design," more or less creating spaces of easy surveillance to ensure only certain people can gain access, see Oscar Newman, *Defensible Space: Crime Prevention through Urban Design* (New York: MacMillan, 1972). Seattle Freeway Park, another Halprin-designed space near downtown, has been consistently threatened with redesign in order to clear the space of the populations the city wants to remove.

19 While permanent supportive housing (PSH) is the end goal to address the humanitarian crisis of L.A. County's 66,000+ unhoused individuals, the reality is that goal is taking time to achieve. Thus, supportive interim strategies must continue to be developed. These conversations were translated into a USC design studio by instructors Lauren Elachi, Jerome Chou, Hyunch Sung and Katy Foley. For more on the studio and its extension through the Landscape Justice Initiative I created at USC, see https://sites.usc.edu/landscape-justice-initiative/projects/ (accessed May 30, 2024).

20 Billy Fleming, "Design and the Green New Deal," *Places Journal* (2019), https://places-journal.org/article/design-and-the-green-new-deal/ (accessed July 16, 2021).

21 Teardrop Park was designed by Michael Van Valkenburgh Associates in Battery Park City and opened in 2004. Tanner Springs Park was designed by Atelier Dreiseitl with and GreenWorks, P.C. and opened in 2005.

22 Jane Hutton, *Reciprocal Landscapes: Stories of Material Movements* (New York: Routledge, 2019), 135–136.

23 Jane Hutton, *Reciprocal Landscapes: Stories of Material Movements* (New York: Routledge, 2019), 206.

24 Jane Hutton, *Reciprocal Landscapes: Stories of Material Movements* (New York: Routledge, 2019), 206.

25 I include these names with great hesitance since both Friedberg and Halprin were trying to alleviate or mitigate the damage these policies of clearance had on communities, but funding sources were often the same as those for clearance and redevelopment. See Alison Hirsch, "From 'Open Space' to 'Public Space': Activist Landscape Architects of the 1960s," *Landscape Journal* 33, no. 2 (2014): 173–194.

26 Bryan Lee, Jr., "America's Cities were Designed to Oppress," *Bloomberg/City Lab* (June 3, 2020): https://www.bloomberg.com/news/articles/2020–06-03/how-to-design-justice-into-america-s-cities (accessed January 25, 2023). Further articles related to this racialized access to shade and other human health benefits of planted cityscapes are cited in note 36.

27 Many essays in the book *Black Landscapes Matter*, eds. Walter Hood and Grace Mitchell Tada (Charlottesville: University of Virginia, 2020), begin to take up this analysis with a focus on elevating and celebrating the cultural heritage and design of Black landscapes. Kofi Boone's standalone essay, "Black Landscapes Matter," *Ground Up Journal*, no. 6; https://groundupjournal.org/black-landscapes-matter (accessed May 30, 2024); became a chapter adapted for the aforementioned book, is among the most direct analyses of the racialized dimensions of landscape architecture. In terms of historic analyses of built landscapes, questions of race and space are the focus in a growing body of literature (see, for instance, the work of Dianne Harris and Thaisa Way, current Director of Landscape Studies at Dumbarton Oaks, who are working on programming that focuses on Black landscapes and racialized space). Sara Zewde is a landscape architect and scholar whose forthcoming book on Olmsted will contribute to this racial analysis.

28 For some good examples of such scholarship and applications in disability and the built environment, see Aimi Hamraie, *Building Access: Universal Design and the Politics of Disability* (Minneapolis: University of Minnesota Press, 2017); Sara Hendren, *What Can a Body Do? How We Meet the Built World* (New York: Riverhead Books, 2020); Alexa Vaughn-Brainard, "Design with Disabled People Now," https://www.designwith disabledpeoplenow.com (accessed July 16, 2021).

29 James Corner, "Introduction: Recovering Landscape as a Critical Cultural Practice," in *Recovering Landscape: Essays in Contemporary Landscape Architecture*, ed. James Corner (New York: Princeton Architectural Press, 1999), 3.

30 David Harvey, "Where is the Outrage?," *The Newsletter of the Architectural League of New York* (2005), 4–7.

31 John Beardsley, "Conflict and Erosion: The Contemporary Public Life of Large Parks," *Large Parks*, eds. Julia Czerniak and George Hargreaves (New York: Princeton Architectural Press, 2007), 209.

32 Boone, "Black Landscapes Matter." Relatedly, Diane Jones Allen, in a recent symposium at the University of Texas at Austin titled "Transgressive Practices to Transformative Policies," described in the context of her work on the maroons, or "self-liberated slaves," that Africans were not brought across the Atlantic for labor, but "for knowledge." By this she partly means the technical knowledge and skill needed to live and thrive in a wetland landscape which had been honed by West Africans of the Senegal Valley. Two-thirds of enslaved peoples brought to Louisiana to cultivate rice were from this region in Senegambia she explained here and in her essay, "Living Freedom Through the Maroon Landscape," *Places Journal* (September 2022); https://placesjournal.org/article/the-maroon-communities-and-landscapes-of-louisiana/ (accessed March 18, 2023).

33 Dolores Hayden, "Urban Landscape History: The Sense of Place and the Politics of Space," in *Understanding Ordinary Landscapes,* eds. Paul Groth and Todd Bressi (New Haven, CT: Yale University, 1997), 111–133, 113; see also Dolores Hayden, *The Power of Place* (Cambridge: MIT Press, 1995).

34 Thaisa Way, "Why History for Designers (Part 2)," *Platform* (2020); https://www.platformspace.net/home/why-history-for-designers-part-2 (accessed July 6, 2021). There have been some recent notable projects by landscape architects working with indigenous tribes to restore and safeguard important lifeways, including Kinngaaluk Territorial Park in Nunavut, a Canadian territory returned to Inuit control in 1999. See Brian Barth, "In the Hunt: A New Park in Nunavut, Canada, Is Made to Protect Indigenous Hunting Ground," *Landscape Architecture Magazine* 109, no. 1 (January 2019): 62–83.

35 Leon Battista Alberti, *De re aedificatoria* (completed in 1452, printed in 1485).

36 Julian Raxworthy, *Overgrown: Practices Between Landscape Architecture and Gardening* (Cambridge: MIT Press, 2018), 3.

37 Living in Los Angeles, this is particularly notable in areas away from the coast that are inequitably suffering from extreme heat and shade injustice. See Sam Bloch, "Shade," *Places Journal* (April 2019); https://placesjournal.org/article/shade-an-urban-design-mandate/ (accessed July 2, 2021). See also Brad Plumer, Nadja Popovich and Marion Renault, "How Racist Urban Planning Left Some Neighborhoods to Swelter," *New York Times* (August 26, 2021): https://www.nytimes.com/2020/08/26/climate/racist-urban-planning.html (accessed July 2, 2021).

38 For his impact on the Community Design Movement, see Anna Goodman, "Karl Linn and the Foundations of Community Design: From Progressive Models to the War on Poverty," *Journal of Urban History* 46, no. 4 (2020): 794–815. See Alison Hirsch, "Urban Barnraising: Collective Rituals to Promote *Communitas*," *Landscape Journal* 34, no. 2 (Winter 2015): 113–126.

39 Anne Whiston Spirn, "In Place Over Time: A Case for Longitudinal Action Research," *LA+, "Community,"* no. 13 (2021): 58–65.

40 See many of these practitioners featured in David de la Pena, Diane Jones Allen, Randolph T. Hester, Jeffrey Hou, Laura J. Lawson, Marcia J. McNally, *Design as Democracy: Techniques for Collective Creativity* (Washington, DC: Island Press, 2017).

41 See Landscape Justice Initiative (LJI) https://sites.usc.edu/landscape-justice-initiative/ (accessed May 30, 2024). Current projects include work in the Black Town of Allensworth, CA, founded by Allen Allensworth in 1908 and community monument building in the South Los Angeles neighborhood of Willowbrook, CA, both of which have relied on sustained processes of building relationships and networks of trust to work together toward design development. See https://sites.usc.edu/landscape-justice-initiative/projects/ (accessed May 30, 2024).

42 Billy Fleming, "Design and the Green New Deal," *Places Journal* (2019), https://placesjournal.org/article/design-and-the-green-new-deal/ (accessed July 16, 2021).

43 Kofi Boone, "Design Justice Q+A," *LA+, "Community,"* no. 13 (Spring 2021): 30.

44 Ed Wall, "Post-landscape or the Potential of Other Relations with the Land," *Landscape and Agency: Critical Essays*, eds. Ed Wall and Tim Waterman (New York and Abingdon: Routledge, 2018), 154.

Chapter 9: Is Landscape Just?

Kofi Boone

INTRODUCTION

Landscape is not yet just. Justice, "the quality of being just, impartial, or fair," has many ecological and societal dimensions that affect how we perceive its expression in the landscape. The myriad interpretations of justice in the landscape also confound our collective ability to reach an agreement on what constitutes fairness. It may not be possible to define this in a field defined by fee-for-service work for clients operating within the constraints of private capital and property rights. However, if justice is not framed as a steady state but a continuous process of facing and dismantling injustice, the landscape and landscape architects are situated to contribute to making our decisions and actions align with fairness to Earth and its people.

Injustice in the landscape takes many forms. Design and Protest and the Design Justice Movements[1] have both successfully argued that we inhabit a designed world that spatializes forms of justice and injustice. Everything, from decisions affecting the exposure to toxins and noxious uses to the aesthetics of streets and public plazas, is infused with our sense of belonging, exclusion, and social order.

This chapter focuses on injustice in the landscape and the potential roles for planning and design strategies to address it, using an environmental justice framework to characterize political, economic, and cultural situations that have perpetuated injustice in American landscapes. Additionally, the need for recognition, reconciliation, and reparation is presented as concepts for disciplinary exploration to better integrate the tools and strategies of dismantling injustice in the landscape.

ENVIRONMENTAL JUSTICE FRAMEWORK

The Environmental Justice Movement was the first major environmental movement founded and led by Black and Brown people in the modern era. Its 17 principles reflect decades of multiracial and multi-sector coalition building asserting everyone has a human right to clean air, water, land, and interdependence with nature. Conflicts in environmental values between indigenous people and settler colonial interests trace one of the pillars of the Environmental Justice Movement.

Historically excluded from the annals of the American mainstream environmental movement, the living conditions and ecological harm imposed on Black and Brown communities were not included in the interests and actions of mainstream environmentalists. Whether sanctioned by European ruling classes or incentivized by capitalist land exploitation for raw materials, the result was the imposition of Western European norms on non-European landscapes. This is not to overly romanticize the environmental and social values of indigenous communities. But it is suggested that the violent stamping of morals and ethics that allowed landscape to be considered property, measures of value connected to extraction, and the justification of human subjugation (from indentured servitude to slavery) to increase efficiencies were all disruptive practices that echo into the current day.

RECOGNITION

Indigenous resistance movements occurred throughout the 500 years of conflict with European settler colonialism. These historical and contemporary resistance movements have recently been made more visible to landscape practitioners. From the rising popularity of texts such as *Braiding Sweetgrass: Indigenous Wisdom, Scientific Knowledge, and the Teachings of Plants* to the critical rethinking of American history in *An Indigenous Peoples' History of the United States,* there is increasing access to traditional environmental knowledge that reframes the role of the landscape and our positions within it.[2] Research has forced a revision of the perception of well-regarded mainstream environmental icons, including John Muir. Muir is widely revered for his instrumental role in advocating for the creation of Yosemite National Park and popularizing the benefits of experiencing National Parks. However, his historical role has been revisited to include his indifference to the removal of the indigenous Ahwahneechee or his regressive views of African Americans.[3]

From the fight for indigenous land rights in Oklahoma and the American West and sovereignty in current US territories to the open resistance to the Dakota Access Pipeline at Standing Rock to the proliferation of memorial and monument landscapes dedicated to indigenous people to the increased disciplinary interest in learning, we are currently riding a wave of recognition of the harm done among a growing number of landscape practitioners. Of course, recognition of harm done is necessary but not sufficient to face injustice. But it represents an essential step in acknowledging the inadequacies of our mainstream framing of the extent, meaning, and sense of belonging to the landscape required to engender the formation of broad coalitions affecting environmental and social change.

RECONCILIATION

A hallmark of environmental justice is the deployment of many non-violent, direct-action organizing tactics used by the Modern Civil Rights Movement.

Dr. Martin Luther King, Jr.'s decision to support Black sanitation workers in Memphis is often cited as a precedent for the merger between the social, economic, and environmental rights pursued by Black people. Mainstream accounts of Civil Rights history reify the contributions of individuals, but the real power of the era was based on scores of decentralized community organizers, many of them Black women.[4]

This became more apparent with regard to fighting injustice in the landscape in the late 1970s and early 1980s in the American South. In Houston, Texas, a class action lawsuit accused the city of disproportionately citing waste facilities in Black communities. Although only representing 25% of the city's population, over 80% of waste facilities were in Black communities.[5] Although the lawsuit failed, it introduced a methodology of mapping the correlation between demographics and exposure to toxins that was used broadly. One of the subsequent places using this approach was Warren County, North Carolina. The development of a PCB Landfill in one of North Carolina's poorest and Blackest counties triggered nationally publicized protests. Most notably, the protests used tactics from the Modern Civil Rights Movement and saw the re-emergence of Civil Rights leaders who were inactive for over a decade. Among them was Benjamin Chavis of the United Church of Christ, who worked with researchers to produce *Toxic Wastes and Race* in 1987.[6] The report was a national study with evidence of race being the most significant determinant in one's proximity to toxins. This pattern, coined "environmental racism," was the immediate predecessor to what we now call environmental justice.

Organizing tactics continued to build multiracial coalitions around the world and led to the First National People of Color Environmental Leadership Summit, where participants crafted the 17 principles of the Environmental Justice Movement.[7] This global movement, which comprised a collection of indigenous, Black, and Brown environmental values and aspirations, successfully lobbied the Clinton Administration to enact Executive Order 12898, requiring all projects under NEPA to demonstrate that they are not increasing the burden of toxins on communities of color. The EPA formed the Office of Environmental Justice. Although maligned for being severely under-resourced, it operates as an advocate for environmental justice within federal programs, providing funding and organizing resources for communities across the country. These include the Collaborative Problem-Solving Model (derived from an environmental justice effort in Spartanburg, South Carolina) and EJSCREEN, an open-access online mapping tool visualizing demographic disparities in toxic exposure throughout the United States and its territories. More recently, the Climate Justice Movement has identified the disproportionate impacts of climate change on Black and Brown communities globally, especially those with low carbon footprints and not contributing significantly to greenhouse gas emissions. The Biden Administration's Justice40 Executive Order,[8] challenging all federal agencies to develop strategies for investing 40% of renewable energy dollars in disadvantaged communities with a legacy of toxins associated with fossil fuel extraction, represents the continuation of federal policies instigated by grassroots environmental movements.

From the rise of a global movement conceived and led by people of color to the inculcation of their values into federal policies and the actions of federal agencies, the results of non-violent direct action by people in their landscapes represent a form of reconciliation. Beyond recognition, reconciliation includes actions bringing divergent views together and building functional working relationships. The irony of reconciliation in the context of injustice in the United States is that it attempts to activate a moral response in leaders to respond to the letter of the US Constitution and federal laws that were created by people that seized indigenous land and enslaved people. Additionally, an underlying thesis of the Environmental Justice Movement describes the reality of "residential apartheid." Racial segregation in American communities enables the phenomenon of disproportionate impacts in the first place. Dr. Robert Bullard, one of the founding academic leaders of the Environmental Justice Movement, points to the need for equal protection under the law. In this way, Bullard invokes one of the radical metaphors created by Dr. Martin Luther King, Jr. who proclaimed that equal protection is the debt owed to Black people.

> "We refuse to believe that the bank of justice is bankrupt. We refuse to believe that there are insufficient funds in the great vaults of opportunity of this nation. So we have come to cash this check, a check that will give us upon demand the riches of freedom and the security of justice."[9]

However, incremental steps toward reconciliation through broader access to tools and resources for communities facing environmental injustice are significant. These approaches have impacted landscape practitioners in areas ranging from public health-derived strategies based on social determinants of health to the justification for new public open spaces based on disparities in park access. Generally, reconciliation has been undergirded by the assumptions of market generation. Reduced taxes and a shrinking public appetite for government spending have come with alternative funding strategies that see people less as citizens and more as customers of services. Finding political and economic support for developing a community garden or playground is more palatable than commiserate support for removing toxins and environmental hazards. Focusing on developing more landscapes that are in the wheelhouse of mainstream landscape practitioners, even with just and equitable intentions, does not disrupt the most noxious practices and outcomes that drive the Environmental Justice Movement.

REPARATION

Reparation, making amends for harm done through compensation to the injured party, has been one of the most contentious topics in American history. Especially when the harm done was authorized by the US government. Although granted in discrete instances, including to the families of Japanese Americans wrongfully relocated to internment camps during World War II, reparations in the context of broader landscape injustice remain elusive.[10] Unjust treaties with indigenous groups that resulted in their violent relocation to reservations and enrollment in normal schools represent some of the most egregious harm done in the national interest. The burdens of this centuries-long dislocation are almost incalculable. Many died en route to new locations. Communities lost their ancestral access to sacred lands, social networks, foodways, and other essential components to sustain social cohesion. Languages and spiritual traditions were lost. In some cases, indigenous groups were forcibly relocated to different ecosystems, requiring adaptation to new environments. Systemic harms are plentiful. Indigenous Americans have diminished life outcomes, including low life expectancy, educational outcomes, and opportunities for social mobility.[11]

Treaties between colonists and indigenous communities were often not enforced and honored, especially in relation to mineral and fossil fuel extraction. The violent repression of Indigenous American protests at Standing Rock illustrates the continuing volatility of land rights issues in the United States. Additionally, landscape practitioners are not required to include acknowledgment of land treaties in the context of their projects. Most designers completely ignore the indigenous groups associated with the landscape, let alone the agreements they formed defining suitable uses of treaty lands. There have been steps toward reparation for this work. "No design on stolen land," a movement, including the work of Pierre Belanger, has challenged the profession of landscape architecture to go beyond making land acknowledgment statements and requiring practitioners to research and respond to land treaty agreements as a part of their work.

Land in Oklahoma, Montana, and Hawaii was returned to indigenous sovereign rule, with additional cases being explored in California and other areas.[12] This is significant because it offers an opportunity to reinterpret land-based cultural traditions and sustain land-based practices. However, other than governance, few reparative resources are being provided. Investment in the continued growth of governance and indigenous land planning and design resources is another arena for reparative work. Isle de Jean Charles has been the historical homeland of the Biloxi-Chitimacha-Choctaw Indians for centuries. The narrow slip of land is almost inhabitable due to the rise in sea level. A federally managed retreat process is underway, resulting in the relocation of people to suburban-style settings on higher ground.[13] Even facing the complete loss of their homeland, the people have been openly critical and disruptive to the process primarily because community-authored settlement concepts were ignored in favor of more economically expedient ones more common among white Louisiana residents. This example illustrates that some of the crises over land rights extend to unjust decision-making processes when indigenous communities depend on US resources and are largely indifferent to the need for just landscapes to respond to the cultural traditions of people.

For Black Americans, the fight for reparations for the generational and systemic harms of enslavement is centuries old. One of the earliest cases was filed by Belinda Sutton in 1783.[14] Repeated appeals were made for reparations through the end of the American Civil War, the creation of the 13th, 14th, and 15th Amendments, and most importantly, Reconstruction. Union soldiers occupied the former Confederate States, enforcing voting rights and redistributing resources to formerly enslaved Black people. Foremost among this was the War Department seizing land from former Confederate soldiers for redistribution to Black people. Forty acres and a mule were more euphemism than real and only a small percentage of Black people received any land.[15] What land was received was often difficult to farm without capital and other resources. This flawed land redistribution program ended with the compromise of 1877 when President Johnson overruled the decision to seize land from former Confederates, remove Union troops from the South, and no longer enforce protections for Black people in trade for election votes. This decision is often cited as the beginning of the Jim Crow Era. The unrealized promise of land as reparations is the foundation of all subsequent advocacy for reparations.

A century later, Black farmland ownership shrank by 98%, representing nearly one trillion dollars in lost wealth.[16] Hundreds of thousands of Black people left their land, moving north to escape the oppression of the Jim Crow South in the Great Migration. Black land loss was also the result of racist Department of Agriculture policies that misinformed Black rural landowners and excluded them from support due to delegation of granting to local (and racist) farm boards. The *Pigford v Glickman* case authorized the largest distribution of wealth and resources for Black farmers to retain and reclaim their land.[17] However, conservative congressional resistance has stymied the implementation of the Pigford decision and recent American Rescue Plan resources. To date, HR 40, the bill authorizing Congressional study of reparations, has not been passed.[18]

Mindy Fullilove Thompson notes that Black people have experienced serial displacement from the time of enslavement to the present.[19] "Redlining," the result of a federal program assessing suitability for home and business improvement loans, excluded Black communities from reinvestment.[20] A generation later, urban renewal and freeway construction destroyed Black communities across the nation. The "War on Drugs" and increasingly militarized policing ushered in an era defined by mass incarceration where nearly one in five Black men over 25 years old have some connection to the criminal justice system.[21] Insurrections around the world in the aftermath of the police murder of George Floyd visualized the rage and frustrations associated with generational trauma.

The Biden Infrastructure Bill included resources to fund the removal of federal highways. Downgrading them into at-grade boulevards has been proposed in many Black communities damaged by freeways. However, in the face of mounting displacement due to the ways public infrastructure improvements trigger increased adjacent property values, there is broad concern that the benefits of freeway removal will also put low-wealth homes and businesses at risk of loss and relocation.

Concurrent to centuries of oppression, Black communities engaged in many activities claiming landscapes to enable some measure of cultural, social, and economic empowerment. J. T. Roane discussed the importance of Yam grounds, subsistence plots on plantations that served as social and cultural exchange places for enslaved Africans.[22] Black towns, including Mound Bayou, Eatonville, and Princeville, emerged as landscapes where people owned homes, held businesses, and persevered amid extreme oppression. Cooperative structures of ownership and management, championed by W.E.B. DuBois and others,[23] enabled the development of large-scale community investment and even seed funds for the Greenwood District of Tulsa, also known as "Black Wall Street."

In the twilight of the Modern Civil Rights Movement, Community Land Trusts (CLTs) were formalized and popularized as a means for rural Black people without large amounts of wealth to gain collective and stable control of land and development.[24] CLTs have only increased and expanded in influence in rural and urban areas to protect housing affordability and retail space access, as well as promote community stability. Through partnerships between governments and nonprofits in cities, including Atlanta, Durham, and Boston, city property is becoming more available to support CLT and cooperative development strategies. The Schumacher Center, the group that pioneered the policies leading to CLTs, promotes the idea of "A Black Commons": a common pool of non-contiguous land trusts to bring wealth formation and stability strategies to the scale of national disparities faced by Black communities.

Alternatively, reparations movements have also included decentralization and networks leveraging alternative forms of wealth operating outside of capitalist and land-based systems. Emergent Strategy, a leadership approach conceived by Adrienne Maree Brown, proposes incremental organizational development and change, beginning with personal transformative change.[25] The influence of this decentralized and personal investment approach has translated into many incremental approaches to reparations, ranging from universal basic income to home improvement resources available to those harmed by systemic injustice. It also mirrors the Capability Approach proposed by Amartya Sen, which focuses on investing in people to help them pursue lives that have value for them.[26] This approach, highly influential to the United Nations Sustainable Development Goal formation process, takes the emphasis off of the restoration of lost resources (a focus of traditional reparations) and instead focuses on investments to increase and pass along investments in people.

Disassociating reparations and landscape justice from solely the restoration of land and property due to systemic oppression has many challenges. Namely, it is outside the historical and cultural narrative associated with resistance to settler colonialism. Massive and long-term harm in many ways begs an equal and opposite response focused on repair. However, in the face of extreme political polarization and a general lack of appetite for large economic and political strategies, adding strategies aligned with building human capabilities may augment the expanding toolkit for pursuing spatial justice. Investment in people and not just place was critical to all of the component elements of the Environmental Justice Movement.

Resistance to spatialized injustice in the landscape offers landscape practitioners many opportunities to extend and deepen the reach of landscape strategies for environmental change. How can we recognize the harm caused to communities and begin the process of building trust? How can we reconcile with harmed communities and promote collaboration and co-creation of sustainable approaches? And how can we continue to repair the harm done through advocacy for community control of land and resources that can build each person's capabilities to pursue the lives they value? These are not traditional questions for those engaged in landscapes of the past. But they are critical to a future where landscapes are, in fact, just.

NOTES

1   Design as Protest and the Design Justice Movement emerged as means of dismantling White Supremacy through the tools used in the built environment professions. They couple non-violent direct-action techniques with the tools of architecture and graphic design in the interest of justice.
2   See Robin Wall Kimmerer, *Braiding Sweetgrass: Indigenous Wisdom, Scientific Knowledge, and the Teachings of Plants* (Minneapolis, MN: Milkweed Editions, 2013); and Roxanne Dunbar-Ortiz, *An Indigenous Peoples' History of the United States* (Boston, MA: Beacon Press, 2014).
3   See John Muir, "Chapter 5: The Passes," *The Mountains of California* (New York: The Century Co., 1894), 74–96. Muir referred to African Americans as "Dirty and Lazy."
4   See Aldon D. Morris, *The Origins of the Civil Rights Movement: Black Communities Organizing for Change* (New York: Free Press, 1984).
5   See Robert D. Bullard, "Solid Waste Sites and the Black Houston Community," *Sociological Inquiry* 53, no. 2/3 (April 1983): 273–288.
6   Commission for Racial Justice, *Toxic Wastes and Race in the United States: A National Report on the Racial and Socio-Economic Characteristics of Communities and Hazardous Waste Sites* (New York: United Church of Christ, 1987); https://www.nrc.gov/docs/ML1310/ML13109A339.pdf (accessed May 28, 2024).
7   First National People of Color Environmental Leadership Summit, October 24–27, 1991, Washington, DC; https://www.ejnet.org/ej/principles.html#:~:text=Delegates%20to%20the%20First%20National,17%20principles%20of%20Environmental%20Justice (accessed May 28, 2024).
8   US Executive Order No. 14096, *"Revitalizing Our Nation's Commitment to Environmental Justice for All,"* April 21, 2023; https://www.whitehouse.gov/briefing-room/presidential-actions/2023/04/21/executive-order-on-revitalizing-our-nations-commitment-to-environmental-justice-for-all/ (accessed May 30, 2024).
9   Martin Luther King, Jr., "I Have A Dream," August 28, 1963.
10 US HR 442, Civil Liberties Act of 1988, August 10, 1988.
11 See Elizabeth D. Hathaway, "American Indian and Alaska Native People: Social Vulnerability and COVID-19," *Rural Health* 37, no. 1 (Winter 2021): 256–259.
12 Claire Elise Thompson, "Returning the Land," *The Grist* (November 25, 2020).
13 Coral Devenport and Campbell Robertson, "Resettling the First American Climate Refugees," *The New York Times* (May 2, 2016).

14 Belinda Sutton, "Petition to the Massachusetts General Court February 14, 1783," *Royal House & Slave Quarters*; https://royallhouse.org/slavery/belinda-sutton-and-her-petitions/ (accessed May 30, 2024).

15 See William Darity, "Forty Acres and a Mule in the 21st Century," *Social Science Quarterly* 89, no. 3 (September 2008): 656–664; and William A. Darity and A. Kirsten Mullen, *From Here to Equality: Reparations for Black Americans in the Twenty-First Century* (Chapel Hill: University of North Carolina Press, 2020).

16 Nathan Rosenberg and Bryce Wilson Stucki, "How USDA Distorted Data to Conceal Decades of Discrimination Against Black Farmers," *The Counter* (June 26, 2019).

17 *Timothy Pigford, et al., v. Dan Glickman, Secretary, United States Department of Agriculture*, US District Court for the District of Columbia, Civil Action No. 97–1978 (PLF). Paul L. Friedman, U.S. District Judge.

18 US HR 40, Commission to Study and Develop Reparation Proposals for African Americans Act, January 24, 2023.

19 Mindy Thompson Fullilove, *Root Shock: How Tearing Up City Neighborhoods Hurts America, and What We Can Do About It* (New York: New Village Press, 2016).

20 Richard Rothstein, *The Color of Law* (New York: Liveright, 2017).

21 Nazgol Ghandnoosh, "One in Five: Ending Racial Inequity in Incarceration," *The Sentencing Project* (October 11, 2023); https://www.sentencingproject.org/reports/one-in-five-ending-racial-inequity-in-incarceration/ (accessed May 30, 2024).

22 J. T. Roane, "Plotting the Black Commons," *Souls: A Critical Journal of Black Politics, Culture, and Society* 20, no. 3 (2018): 239–266.

23 Jessica Gordon Nembhard, *Collective Courage: A History of African American Cooperative Economic Thought and Practice* (University Park: Penn State University Press, 2014).

24 Schumacher Center for New Economics, "Background and History," Community Land Trust Program; https://centerforneweconomics.org/apply/community-land-trust-program/ (accessed May 30, 2024).

25 Adrienne Maree Brown, *Emergent Strategy: Shaping Change, Changing Worlds* (Chico, CA: AK Press, 2017).

26 Ingrid Robeyns and Morten Fibieger Byskov, "The Capability Approach," in *The Stanford Encyclopedia of Philosophy* (Summer 2023 Edition), eds. Edward N. Zalta and Uri Nodelman; https://plato.stanford.edu/archives/sum2023/entries/capability-approach/ (accessed May 30, 2024).

Chapter 10: Is Landscape Labor?

Danielle Narae Choi

There are two established ways of defining landscape as labor; they are distinct but interrelated. In the most widely used meaning, understood across design and the natural sciences, landscape is a protagonist in the story of natural processes. Certain landscapes aid in converting natural matter into human provisions (e.g., food, fiber, fuel) and, as subspaces of ecosystems, they locally maintain environmental life support systems (e.g., filtering water, oxygenating the atmosphere, regulating temperature). These landscapes—sometimes explicitly called "working landscapes," "productive landscapes," or providing "ecosystem services"—sustain human life. The second definition concerns social processes: human labor is embedded in landscapes. Landscapes are labor because they are not naturally occurring; they themselves are made and managed over time. This definition encompasses the intentional and coordinated human activity that produces landscapes: the conscious articulation of form by trained designers and non-professionals alike, the physical construction of a site, and the ongoing maintenance of these places. The first definition—the working landscape—allows humankind to make a living on the Earth. The second definition—the worked landscape—is one way of making a life through an ever-changing (and always contested) social agreement on the continuous process of building the environment.

To define landscape as labor is to define the synthesis of making a living and making a life—and to do so to distinguish landscape from nature as a scale of authorship, interpretation, and representation. The meaning of the "working landscape" vis-a-vis landscape architectural theory and praxis has evolved through time to parallel a shift from landscape as process toward landscape as a co-worker. In "worked" landscapes, designed sites of direct human intervention, the construction of the immediate social milieu begins during the construction of the physical milieu. As the pastoral is still a dominant aesthetic in Western landscape architecture, the metabolic and economic processes of making and maintaining a landscape are often concealed by an inherited affect of restrained productivity in the public realm. As a framework for decoupling economic activity in the landscape from productive social activity, philosopher Hannah Arendt's concepts of *labor*, *work*, and *vita activa*—and more recent scholarship across ecofeminism, political ecology, and Black ecology—offer new interpretations of labor as

DOI: 10.4324/9781003148142-12

an equalizer when it intersects with other forms of kinship and solidarity. A theory of *landscape as labor* discerns the interwoven productive activity of natural processes and human actors, de-universalizes what is natural and what it means to be human, and perceives the social worlds created by these relationships.

THE DEVOURING PROCESS: DEFINING LABOR

In its earliest usage in the English language, "labor" described physical exertion that had a habitual association with difficulty, sorrow, and pain, with possible origins related to "slipping or staggering under a burden."[1] By the early eighteenth century, labor had acquired its more abstract meaning as a social activity with physical requirements; labor was not inherently about a specific individual but a pool of individuals (and their actions) with a subjective economic cost.[2] In *The Human Condition,* Hannah Arendt reclaims the fleshy, corporeal aspect of labor as a precondition to the societal; this metabolic specificity has the greatest relevance to defining landscape as labor. To Arendt, *labor* sustains biological life and is most directly related to the cycles of energy that bind human existence to the greater natural world. Labor is an incessant "devouring process"—a cycle of bodily needs and the fulfillment of these needs through transforming material for consumption.[3] It is, essentially, a private activity of elemental human sustenance made public in the modern age through markets and institutions. *Work*, by contrast, constitutes social worlds. Arendt claims, "From the viewpoint of nature, it is work rather than labor that is destructive, since the work process takes matter out of nature's hands without giving it back to her in the swift course of the natural metabolism of the living body."[4] Put crudely—to eat, shit, and die is a process of labor; the human as *animal laborans* merely "mixes with" nature.[5] The working human, *homo faber,* "works upon" nature.[6] Arresting matter into durable forms (such as buildings and furniture) offers the potential for ingenuity, creativity, and craft.[7] It is important to note that Arendt's argument does not aim to demean the work of "laborers" as somehow less-than-human. Rather, it identifies the kind of society that such a being inhabits. By her account, the foodstuffs produced by advanced manufacturing processes could be part of a laboring society without meaning.[8] The *vita activa*, never fully realized in Arendt's view, is the active life "devoted to public-political matters," which reproduces the shared interests of humankind.[9]

Although Arendt does not discuss "landscape" as such, she is perplexed by a paradoxical aspect of human intervention on the land as a project of persistence without durability.

> Cultivated land is not, properly speaking, a use object. . . the tilled soil, if it is to remain cultivated, needs to be labored upon time and again. A true reification in which the produced thing in its existence is secured once and for all has never come to pass; it needs to be reproduced again and again in order to remain within the human world at all.[10]

Landscapes (unlike chairs or tables) demand the ongoing attention of labor *and* work, of being bound to basic metabolic cycles *and* the stability granted by deliberate and strategic creativity. Arendt's description is of a nineteenth-century understanding of landscape, wherein maintaining a productive agricultural field is a local affair. In the twenty-first century, the duality of labor and work necessary to maintain a landscape has global implications; it is central to questions of scale, scope, and complexity in landscape architecture and the discipline's relevance in the face of climate change. The climate crisis is undoubtedly one cause of the present-day enthusiasm for new materialist theory in the design disciplines, wherein the disposition of nonhuman or more-than-human entities and actors can have profound societal effects—and, in many ways, constitute society itself.[11] For designers, it is important to parse the differences between "nature" and "landscape" as nonhuman entities in applying such theory; these distinctions matter if there are to be any political or ethical conclusions drawn from knowing the interconnectedness of things. To ask the question, "What kind of person makes a landscape. . . a landscape?" is to suggest the constitution of new human subjects beyond *animal laborans* and *homo faber*.

CO-OPTING NATURAL CAPITAL

In classical economics, "land"—broadly defined to include land and sea—and the stuff that can be extracted from it was considered capital, the primary element of production beyond what is necessary for sustenance alone.[12] Land is material (tangible) and alienable (transferable), but natural processes—ecosystem services—were taken for granted as always available. As early as the mid-nineteenth century, amidst the polluted air and water of industrial cities and depletion of soil fertility in rural areas, the concept of "natural capital" arose to describe "natural-material use values constituting *real wealth*," to be seen in opposition to a capitalist system concerned only with exchange value.[13] Natural capital—inclusive of place and processes—played an exceptional role in creating monetary wealth because it was underestimated in all forms of accounting. These concerns were resurrected in the mid-twentieth century by British economist E.F. Schumacher in his 1973 *Small is Beautiful: Economics as if People Mattered* (Figure 10.1):

> To measure the immeasurable is absurd and constitutes [on the part of the economist] but an elaborate method of moving from preconceived notions to foregone conclusions: all that one has to do to obtain the desired results is to impute suitable values to the immeasurable costs and benefits of nature. The only real result of such an endeavor was to perpetuate the myth that "everything has a price," or, in other words, that money is the highest of all values.[14]

For Schumacher, natural capital is unique in that the consequences of its mismanagement result not only as a threat to civilization but also to human existence.[15] Presciently, Schumacher used the concept of natural capital to critique

Figure 10.1 British economist E.F. Schumacher resurrected the term "natural capital" in the twentieth century as a rhetorical device, not economic theory. *Small Is Beautiful* by E. F. Schumacher. Copyright (c) 1973 by E.F. Schumacher. Used by permission of HarperCollins Publishers.

the economic presumption of fossil fuels as a sort of business income, an expendable resource.

> If we treated them [fossil fuels] as capital items, we should be concerned with conservation. . . money obtained from the realization of these assets—these irreplaceable assets—must be placed into a special fund to be devoted exclusively to the evolution of patterns of living which do not depend on fossil fuels at all.[16]

Schumacher was not calling for a greener capitalist system and was critical of socialism's attempts to "out-capitalize the capitalists" through an abiding faith in technology.[17] Rather, the accelerated growth of Western institutions through the consumption of finite resources was highly irrational by its own logic. The ongoing survival of the species required that the frameworks established by the exploitation of nature undergo a sort of controlled demolition: "What is at stake is not economics but culture; not the standard of living but the quality of life."[18] For Schumacher, the rendition of ecological crisis into economic terms was a rhetorical device, not economic theory. By the late twentieth century, with oft-cited estimations of the value of natural capital (and natural processes) in the trillions of US dollars, these concepts lost their non-conformist associations as they were absorbed into mainstream policy literature.[19]

FROM THE "WORKMAN'S CODE" TO ECOSYSTEM SERVICES

In Karl Marx's labor theory of value, labor (rather than capital) is the primary source of wealth.[20] For instance, the labor of weaving linen converts the natural matter of raw flax into its "general social form" of cloth.[21] This transformation of flax into cloth—labor in its "coagulated state"—renders a socially recognizable material ready for exchange: "As a commodity, it is a citizen of the world."[22] In her critique of Marx, Hannah Arendt asserts that material provisioning for an individual person can be decoupled from the reproduction of society; the expenditure of natural resources through industrial capitalism is not the only possible trajectory for modern life. For Arendt, in the "unutopian ideal that guides Marx's theories . . . all things would be understood not in their worldly, objective quality, but as results of living labor power and functions of the life process."[23] The durability of a linen coat indeed *should matter*, not only for its exchange value or its use value of protecting a body from the elements but for the customs and habits that arise from its existence. Does it swaddle the wearer? Accentuate the human form? Is it adorned with embroidery, logos, or sequins? Or are the details of its construction simple? A coat does not merely exist in the world; its endurance offers stability and meaning to a specific world of human beings.

Arendt's distinctions between the products of labor and work—for cyclical consumption and a durable environment of things, respectively—can be found to coexist in conceptions of landscape in the twentieth century. Throughout various elements of Western landscape environmentalism, those more pragmatic than transcendental in their outlook, nature itself is identified as the laborer in alliance with human workers. George Perkins Marsh, in his 1864 *Man and Nature,* calls for "reclaiming and reoccupying lands laid waste by human improvidence. . . . The task is to become a co-worker with nature in the reconstruction of the damaged fabric."[24] Anne Whiston Spirn describes how Frederick Law Olmsted likewise employed "natural and cultural processes as 'co-workers'" in the sanitary engineering of the American metropolis and represented a middle ground "between John Muir's idea of nature as 'temple' and Gifford Pinchot's idea of nature as 'workshop.'"[25] Humans are no longer directly integrated into the metabolic processes of landscape—Arendt's "swift" return of nutrients through night soil, cadavers, or replanting crops—but the deleterious effects of human consumption and waste could be remediated through urban development (Figure 10.2). During the nineteenth century, this scale held the potential for reciprocity among the primary biological appetites of humans, the human conversion of natural matter to satisfy these appetites, and the ongoing commitments to landscapes needed for repair and regeneration. However, with the expansion of industrial capitalism and globalization and increasing scientific knowledge about ecosystems and planetary systems, the predominantly local concerns of landscape labor and the increasingly complex regenerative requirements of landscape work became too convoluted to understand within a single site.

Landscape planners and designers of the mid-twentieth century, notably Ian McHarg, sought systematic methods of understanding human and

Figure 10.2 Nineteenth-century designers and engineers confronted lands "laid waste" by human development. Chicago Daily News, Inc. *Chicken standing on crusted sewage on Bubbly Creek at Morgan Street.* April 1911. Photograph. DN-0056899, Chicago Daily News collection, Chicago History Museum.

nonhuman metabolic processes at multiple scales. Throughout *Design with Nature,* McHarg foregrounds the "natural processes that perform work for man . . . [such as] natural water purification, atmospheric pollution dispersal, climatic amelioration, water storage, flood, drought and erosion control, topsoil accumulation, forest, and wildlife inventory increase."[26] McHarg's "workman's code" calls for an ecological mandate for design that is grounded in the scientific knowledge of the period.[27] Although the term "ecosystem services" never appears explicitly in *Design with Nature*, it emerges from the same arena of ecology legitimized as a quantitative science in the late 1960s and early 1970s.[28] A "strong form" of science-based utilitarianism, one that systematically recognizes the benefits that ecosystems offer to humankind, was considered a safeguard against weaker, erratic or poorly informed kinds of humanistic utilitarianism[29] (Figure 10.3).

The concept of ecosystem services, as it emerged during this period, was not concerned with intrinsic values in the moral sense, nor, despite the transactional inflection of terms like "inventory," was it initially concerned with economic value. It demonstrated that the ecological processes that sustain human life are nested within larger systems. The 1970 *Study of Critical Environmental Problems (SCEP)* was a landmark interdisciplinary conference convened by MIT professor of

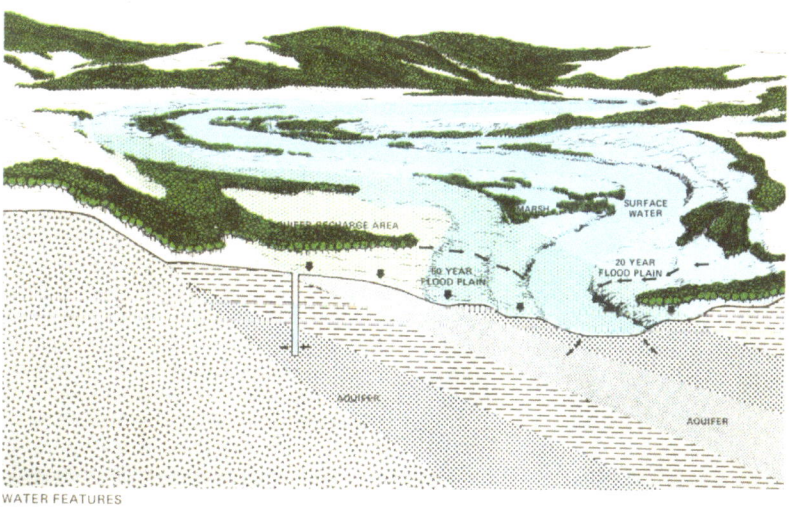

Figure 10.3 Ian McHarg's 1969 study of the Philadelphia region in *Design with Nature* enumerates "natural processes that performed work for man." Ian L. McHarg, *Design with Nature* (Garden City, NY: Doubleday & Company, 1971), 59.

management Carroll L. Wilson. In the conference report, the Work Group on Eco-logical Effects (chaired by biologist Frederick E. Smith, then faculty in the depart-ment of landscape architecture at the Harvard Graduate School of Design) lists nine threatened "environmental services" to humankind, ranging from insect polli-nation to flood control. The authors of this section wryly note, "It is a mark of our time, and a signal of the degree to which man is ecologically disconnected, that the benefits of nature need to be enumerated."[30] In fact, the report does not attempt to calculate any environmental services or their replacement costs. Instead, it offers detailed summaries of the threats to their efficacy by industrial waste, urban runoff, agricultural pesticides, and the combustion of fossil fuels.

By the time that "ecosystem services" appeared in print in 1981, from con-servation biologist Paul R. Ehrlich, it was explicitly opposed to any affiliation with economic logic: "In their [economists'] view, environmentalism is simply a demand for more goods and services (clean air, water, and so forth) . . . Similarly, other species are commodities that society can value or not value, depending on its desires."[31] As scientists like Ehrlich continued to advance knowledge of the complexity of ecosystems, they bristled against efforts to align natural and con-structed systems: "technological substitutes for ecosystem services are no more than partially successful in most cases. Nature nearly always does it better. When society sacrifices natural services for some other gain. . . it must pay the costs of substitution."[32] A deeper scientific understanding of ecosystems offered mount-ing concrete evidence for their irreplaceability. Like economist E.F. Schumacher's efforts to reframe natural resources as natural capital, the term "ecosystem ser-vices" emerged to conceptualize natural processes to the broader non-scientific community that directed economic and land-use policy. However, by the late twentieth century, the term's usage shifted from the conceptual toward the quantitative; scientists were developing new techniques of calculating ecosystem services to understand the "replacement costs" of nature to humankind.[33] In the face of dramatic environmental degradation, nature was no longer a reliable "co-worker" in the arena of biological exchange. The concept of ecosystem ser-vices, which was initially proposed as a governor on the work of productive human activity, could now be reinterpreted as an allowance for new develop-ment or the creation of new markets.

UNDERCOUNTING ECOSYSTEM SERVICES

"Ecosystem services" is now the mainstream term in landscape architecture for identifying the benefits provided to humans by natural processes. The concept is central to the Sustainable SITES Initiative, the landscape complement to the US Green Building Council's LEED rating system.[34] Through the evaluation of individ-ual projects, the program is designed to "distinguish sustainable landscapes, measure their performance, and elevate their value."[35] The rating scorecard, in which points are accumulated (but never deducted for ecosystem disservices), lists goals ranging from "Conserve and use native plants" to "Divert reusable vegetation, rocks, and soil from disposal."[36] Under this rubric, constructed land-scapes are agents of ecological harm reduction, reflecting a general shift in the American discipline from Ian McHarg's suitability studies toward sustainable development in the latter half of the twentieth century.[37]

A recent survey of global assessments of ecosystem services claims that most assessments *exclude* cultivated and urban areas because they are human-dominated environments.[38] Even with more advanced modeling tech-niques than those available to scientists like Paul R. Ehrlich in the 1980s, the functions of anthropized ecosystems are considered too difficult to calculate or too compromised by the concentration of global resources relative to their foot-print. For instance, the ecosystem services provided by a green roof are heavily subsidized by non-ecological processes such as fossil fuel extraction. A certain generation of Western landscape architects—those who "identified as advocates for nature"[39]—may be chastened by the invisibility of the discipline to main-stream scientific discourse on climate change and the pairing of conservation ecology with the engineering sector as the most prominent interdisciplinary alli-ance toward climate action.[40] If the global scientific community does not recog-nize the ecosystem services offered by urban landscapes as a net benefit to planetary systems, then one liberatory response is to shift the focus from land-scape processes alone toward new hybrids of landscape labor (ecological and metabolic) and landscape work (sociocultural).[41] Such a shift resonates with the trajectory of the mainstream discipline from sustainable development toward landscapes of resilience and adaptation.[42] Crucially, the making of a landscape through both labor and work does not merely provide a "service" to a pre-determined public in the manner of landscape construction followed by mainte-nance. It holds the potential for ongoing co-authorship between the metabolic processes harnessed by a landscape and a new public that is produced.

MORE-THAN-HUMAN, MORE-THAN-LANDSCAPE

The concept of the "more-than-human" emerged in the early twenty-first century as a new paradigm for humanity's relationship to nature. The concept circulates across the humanities, art, and design, amalgamating deep ecology's moral unification of human life and all other forms of life, actor-network-theory's assignment of social agency to nonhuman things and beings, and new material-ism's messy, ever-unfolding pluralism. In contemporary design, just a few recent expressions include the theme of the Venice Architecture Biennale 2021, *How Will We Live Together?*; the 2021 *Broken Nature* exhibition at MoMA in New York and the Triennale di Milano; the 2020 issue of *LA+*, "Creature"; and the 2021 *Feral Atlas: The More-Than-Human Anthropocene* (Figure 10.4).[43] Follow-ing the conceptual shifts from "nature as co-worker" to "nature as governor" to "nature/landscape as co-author," this sensitivity to the more-than-human is unsurprising, given the environmental volatility of the present-day and unknown future effects of climate change.

However, some aspects of the more-than-human have met friction from different quarters of feminism, queer theory, and Black ecology, where there can be no universal locus of who precisely is the "human." In climates past and climates future, "we" have never been in this together. Philosopher Rosi Braidotti asserts that the humanities have always been androcentric, excluding not only women but a plurality of genders.[44] In refiguring the relationship of Blackness to animal life and labor, the writer Joshua Bennett asks, "If we are willing to militate toward the abolition of the genre of Man and think companionship anew. . . What rises to the fore in the wake? What beauty? What unthinkable terror?"[45] Earlier scholars explored a more direct accounting of labor to critique the universal "man" in relation to nature. In demanding wages for housework in the 1970s, Silvia Federici and Nicole Cox identified the disproportionality of socially necessary domestic labor to monetary compensation. Importantly, Federici and James were not asking for recognition or sympathy—they were declaring the power of women's work and making real the consequences of domestic strikes (Figure 10.5).[46] In a similar vein, philosopher Val Plumwood identifies naturalization as a means of devaluing work: "Those areas previously excluded as nature— the nonhuman, the reproductive and bodily sphere, the labour of those colonized as nature are treated as invisible inputs to the rational economy."[47] Ecosystem services are just one form of hidden labor among many that have been historically undervalued. Across the work of all these scholars is a demand for the moral and ethical claims of human kinship with the nonhuman to lead to radical transformations of material relations. For Braidotti and Bennett, this may take the form of kinship that can only develop through mutual senses and sensations. Living and breathing bodies may find unanticipated commonalities in how they are affected by the specific landscapes they inhabit. For Federici, Cox, and Plumwood, the adverse affiliation of women with nature can yield a forceful politics of alienation; reproductive work can be celebrated, refused, or substituted

Figure 10.5 In the 1970s, feminist activists' refusal to naturalize reproductive work led to a forceful politics of alienation. Jacquie Ursula Caldwell. *Wages for Housework*, Poster, ca. 1974.

on its own terms. None of these examples are stifled by static representations of identity or demographics passing through a preexisting "public realm"; the plurality of what it is to be human can change in time and space.

LANDSCAPE AND THE PLANTATIONOCENE

The roots of the discipline of landscape architecture in North America are entangled with the unequal recognition of humanness and human labor in the landscape. The plantation and the pastoral both represent a synthesis of productivity and a particular aesthetic mode. The emblematic American plantation is

Monticello, designed by Thomas Jefferson, arguably one of America's earliest landscape architects. Over 600 enslaved people cultivated and lived in this landscape, which persists today as a tableau of subdued abundance.[48] Landscape architect Kofi Boone, writing about the Middleton Place plantation, notes the invisibility of Black labor in the field at large. He argues that the sophisticated landforms and waterworks were evidence of a "high level of talent and ingenuity, even under extreme duress," and that "by any other name, the Wolof people who built Middleton Place. . .were landscape architects" (Figure 10.6).[49] The hidden labor of the American pastoral landscape, as exemplified by Central Park, assumes a related but rather different form. The flock of grazing sheep, symbols of genteel productivity, persisted into the 1930s despite increasing difficulty maintaining their well-being.[50] Their presence belied the tremendous human effort required to blast underlying bedrock, regrade the land, and construct the vast subterranean drainage network.[51]

In 2014, a group of social scientists collectively coined the term "Plantationocene" to join the scene of the "cenes"—Anthropocene, Capitalocene—that attempt to define the indelible human mark on the geologic record. In this conversation between Donna Haraway, Noboru Ishikawa, Anna Tsing, and others, the Plantationocene, as a concept, indicts a particular ethos of transforming the land through concentrated practices of relocation.[52] Tsing and Ishikawa note the "long-distance simplification of landscapes" propped up by slavery and indentured servitude. The plantation requires "the historical relocations of the substances of living and dying around the Earth as necessary to

Figure 10.6 The plantation aesthetic relies on the "long-distance simplification of landscapes" and the subdued presence of labor. Frances Benjamin Johnston. *Middleton Place, Ashley River vic., Dorchester* County, South Carolina, 1938. Photograph. Frances Benjamin Johnston Photograph Collection, Library of Congress, Prints & Photographs Division, LC-DIG-csas-03840.

their extraction. . . These processes depend on the relocation of the generative units: plants, animals, microbes, people."[53] The plantation and the pastoral choreograph the relocations and transformations of matter as a landscape aesthetic, yet, as noted by Raymond Williams, "the meanness of a shepherd's life" is never on display.[54]

Evidence of the Plantationocene's more recent relocations can be found in present-day parks. Post-industrial landscapes—and the people who once labored in them—were left behind across North America because of major geopolitical realignments. These post-industrial landscapes are not, in fact, "post-" anything; the fuel, food, and fibers that support everyday life and habitation have, over many decades, been offshored by the economic logic of "fossil capital," thus tightening the knot between globalization and global warming.[55] For the last several decades in landscape architecture, the repurposing of these sites into parks and cultural complexes is more aligned with the rise of the service economy, which is well established as a mode of landscape production possessing its own aesthetic. These projects, such as Freshkills Park in New York, are forms of Leo Marx's "complex pastoralism" (as opposed to sentimental pastoralism), which synthesizes "moral ambiguity, the intertwining of constructive and destructive consequences generated by technological progress" (Figure 10.7).[56] The genuine threats to human health—belching furnaces, polluted water—now exist elsewhere; relics of industrial activity are integrated into park features, embedded in thriving urban ecosystems.

Figure 10.7 "Complex pastoralism" is morally ambiguous, embedding the evidence of constructive and destructive forces into landscape features. Copyright 2019, Jade Doskow. Jade Doskow, *East Mound Low Road, Iron Stained Rainy Seeping, Gabion Walls, and Phragmites*, 2019. Photograph. Copyright 2019, Jade Doskow.

LANDSCAPE AS A WORKPLACE

Landscape labor and landscape work—to harness natural processes and to con-tinuously renew these relationships through deliberate human action—create social worlds. The concepts of natural capital and ecosystem services once described the notional limits of nature's resources; by the early twenty-first century, their meanings had diversified, drifting toward accumulation and exchange. Theories of the more-than-human attempt to redress this shift, but to further complicate things, the universality of the "human" in private and public realms has been rightfully challenged. What is it like today to work in a land-scape as it is created and managed, with nature as an aloof (and periodically hostile) co-worker?

From modernism onward, architecture has engaged workplace design as a coherent design exercise that organizes the programs and productive activities of an office building, hotel, factory, etc., and solidifies the social protocols specific to an organization in space. Designed landscapes are not typically sites of pro-ductive economic activity. However, they are workplaces for the people who build, maintain, and manage them through landscape labor and landscape work. The chairs in an office are not rebuilt every year, but ecological urban landscapes require a suite of episodic and recurring activities to respond to dynamic environ-mental conditions. The outcomes of these actions—managing aggressive invasive species, reconciling dead wood with public safety, periodic replanting—are more deeply integrated into the spatial constitution of a place for landscape than for architecture. The design of the spaces in which this recurring physical labor takes place is seldom thought of as workplace design, reinforcing hierarchies between the intellectual work that takes place in the professional design studio and the manual labor that takes place in the field. More attention has been given to this in scholarly and professional contexts in recent years, with studios such as Los Angeles-based Terremoto attempting to make structural adjustments to the frameworks of compensation, social standing, and security that distinguish manual and non-manual labor in a project with normative methods of delivery.[57] Furthermore, these efforts recognize workers' deep knowledge and creativity in the field as a continuum of the design process; the flows of information and influence should move in multiple directions, from abstraction to materiality, from the drawing set to the site. To take the design of the landscape seriously as a workplace in this way, as a reformist social project, can have tremendous cumulative effects.

The grids, modules, and hierarchies of the modern factory and office were mutually developed with the regimentation of the workday and the translation of manufacturing processes into spatial relations; they continue to evolve with emerging forms of automation, modes of delivery in manufacturing, and the pre-carity/mobility of a white-collar workforce. For the landscape labor and landscape work of designed sites, the relationships between spatial organization and tem-poral occupation are less well-defined—and thus open to greater reinvention.

Figure 10.8 Operation by women of a new sawmill at Turkey Pond, NH, 1942. Records of the Forest Service 1870–2022, Records Relating to Timber Salvage. National Archives at Boston. Courtesy: United States Department of Agriculture, Forest Service.

However, the nineteenth-century notion of large parks as places for the private body and mind to recharge the capacity to work still lingers. Jack Halberstam's concept of "queer time" asserts the possibility of creating alternative worlds through "strange temporalities, imaginative life schedules, and eccentric economic practices" that are less about sexual identity than a challenge to traditional frameworks of family and work (Figure 10.8).[58] As potentially applied to landscape, temporal considerations—such as seasonality and diurnal cycles, human and animal migrations, and the distortions caused by climate change—may engender novel possibilities for form and occupation. If designers can consider the possibility that hedonism and productivity are not mutually exclusive, as Kenneth Frampton asserts, they may also consider with greater intensity the design of landscape as a workplace for nature and people—perhaps irrespective of the program as a park, campus, farm, etc.[59] Can the labor of landscape management, research, and renewal be brought into the public realm without resorting to self-conscious didacticism or the fuzzy romanticism of the pastoral? The "long distance simplification" of landscapes has not only distorted the true accounting of material flows but prolonged the pastoral as a dominant aesthetic mode that represents the "work" of landscape and people (often historicized or hidden), setting up a false opposition between a seemingly self-sufficient natural environment and the complex economic activity of the city.

Natality and Design

Hannah Arendt critiques the obsession with death in Western human relations made possible by the undervaluation of natality, a notion that is less about childbirth or child rearing than the latent capacity of every individual to create a new polis in coordination with others. Labor and work are rooted in natality as they must "provide and preserve the world for, to foresee and reckon with, the constant influx of newcomers who are born into the world as strangers."[60] Natality is an inchoate capacity for productive activity beyond economic impulse. As Virginia Tassinari and Eduardo Staszowski interpret, "Seen in this light, designing is a moment of the natalic capacity for beginning. . . It is intervention carried through as the affirmative but always propositional negotiation of incommensurability inherent in every human situation and condition."[61] Natality is a form of participation that *does not* express preexisting politics but is the source of a new set of relations. Landscape labor and landscape work present a special case because the durability of the human artifact is dependent on (rather than compromised by) ongoing natural processes. As a framework for understanding landscape design, it is to propose new measures for productive activity that can change over time.

Natality in the landscape, in a flexible form, may resemble landscape architect Jill Desimini's concept of the "fallowscape"; these sites are freed from pressures to produce ecosystem services, generate income, or create jobs.[62] In Desimini's scenarios for urban lands—ranging from nurseries to circuses—fallowness is a precondition for shared delight and regeneration (Figure 10.9). In Newark, New Jersey, anthropologist Kessie Alexandre documents different strands of resistance to city-sanctioned green stormwater infrastructure within Black community gardens. The relative ecological value of water management versus food production cannot be understood apart from a context of massive disinvestment in civic water infrastructure; despite rain gardens being a quantifiable form of water management and climate resilience, their worth could ultimately only be determined collectively by the locals who tended the land.[63] Recent work by designer Kira Clingen examines the spaces and forms of knowledge exchange of twenty-first-century "preppers." Despite wildly divergent ideological beliefs, preppers, perhaps by definition, invent ways of provisioning a world that has not yet come into being.[64] An example of natality in a concentrated form was introduced within Nelson Byrd Woltz's design for Memorial Park in Houston, Texas. A 65-acre grove of pine trees, arranged in grids, marks the location of a World War I training camp and soldiers marching in formation.[65] (It goes unnoted in the 2015 Master Plan that the site, Camp Logan, was home to the all-Black 24th Infantry and a deadly race riot in 1917). The designers proposed that, upon reaching maturity, certain groves could be felled and milled into lumber to construct affordable housing throughout the city (Figure 10.10).[66] This aspect of the proposal does not appear to have been integrated into subsequent design phases, and one could mount a plausible critique that it does little to redress the concentration of public and philanthropic funds; as a productive landscape within a high-profile large park, it is a fragile proposition that makes little ecological and economic sense. However, it is also an unexpected expression of natality in contemporary large park design. It puts faith in future generations to realize a material and social relationship conceived decades prior.

Figure 10.9 Fallowscapes are lands freed from the pressure to produce ecosystem services, generate income, or create jobs. Jill Desimini, Daniel D'Oca, and Julia Czerniak, *From Fallow: 100 Ideas for Abandoned Urban Landscapes*, First edition (Novato, CA: ORO Editions, 2019), 346–347. Copyright 2019, Jill Desimini.

LANDSCAPE IS LABOR!

A theory of landscape as labor is an accounting of the interlinked productive activity of natural processes and human actors. By distinguishing *landscape labor* (harnessing biological processes) from *landscape work* (renewing actions required for durability), it recognizes uncompensated exertion and claims a politics by requiring the definition of whose needs are met and how; it suggests a scale and dominion of exchange. The concepts of natural capital and ecosystem services assign value to the work that landscapes, as subspaces of larger systems, perform for humanity; a shift from "landscape as process" toward "landscape as labor" replaces a framework of distancing with a framework of mutuality. Landscape as labor also encompasses the landscape as a workplace. As a social and aesthetic project, the design of the landscape workplace is inhibited by the legacies of the pastoral and the plantation and underexplored in the discipline. A fuller, more inclusive understanding of the different forms of human labor required to produce a landscape at different

Figure 10.10 Fragile natality: Nelson Byrd Woltz's proposal for Memorial Park in Houston, Texas. Image courtesy: Mir for Nelson Byrd Woltz Landscape Architects. MIR for Nelson Byrd Woltz Landscape Architects, Memorial Park, Houston, TX, 2015.

moments in time must de-universalize, rather than merely de-center, what it means to be "human." Furthermore, labor, as a concept, need not be bound to wage labor and economic activity. To define landscape as labor is to articulate the constant interchange of natural processes and human action, to recognize undervalued or obscured forms of human labor, and to unify making a living and making a life.

Acknowledgments

Thank you to Kareem Rabie and Ed Eigen, who each offered valuable critical insights. Additional thanks to Jacob Cascio, who provided research assistance for this chapter, and to Carson Fisk-Vittori for unearthing archival images of the women of the Turkey Pond sawmill.

NOTES

1 Raymond Williams, *Keywords* (New York: Oxford University Press, 1976), 145.
2 Raymond Williams, *Keywords* (New York: Oxford University Press, 1976), 146.
3 Hannah Arendt, *The Human Condition* (Chicago, IL: University of Chicago Press, 1958), 100.

4   Hannah Arendt, *The Human Condition* (Chicago, IL: University of Chicago Press, 1958), 100.

5   Hannah Arendt, *The Human Condition* (Chicago, IL: University of Chicago Press, 1958), 136.

6   Hannah Arendt, *The Human Condition* (Chicago, IL: University of Chicago Press, 1958), 136.

7   For greater reflection on Hannah Arendt and material practices, see Richard Sennett, *The Craftsman* (New Haven, CT: Yale University Press, 2008).

8   Monica Reinagel, "Soylent vs. Huel: Can Powdered Meals Replace Food?" *Scientific American*, March 31, 2020, https://www.scientificamerican.com/article/soylent-vs-huel-can-powdered-meals-replace-food/.

9   Hannah Arendt, *The Human Condition* (Chicago, IL: University of Chicago Press, 1958), 248.

10 Hannah Arendt, *The Human Condition* (Chicago, IL: University of Chicago Press, 1958), 138.

11 See Elizabeth K. Meyer's references to Donna Haraway and cyborgs in her landmark 1997 essay, "The Expanded Field of Landscape Architecture"; see also public lectures by Jane Bennett, Karen Barad, and Anna Tsing at schools of architecture, respectively at Harvard (2021), Princeton (2018), and Yale (2018), among numerous citations in syllabi and exhibitions.

12 Thomas C. Brown, John C Bergstrom, and John B. Loomis, "Defining, Valuing, and Providing Ecosystem Goods and Services," *Natural Resources Journal* 47, no. 2 (2007): 335.

13 John Bellamy Foster, "Nature as a Mode of Accumulation: Capitalism and the Financialization of the Earth," *Monthly Review* 73, no. 10 (March 1, 2022): 1.

14 E. F. Schumacher, *Small Is Beautiful: A Study of Economics as If People Mattered* (London: Blond & Briggs, 1973), 46.

15 E. F. Schumacher, *Small Is Beautiful: A Study of Economics as If People Mattered* (London: Blond & Briggs, 1973), 17.

16 E. F. Schumacher, Small Is Beautiful: A Study of Economics as If People Mattered (London: Blond & Briggs, 1973), 12.

17 E. F. Schumacher, *Small Is Beautiful: A Study of Economics as If People Mattered* (London: Blond & Briggs, 1973), 261.

18 E. F. Schumacher, *Small Is Beautiful: A Study of Economics as If People Mattered* (London: Blond & Briggs, 1973), 260.

19 "At the current margin, ecosystems provide at least US $33 trillion dollars worth of services annually," Robert Costanza et al., "The Value of the World's Ecosystem Services and Natural Capital," *Nature* 387, no. 6630 (May 1997): 253–260.

20 See Thomas C. Brown, John C Bergstrom, and John B. Loomis, "Defining, Valuing, and Providing Ecosystem Goods and Services," *Natural Resources Journal* 47, no. 2 (2007): 329–376.

21 Karl Marx, *Capital: A Critique of Political Economy, Volume 1*, trans. Ben Fowkes (New York: Penguin Books in association with New Left Review, 1981), 159.

22 Karl Marx, *Capital: A Critique of Political Economy, Volume 1*, trans. Ben Fowkes (New York: Penguin Books in association with New Left Review, 1981), 142, 155.

23 Hannah Arendt, *The Human Condition* (Chicago: University of Chicago Press, 1958), 89.

24 George Perkins Marsh, *Man and Nature*, Weyerhaeuser Environmental Classics (Seattle: University of Washington Press, 2003).

25 Anne Whiston Spirn, "Constructing Nature: The Legacy of Frederick Law Olmsted," in *Uncommon Ground: Rethinking the Human Place in Nature*, ed. William Cronon (New York, NY: Norton, 1996), 112.

26 Ian L. McHarg, *Design with Nature* (Garden City, NY: Doubleday & Company, 1971), 57.

27 Ian L. McHarg, *Design with Nature* (Garden City, NY: Doubleday & Company, 1971), 29.

28 The monthlong 1970 conference at MIT, the *Study of Critical Environmental Problems*, resulted in a report that lists which "environmental services" would decline if there was a decline in "ecosystem function," in Study of Critical Environmental Problems (SCEP), *Man's Impact on the Global Environment* (Cambridge: MIT Press, 1970).

29 Lawrence H. Goulder and Donald Kennedy, "Valuing Ecosystem Services: Philosophical Bases and Empirical Methods," in *Nature's Services: Society Dependence on Natural Ecosystems,* ed. Gretchen C. Daily (Washington, DC: Island Press, 1997), 25.

30 Study of Critical Environmental Problems (SCEP), *Man's Impact on the Global Environment: Assessment and Recommendations for Action* (Cambridge: MIT Press, 1970), 123.

31 Paul R. Ehrlich, *Extinction: The Causes and Consequences of the Disappearance of Species* (New York: Random House, 1981), 245.

32 Paul R. Ehrlich, *Extinction: The Causes and Consequences of the Disappearance of Species* (New York: Random House, 1981), 96.

33 Thomas C. Brown, John C. Bergstrom, and John B. Loomis, "Defining, Valuing, and Providing Ecosystem Goods and Services," *Natural Resources Journal* 47, no. 2 (2007): 350.

34 "About SITES," Green Business Certification Inc., December 27, 2016, https://gbci.org/press-kit-sites.

35 "About SITES," Green Business Certification Inc., December 27, 2016, https://gbci.org/press-kit-sites.

36 Sustainable SITES v2 Scorecard Summary, US Green Building Council 2021.

37 Kristina Hill has recently traced the evolution of environmental concepts in landscape planning. The McHargian concept of spatial suitability that guided planning in most of the twentieth century gave way to "sustainable development," recognizing pre-development conditions such as hydrology and biodiversity. In many planning contexts, sustainability has been replaced by "resilience" and "adaptation" in the face of the effects of climate change. See Kristina Hill, "Climate Change: Implications for the Assumptions, Goals and Methods of Urban Environmental Planning," *Urban Planning* 1, no. 4 (December 29, 2016): 103–113.

38 S. Barot et al., "Ecosystem Services Must Tackle Anthropized Ecosystems and Ecological Engineering," *Ecological Engineering* 99 (February 1, 2017): 486–495.

39 Charles Waldheim, *Landscape as Urbanism: A General Theory* (Princeton, NJ: Princeton University Press, 2016), 53.

40 "It's as if landscape architecture does not exist. . . Is our discipline a necessity? Are we closing the gap between ideals and practice? We are not, I promise, saving the world." See Billy Fleming, "Design and the Green New Deal," *Places Journal*, April 16, 2019.

41 Alissa Battistoni coins the term "hybrid labor" as a useful way of "threading the needle between the categories of intrinsic and instrumental value [for nature]. . . claiming recognition and power on the basis of useful activity while also asserting space for life beyond." It is beyond the scope of her essay to consider the human artifact of landscape architecture. See Alyssa Battistoni, "Bringing in the Work of Nature: From Natural Capital to Hybrid Labor," *Political Theory* 45, no. 1 (February 2017): 7.

42 Kristina Hill, "Climate Change: Implications for the Assumptions, Goals and Methods of Urban Environmental Planning," *Urban Planning* 1, no. 4 (December 29, 2016): 111.

43 Anna L. Tsing et al., *Feral Atlas: The More-Than-Human Anthropocene* (Stanford: Stanford University Press, 2020).

44 Rosi Braidotti, "Posthuman Knowledge" (Harvard Graduate School of Design, Cambridge, MA, March 13, 2019).

45 Joshua Bennett, *Being Property Once Myself: Blackness and the End of Man* (Cambridge, MA: The Belknap Press of Harvard University Press, 2020), 142.

46 Kathi Weeks, *The Problem with Work: Feminism, Marxism, Antiwork Politics, and Postwork Imaginaries* (Durham, NC: Duke University Press, 2011), 122.

47 Val Plumwood, *Feminism and the Mastery of Nature* (London: Routledge, 1993), 141.

48 Lucia C. Stanton, *"Those Who Labor for My Happiness": Slavery at Thomas Jefferson's Monticello* (Charlottesville: University of Virginia Press, 2012), 56.

49 Kofi Boone, "Enabling Connections to Empower Place," in *Black Landscapes Matter*, eds. Walter Hood and Grace Mitchell Tada (Charlottesville: University of Virginia Press, 2020), 58.

50 The sheep of Central Park lived in their eponymous meadow until 1934; their demise may have begun about a decade earlier. "Many [sheep] are expected to die as a result of eating refuse left behind by Memorial Day picknickers. . . Of the herd of ninety, thirty are seriously ill." "PARK LUNCHES POISON SHEEP; 30 of Central Park Herd Ill After Eating Refuse Left by Picnickers," *The New York Times*, June 4, 1921.

51 See Jane Elizabeth Hutton, "Chapter 1: Inexhaustible Terrain," in *Reciprocal Landscapes: Stories in Material Movement* (Milton Park, Abingdon, Oxon and New York: Routledge, 2020).

52 Donna Haraway, Noboru Ishikawa, Scott F. Gilbert, Kenneth Olwig, Anna L. Tsing, and Nils Bubandt, "Anthropologists Are Talking—About the Anthropocene," *Ethnos* 81, no. 3 (2016): 557.

53 Donna Haraway, Noboru Ishikawa, Scott F. Gilbert, Kenneth Olwig, Anna L. Tsing, and Nils Bubandt, "Anthropologists Are Talking—About the Anthropocene," *Ethnos* 81, no. 3 (2016): 557.

54 Raymond Williams, *The Country and the City* (New York: Oxford University Press, 1975), 19.

55 Andreas Malm, *Fossil Capital: The Rise of Steam-Power and the Roots of Global Warming* (New York: Verso, 2016).

56 Leo Marx, "Does Pastoralism Have a Future?" *Studies in the History of Art* 36 (1992): 221.

57 "Landscape Architecture Has a Labor Acknowledgement Problem," *Metropolis* (blog post), April 14, 2021, https://www.metropolismag.com/architecture/landscape/landscape-architecture-labor-terremoto/; see also Michelle Franco, "Landscapes and Invisible Labor," Lecture, Council of Educators in Landscape Architecture Annual Conference Online Conference, April 21, 2021.

58 Jack Halberstam, *In a Queer Time and Place: Transgender Bodies, Subcultural Lives, Sexual Cultures* (New York: New York University Press, 2005), 1.

59 Kenneth Frampton, "The Volvo Case," in *Labour, Work and Architecture: Collected Essays on Architecture and Design* (London and New York: Phaidon Press, 2002), 74.

60 Hannah Arendt, *The Human Condition* (Chicago, IL: University of Chicago Press, 1958), 8–9.

61 Virginia Tassinari and Eduardo Staszowski, eds., *Designing in Dark Times: An Arendtian Lexicon* (London, New York, Oxford, New Delhi, and Sydney: Bloomsbury Visual Arts, 2021), 222.

62 See Jill Desimini, Daniel D'Oca, and Julia Czerniak, *From Fallow: 100 Ideas for Abandoned Urban Landscapes*, First edition (Novato, CA: ORO Editions, 2019).

63 Kessie Alexandre, "When It Rains: Stormwater Management, Redevelopment, and Chronologies of Infrastructure," *Geoforum* 97 (December 1, 2018): 66–72.

64 Kira Clingen, "Preppercraft: Speculative Fiction from the Usable Past," *LUNCH* 17 (2022).

65 Nelson Byrd Woltz Landscape Architects, *Memorial Park Master Plan 2015* (Houston Parks and Recreation Department, Memorial Park Conservancy, Uptown Houston), 126.

66 Kath Hudson, "Interview with Thomas Woltz," *CLAD Magazine* 4 (2019); see also Thomas L. Woltz, "Memorial Park Master Plan: Building a Resilient Future on the Foundations of the Past," (Presentation, The Cultural Landscape Foundation, Houston Museum of Fine Arts, April 25, 2016).

Chapter 11: Is Landscape Public?

Ed Wall

As I walked from Westminster Bridge, through Parliament Square, and up Regent Street to Oxford Circus, the center of London felt as if it was occupied by an unusual summer festival. Stages, speeches, performances, installations, and spontaneous applause resonated through the city from early morning to late evening. Across streets, bridges, and squares, there was an abiding generosity and enthusiasm, while all around were bright signs, stickers, flags, posters, and graffiti—many with a distinctive hourglass logo. Having lived in London for over ten years, I was familiar with these central London locations, structured around a steady flow of cars and buses. I had also experienced them closed down for cultural and commercial events, like the city's annual marathon. But during these warm April days in 2019, central London was slowed by a massive climate protest that successfully blocked bridges, roads, street intersections, and building entrances. The aim of what had been termed a "rebellion" was to disrupt vehicular traffic, oil company headquarters, airports, the London Stock Exchange, and other sites that the organizers associated with the climate crisis. For over ten days, there were no cars, with the streets pedestrianized by direct action.[1]

Part of a climate movement that has grown since the 1990s, the occupation claimed and formed public spaces around concerns for the lives and landscapes of the climate crisis—from sites of deforestation, pollution, resource exploitation, and wildlife destruction that accelerate climate change, to places of heatwaves, forest fires, coastal flooding, food shortages, and mass displacement caused by global warming. In 2019, among many landscapes impacted by rising planetary temperatures, Australia experienced devastating fires as it recorded its hottest and driest year on record (Figure 11.1).[2] In April 2019, in the same month as the climate protests took hold in London, over one million people were evacuated from their homes in India and Bangladesh as Cyclone Fani approached—a cyclone that formed over the Bay of Bengal while it was over a degree warmer than usual.[3] The lives lost and landscapes destroyed due to climate warming occurred in the context of global carbon dioxide emissions reaching a record high in 2019.

Figure 11.1 London's Parliament Square appropriated by climate protestors aiming to make visible the damage of global warming, 2023. Drawing courtesy: The author.

INTRODUCTION

In this chapter, I explore relations between landscapes and public spaces, specifically between landscapes of climate change and public sites where the future trajectories of global warming are contested. I attempt to make sense of their commonalities and differences across multiple scales, from material spaces to lived concerns, including conflicts between how landscapes and public spaces are claimed, defined, and produced. I map out overlapping and entangled ways that landscapes and public spaces can productively inform each other. I take the question that geographer Don Mitchell poses: "Landscape or Public Space?"[4] as a point of departure that highlights frictions between Western traditions of landscape and political ideals of public space. Mitchell criticizes centuries-old European practices of landscape that enforce "a particular way of seeing the world, one in which order and control over surroundings takes precedence over the messy realities of everyday life."[5] He argues that such landscape practices undermine public spaces as situated contested public spheres. However, while I acknowledge the frictions between landscape and public space that Mitchell describes, especially where terms of access and inclusion to urban public spaces are undermined, this chapter is focused on more productive relations. In this chapter, I focus on the affinities between landscape and public space, how they can inform each other, and the collective agency that can be claimed. By discussing a range of definitions of landscape and public space across professional terms and academic disciplines, landscape relations that people have with their worlds and public spaces that are reconstructed around issues of concern become entangled.

Having trained as a landscape architect and being drawn to urban public spaces as the focus of studio-based design processes, I recognize landscape practices in the material reconfiguration of public spaces. However, by looking beyond professional frames and accepting contrasting disciplinary perspectives and cultural histories of public spaces, this chapter reveals complex tensions, contradictions, and exchanges between the two. More than a question of landscape *or* public space, I draw associations and highlight reciprocities between protests, demonstrations, and strikes in London's urban spaces and the future of places impacted by global warming. I form a conceptual framework where processes of landscapes (as more-than-human entities) and practices of making public space are connected through public actions (and inactions). The emphasis on *processes* of landscapes and *practices* of making public spaces requires sites of climate change and accessible places of politics to be considered as both social and spatial constructions. I reveal how relations between people and places are produced and how public movements that reconfigure urban space are formed, practices that I recorded during field observations (2019), understood through online accounts, and read through direct action manifestos.

The chapter comprises three sections: I first discuss relations between landscape and public space, drawing on the overlap between Barbara Bender's framing of landscape[6] and Doreen Massey's definitions of public space as constantly in process.[7] Second, I reflect on the potential that places of protest have for reconceiving terms of landscape. Referring to fieldwork undertaken in 2019, I investigate the construction of transnational publics around concerns for the climate crisis, particularly the sites and networks formed by Extinction Rebellion. In the third section, I discuss landscapes that comprise contrasting scales from material lives and political actions to regional coastlines and planetary entities impacted by climate change. While often framed globally, these are partial, uneven, and unequal planetary landscapes that affect people and places differently. I argue that in addition to the formation of public spaces informing landscapes of the climate crisis, the notion of landscapes as plural and reciprocal across contrasting scales can also be employed to make sense of and situate these transnational publics.

PUBLIC ACTIONS, CLIMATE LANDSCAPES

Definitions of public spaces vary, from architectural designs for squares and streets owned and managed by the state[8] to descriptions in social sciences that portray sites of politics—from physical to digital and from newspapers to coffee houses.[9] The spatial definition of the former is prioritized in planning and architectural design, even as spaces are planned, used, owned, managed, and policed, while the latter points toward a spatialization of public spheres where the production and exchange of public discourses is of primary concern.[10] Precise definitions are necessary to avoid conflating spatial notions of public space with other collective, common, or state-mandated places and to navigate social terms of public spaces that differ from definitions of public spheres and publics that can be less grounded in specific sites. Concerns for public places, especially how terms of access are limited and how some activities and people are excluded, have also led to public space discourses focusing on contracts of ownership and relationships to private property. While recognizing that "'Public space' has very different meanings in different societies, places, and times," Neil Smith and Setha Low claim, "It is impossible to conceive of public space today outside the social generalization of private space and its full development as a product of modern capitalist society."[11]

However, by focusing on singular terms of public spaces, such as ownership—or relying on dualities with private space—places of public action that are contested and reconfigured around issues of concern can be overlooked.[12] Working across both the spatial and social dimensions of public space, Fran Tonkiss provides a useful tripart sketch, including squares, cafés, and streets: where the square is a "site of collective belonging," cafés are "sites of sociability, exchange, and encounter," and streets are "mundane spaces of communal encounter."[13] However, while Tonkiss provides an intricate and detailed account of these ideal types, the descriptions tend toward either the use of space for social interactions or more abstract, less situated public spheres. In contrast, Mitchell draws these two frames of public space together when claiming that public spaces must be taken and made public through collective actions.[14] He highlights the need to consider both the shared concerns around which public spaces are formed and the activities that produce them.

By considering practices of making public spaces—how, where, and why people gather around issues of concern—it is possible to embrace complex relations of public spaces, public spheres, and publics and the issues at stake as they are formed. As Tonkiss emphasizes: "[Public space] is a primary instance of the way that categories of thought can be seen to extend across and become 'real' in space."[15] The inclusivity of ways of making, in addition to the accessibility of the places produced, can point to the degree to which public spaces become public.[16] In contrast to dichotomies of public and private that tend toward discussions of the privatization of urban spaces,[17] and even the end of public space,[18] by emphasizing ways that public spaces are made—that range from design and planning to occupation and use and from adaptation and repair to cleaning and policing—spatial notions of public spaces as streets, civic squares, and coffee shops and social dimensions of public spheres as sites of public discourses become inextricably bound.[19] In *For Space* (2005), Massey argues that the public nature of space needs to be held up to greater scrutiny. Pointing to contradictions in the term open space, Massey proposes interrogating the social relations that could inform other notions of public space. I would go further to argue that it is not sufficient to focus on the social ties that configure public spaces, but there is a need to consider wider landscape relations, including non-human and more-than-human ecologies, and how they form public landscapes. The focus moves from tensions between public and private spaces to open questions of what can be lost and gained through different and collective practices. Examining how and why public spaces are produced can reveal struggles over urban space and broader territories. Furthermore, considering relations of public spaces to include more-than-human entities, such as rivers, tidal zones, oceans, and weather, can illustrate the role of public space in, as well as the political struggles of, issues such as climate justice.

The question of what is at stake in such direct public actions of environmental protests and climate strikes is evident as state and private interests respond to limit economic disruption. While releasing plans for net zero,[20] the UK government has also moved to change laws to limit political protests, especially "eco-protests" (Figure 11.2).[21] Following the demonstrations in 2019 in London and across other sites in the UK, where alternative tactics were adopted that were undeterred by the arrest of large numbers of demonstrators, the UK government changed the legal terms of public action and public space in England and Wales. Part three of *The Police, Crime,* Sentencing, and Courts Bill 2021 grants the police new powers to restrict public protests deemed disruptive. The bill is seen as a direct response to the 2019 climate protests that intentionally blocked movement across large areas of London for several weeks. Unlike day-long demonstrations that march toward the Houses of Parliament, from Trafalgar Square to Parliament Square along Whitehall, Extinction Rebellion adapted the tactics of Occupy[22] to maintain a presence in locations across London. The contentious government bill is a reminder that public spaces are and have always been regulated and that governments and corporations frequently rewrite these rules to manage dissent. The power to remake public space is always contested, and as Smith and Low claim: "Public space, in fact, only comes into its own in the differentiation of a nominally representative state on

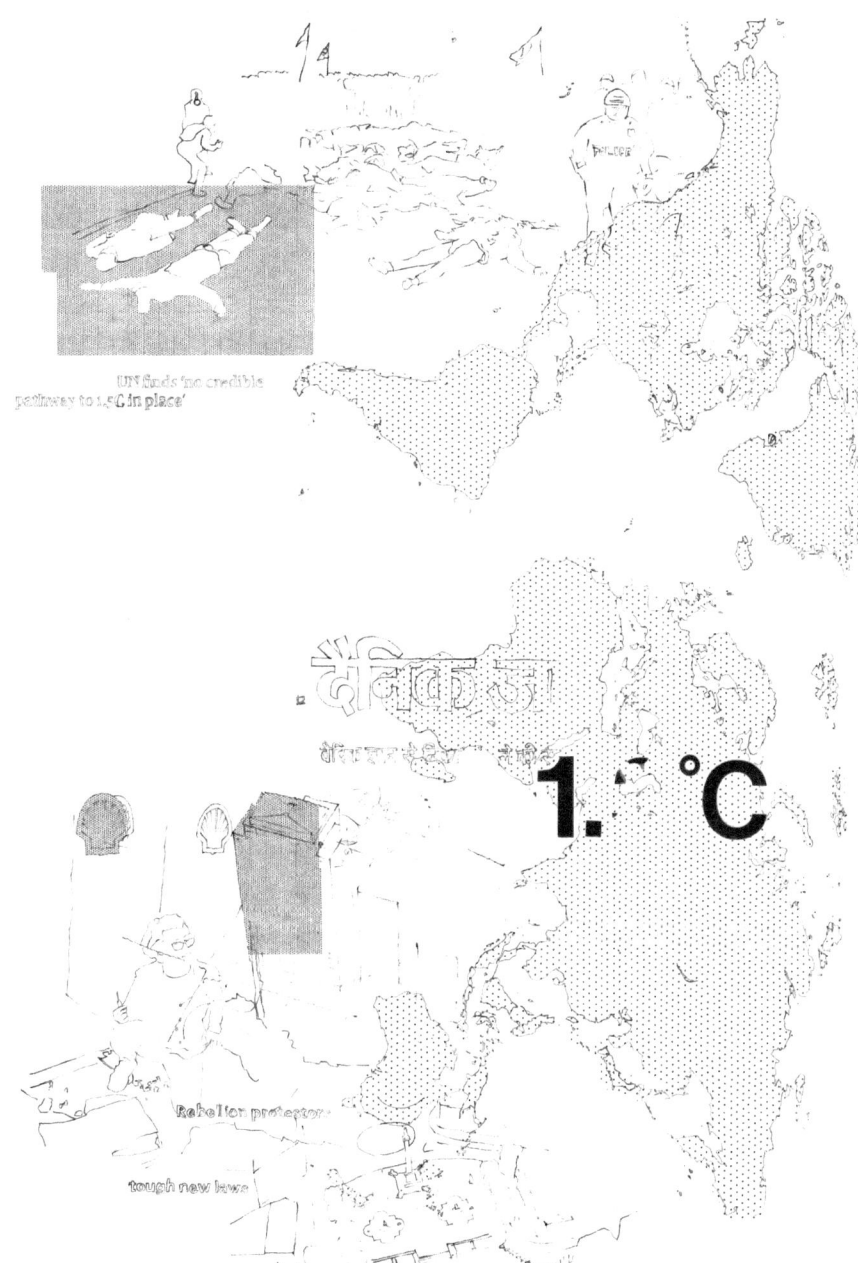

Figure 11.2 The UK government response to climate protests follows a history of laws changed to silence dissent, 2023. Drawing courtesy: The author.

the one side and civil society and the market on the other."[23] The passing of new laws illustrates how public space needs to be challenged, claimed, and made public,[24] how terms of publicness are always in flux, and how the spatial and social relations of such geographies are never settled.

Corresponding with practices of continually making public spaces, Bender describes landscape as always under construction.[25] Landscape practitioners have long recognized material processes of growth, decay, erosion, weathering, movement, and deposition that are essential to all physical landscapes. Recent research into climatic conditions of storm patterns[26] and monsoons[27] highlights further contemporary dynamics inherent to all landscapes. While Henri Lefebvre's often cited description of the "production of space"[28] can be recognized as giving emphasis to processes of making public space,[29] and even Bender's description of landscapes as "in process of construction and reconstruction,"[30] it would seem absurd to ignore wider and more enduring knowledge of landscape processes—whether from agriculture, geography, climatology, or archaeology. Landscapes are not just socially produced but are more-than-human entities that entangle ecologies of maintenance with patterns of weathering and engineering practices with destructive flooding. Landscapes extend from the geologies of the earth to far beyond the satellites of the upper atmosphere. They are material, social, ecological, and political. Projects such as SCAPE's *Public Sediment* and the *Monsoon Assemblages* research led by Lindsay Bremner highlight the complex shared, collective, common, and public nature of such landscape formations. These landscapes reveal places of politics and relational entities formed through intersecting human and non-human processes that are continually bound up with sites and practices of publicness.

Mitchell's concern with landscape as it relates to public space is a critique of landscapes that developed in and expanded from Western Europe in the sixteenth century.[31] From painting landscapes as idealized scenes to reconfiguring the land as reconstructed nature, these concepts of landscape emphasize visual scenes, viewed from static positions of individuals with power and controlled through frames of enclosure—techniques of landscape that continue to be prevalent in many countries and cultures around the world today.[32] If historically such landscape practices are implicated in territories of colonial expansion, appropriations of common land, and continents cleared of Indigenous people, they can be read today in the privatization of public spaces, the development of gated communities, and the gentrification of neighborhoods—as well as territorial disputes between nations and exploitation of mineral landscapes that have continued since the term landscape was coined.[33] Mitchell challenges public spaces made as landscape where the imaging of public space overshadows the realities of daily life—what he describes as "the illusion of control."[34] While visual forms of public spaces are prioritized in competition between cities and markets of rising property values, in what Mitchell terms "landscape's fetishizing agency,"[35] less desirable activities, from homelessness to environmental protests, are restricted. At the same time, more urgent concerns for the climate crisis are denied.

However, as Bender emphasizes, while landscape continues to be employed in such terms—through extractions of fossil fuels and mitigating environmentally destructive developments—there are and have consistently been, "other landscapes" that reveal alternative ways of relating to the land.[36] She states that "Landscapes are polysemic,"[37] pointing to an expanded appreciation of how people have historically interacted and have the potential to relate to the places, beings, and entities around them. Such landscapes may include places where rivers are afforded legal rights, lands not bound by terms of ownership, and forests protected from exploitation. They can also be understood in the actions of nomadic peoples who resist settling and even Western communities who choose to live off-grid from conventional infrastructures.

Like the public spaces of the climate protests, landscapes can be composed of relations with land that are less visual, more collective, and resist conventions of framing. The ways that we relate to land exist, as Bender describes, in a plurality, with each of us holding "many landscapes in tension."[38] What we see with our eyes is one of many landscapes we perceive—along with the relations we draw with places more remote—such as the lands to which we belong, supply chains that sustain our lives, and neighborhoods on which our actions may have an impact. The plurality of landscapes are not just different ways of relating to the same place or the same way or relating to contrasting locations—but politicized entanglements of places and relationships, individual and shared, from geographic and atmospheric relations that span continents to conversations on sidewalks about how to enact change on the street.

SITUATED PUBLICNESS

Relations of landscape are constantly in flux; how people relate to their surroundings is always changing. Recognizing the development of Western notions of landscape in tandem with the advance of merchant capitalism,[39] I previously speculated that if post-capitalist societies were to be constructed,[40] then corresponding "post-landscapes" could also be realized.[41] Reflecting on fieldwork in 2011 at the site of Occupy London Stock Exchange (Occupy LSX)—global events that drew commentators and researchers to imagine post-capitalist futures—I recognized that protest encampments of Occupy formed landscapes that were constructed with urgent concern for political and economic change and less for visual images. I became interested in whether future landscape practices could be informed by situated public protests. In Shelly Egoz's exploration of direct action in public space and the formation of what she terms "landscape democracy," she states: "Another critical function of public space that is fundamental to landscape democracy today is its role as the place for public protests and demonstrations, such as the *Arab Spring* and *Occupy Wall Street*."[42] However, rather than public space providing the *grounds* for emerging landscape democracy, I argued that landscape can employ principles of public protest. Instead of landscapes that favor visual images, static, singular positions of power, and controlling frames of ownership and policing, there is potential for other less-visually dominated landscapes formed through collective actions that reveal the politics from which they are constructed and challenge conventions of framing and enclosure.

This framing of post-landscapes draws on Mitchell's question of "Landscape or public space?"[43] to speculate whether future landscapes could borrow techniques of public protest. It also echoes Mitchell's contribution to discourses of landscape democracy, where he states: "Any movement or struggle to create an alternative spatial organisation of society must necessarily take and produce new spaces."[44] In turn, referring to post-landscapes, while describing gatherings in Union Square after 9/11 and Occupy protests in London, Mitchell suggests we should look "toward a new kind of post-landscape order, one being worked out on the ground."[45] However, while the protests of Occupy were situated primarily in cities such as London, New York, and Hong Kong, the movements' concerns for injustice associated with global processes of capital were more difficult to locate in specific places. Policies of governments, practices of corporations, and the roles of global banks tend first to impact individuals and communities in financial terms, even if this can be later read through neighborhoods devastated by foreclosures, schools underfunded, or public parks sold off to balance state deficits.

In contrast to Occupy, the London climate protests involved direct actions in public sites and made landscape concerns of environmental justice visible and situated. Firstly, the protests targeted industries and politicians facilitating ecological destruction, from the landscapes of fossil fuel corporations to the debating chambers of national governments and from the printing presses of media conglomerations to the marketplaces of global banks; secondly, the climate movement endeavored to make palpable lives and landscapes most impacted by global warming, including those less stable due to their coastal locations, lack of water, sensitivity to temperature change, or increasing frequency of storms; thirdly, at a more localized scale, the disruption to traffic caused by the protests reduced the volume of motor vehicles driven in urban centers and associated air pollution—even if only for 11 days. In contrast to many social movements, climate protests make landscapes more tangible. Caught up in the days of direct action in April 2019, processes of landscapes and practices of public spaces connected, interacted, and made an impact on conversations about the climate crisis in the UK and around the world. The Mayor of London, Sadiq Kahn, stated in November 2019: "I declared a climate emergency in London last year, and I agree with the protesters' view that the Government needs to stop ignoring the climate emergency and immediately deliver meaningful action."[46] The climate protests made landscapes visible as sites of political struggle and more-than-human processes that are public and contested (Figure 11.3).

The actions in London that were followed in September of the same year with a series of "global rebellions" closed so much of central London that the *Metro* newspaper declared, "The best advice for anyone driving in London today would be to stay well away."[47] Gatherings, occupations, demonstrations, talks,

Figure 11.3 Climate protestors highlighting the connection between the actions of oil industries and warming oceans, 2023. Drawing courtesy: The author.

and rallies focused on Waterloo Bridge and Westminster Bridge; the procession of streets and squares popular for mass demonstrations, including the Strand, Whitehall, Parliament Street, Parliament Square, Abingdon Street, and Millbank; The Mall leading to Buckingham Palace; and many adjacent streets. Additionally, localized actions that had the potential to disrupt while simultaneously garnering publicity, most notably the fixing of a large pink yacht at the intersection of Oxford Circus, occurred throughout the 11 days of protests. The movement had three demands for the UK government: (1) to tell the truth about the climate and ecological emergency, (2) reduce carbon emissions and halt biodiversity loss, and (3) create citizen assemblies that would lead decisions on climate and ecological justice. Public spaces were made and remade following strategies planned in advance and through tactical responses to less expected circumstances. They were public spaces defined through the coming together of different publics, fleeting spatializations of public spheres that formed one day to be fought on other streets the next.

Such protest sites contribute to networks of public spaces, including infrastructures of streets, squares, and parks that can accommodate political action, communication networks, and media, links between community and campaigning organizations, and settings conducive to the construction of interest groups. These complex formations of political action also exist in relation to the policies and proposals they aim to inform and in tension with global agendas with which they take issue: "[But] urban protests represent situated moments in much larger politics of space."[48] Sites of localized community organizing (such as the Huaorani peoples of the Ecuadorian Amazon) and think tanks (such as the New Economics Foundation), national governments, political unions (such as the European Union), intergovernmental organizations (such as the United Nations) where climate policies (such as the Climate Change Act 2008) are debated, policy initiatives (such as Green New Deal) are agreed, and global agendas (such as the UNDP's SDGs) are ratified are also political spheres and public spaces that intersect with and define landscapes of planetary climate change, concern, and action: "The sites of urban protest in this way are politicized in terms of a much larger spatial system of global relations and inequities."[49] While Smith and Low also extend their definition of public spaces to such broad geographies,[50] including spaces of the media and online forums, the exclusivity of some public spheres contrasts with the openness of some urban streets and squares of civil protest. Through situated public action, climate movements seek to make visible global connections between landscapes impacted by climate change and ecological damage with fossil fuel industries, inaction of national governments, and obfuscation by global media corporations. Landscapes related to the causes and impacts of global warming are entangled with an array of publics and public spaces, creating what we could term climates of publicness.

PARTIAL, PLANETARY, PUBLIC

Recognizing public climate geographies as inextricably connected, such entanglements of networks, beings, things, and places can also be claimed *as* landscapes. Research that maps landscape relations, from interactions between sites of mineral extraction, material processing, logistics corridors, and exchange to their use, reuse, and refuse, provides clues to such formations.[51] In *The Flip-Flop Trail*,[52] sociologist Caroline Knowles traces the life of a pair of flip-flops, following the intersecting lives and landscapes of global supply chains. Jane Hutton similarly draws connections between distant landscapes, from places of resource extraction to locations where the extracted materials, such as timber and stone, are put to work in creating renowned public landscape architecture projects.[53] These are not merely landscapes of flows; they are instead landscapes formed through intersections and tensions between many different people and places (Figure 11.4).

Mapping landscapes that can be simultaneously read from the intimately small to planetary scales is also an endeavor for practices confronting the climate crisis. Contrasting descriptions of climate landscapes, such as how IPCC (Intergovernmental Panel on Climate Change) reports struggle to represent local experiences and knowledge, highlight complexities in defining such distributed subjects across contrasting scales. Situating the impacts of global warming in specific communities, not just in civic squares that may be thousands of kilometers away, recognizes the many and long struggles for environmental justice. In "The Perils of Climate Activism," Nityanand Jayaraman also cautions against viewing the global ecological crisis merely through that of climate change: "With eyes turned skywards, it is easy to lose sight of the multiple causes of our planetary disease which are all rooted in land."[54] Similarly, the anthropologist Kirsten Hastrup recognizes, "an objectified notion of climate is present as a backdrop for most, localized perceptions of environmental change; people do not reason against such a notion but live and think with it, in their own creative ways."[55] As Hastrup proposes for anthropology, I argue that such distributed and entangled landscapes can reveal a "renewed understanding of the multiplicities of scale inherent in localized knowledge"[56] as well as making sense of local ecological injustice through the lens of planetary climate change.

These landscapes must also be claimed as public spaces. The situated actions of climate protests reveal landscapes of shared, collective, public concern—planetary landscapes of polluting industries, governments failing to regulate, newspapers inadequately reporting, coastlines destroyed by rising sea levels, urban centers choked by pollution, and the public demanding action. Smith and Low claim: "'Public Space' envelopes the palpable tension between place, experienced at all scales in daily life, and the seemingly spacelessness of the Internet, popular opinion, and global institutions and economy."[57] Their writings on public space reflect Nancy Fraser's concerns for transnational public spheres, where public concerns for climate justice and globalization transcend national borders:

> Whether the issue of global warming or immigration, women's rights or the terms of trade, unemployment or the 'war on terror,' current mobilizations of public opinion seldom stop at the borders of territorial states.[58]

Referring to what she terms the "politics of reframing,"[59] Fraser describes that the "Westphalian frame,"[60] a principle of nation-states and territorial sovereignty, is challenged by public spheres that form associations regardless of national boundaries. Traditional national frameworks are insufficient on their own to address transnational public concerns. "In such cases [of global warming], we have no choice but to look to other frames, including for some issues, global frames."[61] When the climate crisis is effectively reframed as a

Figure 11.4 The entanglement of climates of publicness—landscape processes and practices of public spaces public actions connected through public actions, 2023. Drawing courtesy: The author.

global concern, we can recognize that the geographies produced through this global public sphere form larger landscapes of public space.

The question of scale, as it pertains to public space, is raised by Smith and Low, where public spaces include "recognizable geographies of daily movement, which may be local, regional, or global."[62] While acknowledging that over recent decades, increased attention has been paid to urban scales of public space, Smith and Low highlight that growing global concerns have undermined both the power of national governments and the conceptual frameworks of public space: "Public space literature needs to nest its traditional concerns with the urban scale in a wider field of transformations spanning from the body to the global and even now supraplanetary."[63] These are challenges for researchers that Hastrup identifies for anthropology: "to come to terms with an entangled reality that defies conventional understandings of holism and of causation, and calls for a new understanding of scale."[64] Landscapes of the climate crisis draw closer public gatherings in London's urban streets to wildfires in California, school strikes around the world to polluting oil fields in the Gulf of Mexico, and environmental activists in Manila to rainforests in Brazil. The reframing that Fraser advocates provides a basis from which global public concerns can be investigated, and global action can be demanded. Furthermore, the positions from which this reframing occurs reveal prevailing structures of power over places that must also be challenged. Reflecting Donna Haraway's critique of the god-like view from above and her argument for adopting the position of the subjugated,[65] public discourses must keep in focus the precarious lives and landscapes threatened by the climate crisis and the marginalized voices of environmental struggles, frequently in the Global South.

These planetary-scale landscapes of public space—which are always in flux—are also uneven. Despite global warming being experienced across the entire planet, increasing planetary temperatures impact some communities and some landscapes with much greater force. From individuals with resources that enable them to move and isolate themselves to financiers who gain by speculating on the instabilities inherent to a warming planet, landscapes of the climate crisis are unequally distributed and unjustly experienced. The inequities of global warming are also reflected in the politics of climate movements, such as the protests that occupied London's public realm in 2019 that were considered to have ignored other and longer struggles for environmental justice and were condemned for being less inclusive of marginalized communities.[66] These concerns highlight the difficulties of making sense of climate injustice across contrasting scales, communities, and cultures. Such planetary landscapes of public space must strive—even while recognizing the limits of what is possible—to make accessible, inclusive public processes. Fraser explains:

> Public opinion is only legitimate if and only if it results from a communicative process
> in which all who are jointly subjected to the relevant governance structure(s) can
> participate as peers, *regardless of political citizenship*.[67]

To effectively address the climate crisis, focusing action on specific industries, countries, and organizations is necessary. This requires closer scrutiny of the nature of a global frame, regarding both who is included within public discourses and their visibility and presence within these contested landscapes.

The climate protests in London and other cities around the world aimed to make visible connections between global banks funding oil companies that drive ecologically destructive industries and national governments who facilitate these corporations while ignoring their role in rising sea levels, airborne pollution, destructive storms, and polluted oceans. 'Tell the truth' has been the first demand of this climate movement (Figure 11.3),[68] reflecting Mitchell's claim that "to be effective politics must be made visible in public space."[69] The performative blockading of Westminster Bridge, Whitehall, and Parliament Square, adjacent to the seat of the UK Government, as well as the rallies held within Parliament Square that provided a platform for UK politicians such as Caroline Lucas and activists like Greta Thunberg, aimed for the movement's demands to be heard. In contrast to traditions of landscape that have denied their politics, concealed disputed lands, and denied the presence of marginalized communities, landscapes of public space need to be concerned with making explicit tensions between governments, businesses, communities, and their environments. They are the landscapes that Hastrup describes, involving "redistribution of cause and effect across time and space," where "new questions of culpability and justice" can also be posed.[70]

CONCLUSIONS

I find that by connecting landscape processes with practices of public spaces through direct action, planetary-scale landscapes of the climate crisis can be revealed. Rather than maintaining landscape and public space in positions of conflict, reflecting frictions between traditional landscape practices and democratic forms of public space, I aim to make sense of climate protest in streets and squares that draw attention to landscapes endangered by global warming. Building on the argument of post-landscapes, where landscape is challenged to reflect the open, collective, grounded conditions of public spaces of political protest, I propose that transnational public spaces can be claimed and represented by adopting techniques of planetary landscapes. This reciprocity between landscape and public space requires that bounded definitions and spaces of both landscapes and public spaces be reframed through an emphasis on processes and practices of continual remaking. It also demands that the agency of non-human and more-than-human entities, whether rivers and oceans or the Internet and global warming, are accepted as inherent components of these other landscapes and essential considerations in practices of public spaces. While recognizing the extent of landscapes that span from streets in London to oil platforms in North America and from the debating chambers of national governments to Pacific islands inundated by rising oceans, I also underline that these are partially composed and constantly contested forms that hold in tension material and social actions with global concerns and politics. By making explicit relations between environmental movements and the visibility of climate-impacted landscapes through direct actions in specific geographic locations, the capacity to collectively inform new planetary-scale responses to the climate crisis may be found.

ACKNOWLEDGMENTS

This chapter is developed from a keynote address to the International Association of Landscape Ecology (I.A.L.E.) annual conference in Milan (2019) and subsequently developed as a keynote lecture at the Architectural Association (A.A.) in London (2021). Many thanks to Antonella Contin, David Grahame Shane and the organizers of I.A.L.E. and to Alfredo Ramirez at the A.A. for the invitations. I am also grateful to Gareth Doherty and Charles Waldheim for the opportunity to develop the research further for this book.

NOTES

1   The Extinction Rebellion protests of April 2019 continued in London for 11 days, beginning in April.
2   Jeff Masters, "The Top 10 Weather and Climate Stories of 2019," *Scientific American*, https://blogs.scientificamerican.com/eye-of-the-storm/the-top-10-weather-and-climate-stories-of-2019/ (accessed February 1, 2023).
3   Mercy Corps, "Last Year's Natural Disasters Prove that the Climate Crisis is Only Getting Worse," *Global Citizen*, https://www.globalcitizen.org/en/content/climate-disasters-2019/ (accessed February 1, 2023).
4   Don Mitchell, *Right to the City: Social Justice and the Fight for Public Space* (New York: The Guilford Press, 2003), 184.
5   Don Mitchell, *Right to the City: Social Justice and the Fight for Public Space* (New York: The Guilford Press, 2003), 186.
6   Barbara Bender, *Landscape: Politics and Perspectives* (Oxford: Berg, 1993).
7   Doreen Massey, *For Space* (London: Sage, 2005).
8   Stephen Carr et al., *Public Space* (Cambridge: Cambridge University Press, 1992).
9   Neil Smith and Setha Low, "Introduction: The Imperative of Public Space," in *The Politics of Public Space*, eds. S. Low and N. Smith (New York: Routledge, 2006).
10 Nancy Fraser, "Rethinking the Public Sphere: A Contribution to the Critique of Actually Existing Democracy," *Social Text* 25/26 (1990): 56–80.
11 Neil Smith and Setha Low, "Introduction: The Imperative of Public Space," in *The Politics of Public Space*, eds. S. Low and N. Smith (New York: Routledge, 2006), 4.
12 Don Mitchell, *Right to the City: Social Justice and the Fight for Public Space* (New York: The Guilford Press, 2003).
13 Fran Tonkiss, *Space, the City and Social Theory: Social Relations and Urban Forms* (Cambridge, UK: Polity Press, 2005), 68.
14 See Don Mitchell, *Right to the City: Social Justice and the Fight for Public Space* (New York: The Guilford Press, 2003), 142.
15 Fran Tonkiss, *Space, the City and Social Theory: Social Relations and Urban Forms* (Cambridge, UK: Polity Press, 2005), 66.
16 Ali Madanipour, ed., *Whose Public Space?* (Oxon: Routledge, 2010).

17 Anna Minton, *Ground Control: Fear and Happiness in the Twenty-first Century City* (London: Penguin, 2012).

18 Michael Sorkin, ed., *Variations on a Theme Park: The New American City and the End of Public Space* (New York: Hill and Wang, 1992).

19 Ed Wall and Tim Waterman, *Landscape and Agency: Critical Essays* (Oxon: Routledge, 2017); Ed Wall, *Contesting Public Spaces: Social Lives of Urban Redevelopment in London* (Oxon: Routledge, 2022).

20 "UK's Path to Net Zero Set Out in Landmark Strategy," U.K. Government, https://www.gov.uk/government/news/uks-path-to-net-zero-set-out-in-landmark-strategy (accessed February 1, 2023).

21 Harry Taylor and Agency, "Priti Patel Vows to Curb Eco Protests and Asylum Appeals in 2022," *The Guardian,* https://www.theguardian.com/politics/2022/jan/01/priti-patel-vows-to-curb-eco-protests-and-asylum-appeals-in-2022 (accessed February 1, 2023).

22 On January 1, 2023, Extinction Rebellion UK announced that they would "temporarily shift away from public disruption as a primary tactic," in "Extinction Rebellion UK's New Year's Resolution: WE QUIT," *Extinction Rebellion UK*, https://extinctionrebellion.uk/2023/01/01/extinction-rebellion-uks-new-years-resolution-we-quit/ (accessed February 1, 2023).

23 Neil Smith and Setha Low, "Introduction: The Imperative of Public Space," in *The Politics of Public Space*, eds. S. Low and N. Smith (New York: Routledge, 2006), 4.

24 See Don Mitchell, *Right to the City: Social Justice and the Fight for Public Space* (New York: The Guilford Press, 2003), 184.

25 Barbara Bender, *Landscape: Politics and Perspectives* (Oxford: Berg, 1993), 3.

26 Guy Nordensen et al., *On the Water: Palisade Bay* (New York: Museum of Modern Art, 2010).

27 Lindsay Bremner, "On Monsoon Assemblages." Interview with Damaso Randulfe, *Migrant* 3 (2017): 90–99; Christina Geros, "Designing Momentums: Site, Practice, Media as Landscape," *Architectural Design* 90, no. 1 (2020): 14–21; Beth Cullen, "Intuiting a Monsoonal Ethnography in Three Bay of Bengal Cities," *GeoHumanities* 7, no. 1 (2021): 148–163.

28 Henri Lefebvre, *The Production of Space*, trans. (Oxford: Basil Blackwell, 1991).

29 Don Mitchell, *Right to the City: Social Justice and the Fight for Public Space* (New York: The Guilford Press, 2003); Doreen Massey, *For Space* (London: Sage, 2005).

30 Barbara Bender, *Landscape: Politics and Perspectives* (Oxford: Berg, 1993), 3.

31 Don Mitchell, *Right to the City: Social Justice and the Fight for Public Space* (New York: The Guilford Press, 2003).

32 Don Mitchell, *Right to the City: Social Justice and the Fight for Public Space* (New York: The Guilford Press, 2003), 2.

33 Don Mitchell, *Right to the City: Social Justice and the Fight for Public Space* (New York: The Guilford Press, 2003), 2.

34 Don Mitchell, *Right to the City: Social Justice and the Fight for Public Space* (New York: The Guilford Press, 2003), 187.

35 Don Mitchell, "Landscape's Agency," in *Landscape and Agency: Critical Essays*, eds. Ed Wall and Tim Waterman (Oxon: Routledge, 2017), 189.

36 Barbara Bender, *Landscape: Politics and Perspectives* (Oxford: Berg, 1993), 2.

37 Barbara Bender, *Landscape: Politics and Perspectives* (Oxford: Berg, 1993), 3.

38 See Barbara Bender, *Landscape: Politics and Perspectives* (Oxford: Berg, 1993), 2.

39 See Barbara Bender, *Landscape: Politics and Perspectives* (Oxford: Berg, 1993), 2; Kenneth Olwig, "Gods and Humans," in Barbara Bender, *Landscape: Politics and Perspectives* (Oxford: Berg, 1993).

40 Mark Fisher, *Capitalist Realism: Is There No Alternative?* (Alresford, Hants: 0 Books, 2009); David Harvey, *Seventeen Contradictions and the End of Capitalism* (New

York: Oxford University Press, 2015); Paul Mason, *Post Capitalism: A Guide to Our Future* (New York: Farrar, Straus and Giroux, 2015).

41 Ed Wall and Tim Waterman, *Landscape and Agency: Critical Essays* (Oxon: Routledge, 2017).

42 Shelly Egoz, Karsten Jorgensen, and Deni Ruggeri, *Defining Landscape Democracy: A Path to Spatial Justice* (Cheltenham: Edward Elgar, 2018), 66.

43 Don Mitchell, *Right to the City: Social Justice and the Fight for Public Space* (New York: The Guilford Press, 2003), 187.

44 Don Mitchell, "Claiming a Right to Place in the Urban Landscape: Planning Resistance and Resisting Planning in Glasgow," in *Hva betyr landskapsdemokrati? Defining Landscape Democracy Conference Reader*, ed. Shelly Egoz (Centre for Landscape Democracy, NMBU), 16–17 (2015): 16.

45 Don Mitchell, "Landscape's Agency," in *Landscape and Agency: Critical Essays*, eds. Ed Wall and Tim Waterman (Oxon: Routledge, 2017), 192.

46 "Questions to Mayor of London," London Assembly, https://www.london.gov.uk/questions/2019/19706 (accessed February 1, 2023).

47 Richard Hartley Parkinson, "Which Westminster Roads have been Closed by Extinction Rebellion Protesters," *Metro*, October 8, 2019, https://metro.co.uk/2019/10/08/extinction-rebellion-google-maps-shows-central-london-road-closures-10879662/.

48 Fran Tonkiss, *Space, the City and Social Theory: Social Relations and Urban Forms* (Cambridge, UK: Polity Press, 2005), 65.

49 Fran Tonkiss, *Space, the City and Social Theory: Social Relations and Urban Forms* (Cambridge, UK: Polity Press, 2005), 66.

50 Neil Smith and Setha Low, "Introduction: The Imperative of Public Space," in *The Politics of Public Space*, eds. S. Low and N. Smith (New York: Routledge, 2006), 5.

51 Jane Hutton, *Reciprocal Landscapes* (Oxon: Routledge, 2020); Caroline Knowles, *Flip-Flop: A Journey Through Globalisation's Backroads* (London: Pluto, 2014).

52 Caroline Knowles, *Flip-Flop: A Journey Through Globalisation's Backroads* (London: Pluto, 2014).

53 Jane Hutton, *Reciprocal Landscapes* (Oxon: Routledge, 2020).

54 Nityanand Jayaraman, "The Perils of Climate Activism," *e-flux Architecture*, July, 2021, https://www.e-flux.com/architecture/survivance/410014/the-perils-of-climate-activism/.

55 Kirsten Hastrup, "Anthropological Contributions to the Study of Climate: Past, Present, Future," *WIREs Climate Change* 4, no. 4 (July 2013): 270.

56 Kirsten Hastrup, "Anthropological Contributions to the Study of Climate: Past, Present, Future," *WIREs Climate Change* 4, no. 4 (July 2013): 278.

57 Neil Smith and Setha Low, "Introduction: The Imperative of Public Space," in *The Politics of Public Space*, eds. S. Low and N. Smith (New York: Routledge, 2006), 3.

58 Nancy Fraser, *Scales of Justice: Reimagining Political Space in a Globalizing World* (Cambridge: Polity, 2008), 85.

59 Nancy Fraser, *Scales of Justice: Reimagining Political Space in a Globalizing World* (Cambridge: Polity, 2008), 154.

60 Nancy Fraser, *Scales of Justice: Reimagining Political Space in a Globalizing World* (Cambridge: Polity, 2008), 85.

61 Nancy Fraser, *Scales of Justice: Reimagining Political Space in a Globalizing World* (Cambridge: Polity, 2008), 149.

62 Neil Smith and Setha Low, "Introduction: The Imperative of Public Space," in *The Politics of Public Space*, eds. S. Low and N. Smith (New York: Routledge, 2006), 3.

63 Neil Smith and Setha Low, "Introduction: The Imperative of Public Space," in *The Politics of Public Space*, eds. S. Low and N. Smith (New York: Routledge, 2006), 3.

64 Kirsten Hastrup, "Anthropological Contributions to the Study of Climate: Past, Present, Future," *WIREs Climate Change* 4, no. 4 (July 2013): 277.

65 Donna Haraway, "Situated Knowledges: The Science Question in Feminism and the Privilege of Partial Perspective," *Feminist Studies* 14, no. 3 (Autumn 1988): 575–599.

66 Karen Bell and Gnisha Bevan, "Beyond Inclusion? Perceptions of the Extent to Which Extinction Rebellion Speaks to, and for, Black, Asia and Minority (BAME) and Working-Class Communities," *Local Environment: The International Journal of Justice and Sustainability* 26, no. 10 (2021): 1205–1220.

67 Nancy Fraser, *Scales of Justice: Reimagining Political Space in a Globalizing World* (Cambridge: Polity, 2008), 96.

68 See "Extinction Rebellion Demands," *Extinction Rebellion*, https://extinctionrebellion. uk/the-truth/demands/ (accessed February 1, 2023).

69 Don Mitchell, *Right to the City: Social Justice and the Fight for Public Space* (New York: The Guilford Press, 2003), 110.

70 Kirsten Hastrup, "Anthropological Contributions to the Study of Climate: Past, Present, Future," *WIREs Climate Change* 4, no. 4 (July 2013): 278.

Chapter 12: Is Landscape Queer?

Kate Thomas

> What is a landscape
> A landscape is what when they that is I
> See and look.
> Gertrude Stein[1]

There is something queer here, as I will argue in this chapter. Stein asks and answers her question in the same breath, not even pausing to use a question mark, her answer formed mostly from a slew of pronouns. As Stein scholar Sarah Posman argues, the thing about this definition "is that it contains everything at once. There is no chronology of emotion, no Aristotelian development that forces you to keep up. . . . Everything is there for you to explore at once, there are as many ins and outs as you want there to be."[2] Elsewhere, Stein expressed admiration for the way that landscape has "an existence in and for itself,"[3] a dictum that sounds almost like queer self-fashioning; after all, *she* is, herself, Stein, she is stone on which we all stand. But to be clear, this is not an argument for queer belonging, not precisely. Stein lived and wrote as an expatriate, a "deterritorialized" subject, an American who had made her home in European landscapes.[4] But if she resisted considering herself marked and defined by man-made territory, she embraced the idea that she—everyone—is produced by terroir: "anybody is as their land and air and water sky and wind and anything else is."[5] Anybody is as land and air: bodies are produced by landscape, just as landscape is made by us, by "what when they that is I/See and Look." This is not the queerness that is other, but the queerness that is all-togetherness in space and time, the queerness that unmakes ontological difference and replaces it with shifting relationality in and of and across landscape.

Stein wrote *Stanzas* in the country house in the Rhône Valley that she and Alice B. Toklas had started renting in the 1920s.[6] This house, set high on a hill in the small village of Bilignin, had formal and vegetable gardens, a terrace that overlooked farmland and woodland, hills with streams and lakes, and beyond that—the Swiss Alps. Nevertheless, *Stanzas* describes neither flora and fauna nor geographic or environmental features. Instead, Stein disarticulates the compound noun "landscape" into a tumble of monosyllabic words that express position, perspective direction, and relation. It is, therefore, perhaps no wonder that the literary form she came to most associate with landscape was drama. Living in

DOI: 10.4324/9781003148142-13

that "landscape that made itself its own landscape," as she described Bilignin, "so completely made a play that I wrote quantities of plays."[7] Both landscape and drama had, she felt, "formations," and both are about "being always in relation." Landscape is always "in relation one thing to the other thing."[8]

Stein's poetics of land-as-relation finds resonance in the etymology of the English word "landscape." The suffix "scape" doesn't derive, as has sometimes been supposed, from "scope," which invokes visual perception, but is rather a cognate of the German "schaft" or Dutch "skip," which means "creation, creature, constitution, condition" and becomes "ship" in English. "Scape" thus operates in the way it does in words like "friendship" or "companionship": it means the "state or condition of being."[9] Kenneth Olwig has shown how cultural geographers of the 1980s, such as Denis Cosgrove, were guided by a belief that "scape" was a visual prompt and therefore tied the concept of landscape to Renaissance perspectival representation and Johnsonian notions of a pictorial scenic. The field of Landscape Architecture has since unearthed older definitions of landscape that emphasized not scenery but instead place and polity. Landscape is not, in other words, the province of painters, architects, and theater designers. Landscape is neither a scenic view nor a backdrop. It is also neither solely nature nor solely culture. It is land in a relationship. As Anne Whiston Spirn points out, *landscape* contains the notion of a "mutual shaping of people and place: people shape the land and the land shapes people."[10] But is this etymology a prompt to understand landscape as a thing in and of itself, in communion with itself, self-constituting, or rather a thing only brought into being by an encounter with a sensate "I"? The latter might emphasize cultivation, the former upon something more like *genius loci*, a spirit of place.

We might therefore begin to answer the question "Is landscape queer?" by noting that when we talk, as we commonly do, of "sexual orientation," we are using a spatial metaphor. The idiom acknowledges that sexuality might be a matter of, as Sara Ahmed writes, "how we inhabit spaces as well as 'who' or 'what' we inhabit spaces with."[11] It is also a matter of how we traverse those spaces. Back in 1993, Eve Kosofsky Sedgwick made the observation that has become foundational to the field of queer theory; that the root meaning of "queer" is "across."[12] "Across" is a preposition, the part of speech that signals direction, time, place, location, or spatial relationship. In what might be called a "prepositional turn," queer theory veered away from nouns and definite articles, declining to define "the homosexual" as a distinct, discreet, and identifiable subject, turning instead toward ideas of relation and, I argue, landscape. Twelve years before Sedgwick showed us queer's affiliation with being "across," "transverse," or "athwart," Michel Foucault had also spoken of how the homosexual lived "slantwise."[13] In this interview, he argues that the "diagonal lines" homosexuality lays out in the social fabric reveal other possible forms and textures of relation: it discloses "the formation of new alliances and the tying together of unforeseen lines of force."[14] Once again, we see (even very early) queer theory declining to define "the homosexual" as a fixed point, a delimited, delimiting noun, and a definite article. Instead of being a bounded subject, the homosexual is, for Foucault, an "occasion" for making manifest many "virtual" kinds of affiliation and relation. The term on which he comes to rest is spatial. "We must think," he concludes, "that what exists is far from filling all possible spaces."

I would propose that Foucault's "all possible spaces" might be glossed as "landscape." After all, they are physical and social spaces that promise dimension, heterogeneity, and interrelation. Foucault's polymorphous relations that can be manifested out of and into "all possible spaces" are reminiscent of the "endless forms" dwelling in a "tangled bank" that heaves into view in the final, deeply poetic paragraph of Charles Darwin's *The Origin of Species* (1859). This bank teems with vegetation and creatures of "many kinds," "various" life forms, all "different from each other" but contiguous, touching, and entangling. Homosexuality, Foucault suggests, is a point from which we might apprehend a full, free-ranging landscape rather than looking out only at sanctioned scenic views with their hierarchies of foreground and background or treading only paved and narrow pathways. By the end of the interview, it's as if homosexuality has been the agent that materialized a full and lush landscape in place of what Foucault calls "a background of emptiness." We can now return to Ahmed, who makes this latent connection between queerness and landscape architecture fully explicit:

> In landscape architecture, they use the term 'desire lines' to describe unofficial paths, those marks left on the ground that show everyday comings and goings, where people deviate from the paths they are supposed to follow. Deviation leaves its own marks on the ground, which can even help generate alternative lines, which cross the ground in unexpected ways. Such lines are indeed traces of desire: where people have taken different routes to get to this point or to that point. It is certainly desire that helps generate a lesbian landscape, a ground that is shaped by the paths that we follow in deviating from the straight line.[15]

Ahmed's reflection on how unsanctioned desires form new pathways, any which way, is a version of literary critic Catherine Belsey's conviction that "Desire. . . can go anywhere."[16] Belsey's "anywhere" suggests that desire can be found in any or every place and also that it can take you to those places. It is omni-locational. It would follow, therefore, that landscape, a "ship" containing and connecting locations, can take you to all desires.

When Ahmed makes the connection between queer theory and landscape architecture so direct, she is, herself, taking a "different route" from most other critics. When queer theory turned its attention to spatiality, it first focused on the way major cities offered refuge to queers migrating from presumptively hostile small towns and rural settings, congregating in urban bars and public spaces and cruising grounds. Metropolitan centers occupied the center ground of queer studies. As Scott Herring observed in 2010, "Much of queer studies wants desperately to be urban planning, even as so much of its theoretical architecture is already urban planned."[17] Herring is referring to epistemologies such as the closet and politics such as "coming out"; these supposedly foundational and universal structures of queer life are, Herring argues, "urban-based."[18] In between the queer theory of the early 1990s and Herring's book, there had, in fact, been a wave of scholarship—much of it by geographers, cultural studies, and literary scholars—that had tried to provide a corrective to the way that queer theory was privileging what David Bell calls "metrosexuality."[19] Coming out, going west, moving on up, following the yellow brick road: the disco choruses of the queer liberation movement had allied queerness with leaving places behind. In the course of this, certain kinds of places with certain kinds of affiliations were left behind. Despite queer studies' intellectual and political allegiance to those who are sexually marginalized, many of its scholars and activists ended up, themselves, marginalizing non-urban spaces: rural, suburban, the heartlands, the farm.[20]

These spatial chauvinisms edged out queer people whose access to metropolitan spaces was more difficult or who were simply more at home in other kinds of communities and landscapes. Leaving a small town might also mean leaving behind working-class culture. Navigating new paths in a new city might be less enticing to those dealing with disability or chronic health conditions. Relocating in order to find a sexual community might produce dislocations of racial and ethnic ties. So those who are sexually othered might nonetheless cherish varying kinds of pleasure that lead them, as Elton John's 1973 "Goodbye Yellow Brick Road" has it, away from the penthouse and back to the plough: "When are you going to land?" the song asks, counterpointing the rustic scapes of the woods and the farm to the rootlessness of disco and emerald cities. Another version of this call to be "close to the land" would be taken up by the Radical Faeries just a few years later; influenced by the hippie, neo-pagan, environmental, and feminist movements, the first Spiritual Conference for Radical Fairies took place in Arizona in September 1979. Lesbian back-to-the-land movements date back even further; the landdyke, or womyn's land movements of the 1970s and 1980s that sought to carve separatist utopias out of a recognition of interdependence were following in the footsteps of nineteenth-century women's suffrage movements that similarly used rural retreats to try and remove themselves from the bounds of patriarchy.[21] Lisa Moore has documented how "back to the land" felt like a move "back to ourselves," creating "a lesbian aesthetic of the domestic outdoors that transcends the traditional garden space."[22] Efforts to find and farm "promised lands," of course, risk reproducing the problems of settler colonialism, whether it be by imposing white and middle-class notions of property ownership, displacing established communities, or "playing Indian" on Indigenous land.[23] Even lesbian-of-color land movements that grew out of resistance to white middle-class dominance struggled to establish strong and lasting roots or, indeed, avoid the fetishization of Indigenous connections to the land.[24] La Luz de la Lucha, for example, which became womyn of color land in the Fall of 1977, was empty and in foreclosure by 1979. Claiming and queering land has been both a powerful and problematic practice.

Critical theory has had a similarly hard time knowing how to put sexuality and landscape into relation with each other. The disciplines of landscape architecture and queer theory have played coy with each other; it is rare to find an entry on sexuality in the indexes of landscape architecture texts, and similarly rare to find queer theory, considering how landscape might shape and be shaped by sexuality. But we have long known that landscapes have an erotics. We might think of the fashion for ribald gardens in the eighteenth century or how, at Versailles, as Marc Treib fetchingly puts it, "amorous trysts occurred in the bosks."[25] Or we could look up the word "sexuality" to find that its first documented use refers to the sexuality of plants; this application comprises the first six entries in the Oxford English Dictionary. Indeed, our very origin stories often site sex in a garden. Literary scholar Lisa Moore points out, "The rich history of associations between transgressive sexual knowledge and the garden goes back to the myth of Eden."[26] Moore's particularization that the sexuality found in gardens skews transgressive is pertinent. E. M. Forster's groundbreaking "gay novel," *Maurice*, is as much a love letter to the "greenwood" of England as it is a love story between two men.[27] The book, written in 1913–1914, revised in 1932, again in 1959–1960, and only published posthumously in 1971, defied literary convention by giving Maurice and his game-keeper lover Alec a happy ending, a plot in which they were "parted no more."[28] Forster writes in the terminal note to the novel, "I was determined that in fiction anyway two men should fall in love and remain in it for the ever and ever that fiction allows."[29] Forster imagines this space of queer eternal felicity very specifically as a landscape: "Maurice and Alec still roam the greenwood," he writes.[30] Forster glosses this greenwood as a pastoral idyll "in which it is still possible to get lost," a place where one can take refuge in a "forest or fell . . . [or] cave."[31] Forster had first-hand experience of this kind of greenwood, an experience that prompted him to write *Maurice*. The novel was, he wrote, a "direct result" of a visiting his friend Edward Carpenter's back-to-the-land gay commune at Millthorpe in Derbyshire. Seeing Carpenter relish what he called "simple living" in his "happy valley" with his lover George Merrill, farming and making sandals was like glimpsing into an Arcadian grove of gay possibility.[32] Forster admired Carpenter for having pushed through the privet hedges of suburbia and cut across the manicured quadrangles of Cambridge to root himself in nature instead of convention. In the novel, Maurice wonders if same-sex relationships will ever be acceptable in England, to which the doctor trying to cure him of his homosexuality says, "I doubt it. England has always been disinclined to accept human nature."[33] Forster fights this naturalization of heterosexuality by arguing for the queerness of landscape. He uses the topos of the greenwood to figure nature as validating human nature's manifold desires. For many gay rights activists emerging from the fin de siècle, the English bucolic—and the cross-winds of Walt Whitman's American eco-erotics—provided a strong defense of homosexuality. As Matt Cook observes, the bucolic is "used by John Addington Symonds, Edward Carpenter, and E. M. Forster to legitimize their queer desires and to show continuity with the natural as it enfolds them."[34] Green, we might say, is gay.

This fin-de-siècle queer embrace of the bucolic tended, however, toward earnest utopianism. And as such, it involved a certain placelessness: "utopia," meaning "nowhere" or "not place." It involved the spatial and temporal displacements of "not here, not now," and because it turned toward the Classical world and its climes, it also involved investment in "over there" or "back then." Symonds and Carpenter both, for example, held passions for all things Greco-Roman, and Carpenter credited "the delightful landscape and climate of Italy" for resuscitating him to a "new life."[35] The fin-de-siècle bucolic romanticized the rural, or Arcadian landscapes, turning its back on others as commercial and crass. But of course, there is no equal sign between "landscape" and "rural." A landscape does not have to be green. Nor does a landscape have to be welcoming or even sustaining. As anthropologist Anna Tsing puts it, "A landscape is a gathering in the making. . . . Landscapes are both imaginative and material; they encompass physical geographies, phenomenologies, and cultural and political commitments."[36] The fin-de-siecle queer writers described and built—whether literally, in Carpenter's case, or literarily in Forster's case—their queer landscapes. But these were gatherings particular to their own queer moment. They were describing—or imagining—the only kinds of landscapes in which they could be queer. The later "Go West" generation ("west" meaning not the high plains and mountains of the phrase's first iteration but the rainbow-flag-draped streets of San Francisco) would invert the paradigm of the bucolic, turning its back on the rural and the suburban and the small town, embracing cities as the only places in which they could live gay lives.

There is, of course, no one site of sexual liberation. No single type of landscape that could house, express, sustain or reflect a queer life. Whitman celebrated the sensuousness of nature in *Leaves of Grass*, but he also celebrated the sensuousness of crowds on trams and buses. Artist, filmmaker, and gay activist Derek Jarman's first muse was London; his earliest films focused on the Docklands. But as a frequent visitor to the queer cruising grounds of Hampstead Heath, he also well understood how any metropolis has bucolic pastoral spaces enfolded within it; when he ventured over "the invisible border [where] your heart beats faster, and the world seems a better place" he found his own Eden, his own Arcadia: "lying in the grass under the stars with some stranger was ecstasy," he wrote.[37] If he could perceive the heterogeneity of landscapes, it was perhaps because he understood himself as an amalgam, an artist who could work with a range of media and materials. In *Modern Nature,* he writes that if "fate had turned out different," he would have been a professional gardener because he was a passionate amateur practitioner of the horticultural arts.[38] He devoted the last years of his life to his beach garden at Prospect Cottage in Dungeness, which he bought in 1987. Having been diagnosed with HIV a year earlier, he was "gardening on borrowed time," as he described it in a scribbled note in a sketchbook. The cottage and garden, which would be his home until his death in 1994, sit on a shingle shore in the shadow of the Dungeness nuclear power station. To Jarman's eye, this landscape was "parched," "bone dry," and "wounded."[39] It expressed, in other words, physical endangerment and a need for care. Jarman took his ailing body to a failed and abandoned landscape, where he became an architect of wonderment and compassion. Garden and artist would salvage each other together. The dominant man-made features of the borrowed scenery of Prospect Cottage invoke both life-saving and peril; it looks out on two lighthouses, two lifeboat stations, and two nuclear power reactors. Jarman called it a "landscape of past endeavours."[40] It's a phrase that conjures up a sense of struggle, possibly futile. These six built structures navigate the divide between salvation and destruction. The lighthouse and lifeboats are civic furniture devoted to safety and rescue but are needed because of the peril of storms and sea. Nuclear power stations might arguably be classed as life-sustaining because they provide energy but given that the Chernobyl disaster had occurred in April 1986, just under a year before Jarman bought the cottage in May of 1987, they would instead have been viewed as ominous. They were ominous and politically odious: an insignia of Margaret Thatcher's pledge to build one nuclear power station every year and a reminder of lesbians at Greenham Common protesting nuclear weapons. But Jarman had a perverse—perhaps compassionate—attraction to the apocalyptic. In *The Last of England* (1987), he'd fantasized about living in a "little lead-lined house," calling it The Villa Chernobyl and furnishing it with a "Geiger-counter in the hall ticking where the grandfather clock used to chime away the hours."[41] Living with nuclear reactors in full view was, of course, a wry metaphor for how his own life had been turned into a half-life by a virus. It also metaphorized not being afraid to stand on the front lines, squarely facing a fucked up world and the forces that power it. Jarman had always, Peake tells us, liked pylons for that reason, too.[42] Rather than harboring a Ruskinian hatred of the ruination of "scenery" or seeking solace in the purely picturesque, Jarman embraced landscapes that were blighted, stigmatized, or abandoned. He was a high priest of queer art's theology of salvaging discarded spaces and materials and of finding beauty in that which others consider ravaged or toxic.

Jarman's biographer, Tony Peake, explains the draw of Dungeness: "Jarman had loved places that were interzonal, that stood between other worlds, or on the fringes of them."[43] Prospect Cottage, which was originally a fisherman's shack, sits lightly and liminal on a stretch of shingle beach where land meets sea. It is an unenclosed plot, fenceless and thus un-English, fully open to the elements, whether gentle or buffeting, witness to the constant mutability and also endurance. For Jarman, gardening supplied a mystical, metaphysical release from what Elizabeth Freeman has called "chrononormativity," the conventional timelines of what Jarman called "heterosoc."[44] Jarman writes:

> The gardener digs in another time, without past or future, beginning or end. A time that does not cleave the day with rush hours, lunch breaks, the last bus home. As you walk in the garden you pass into this time—the moment of entering can never be remembered. Around you the landscape lies transfigured. Here is the Amen beyond the prayer.[45]

The key term in this passage is "transfigured." As a gardener, he is doing the transfiguring—the planting, the pruning, the planning—but he is also profoundly changed by the landscape in which he works. When he broke ground on his Prospect Cottage garden—his first garden of his own—he imported 30 rose bushes from a supplier in Kensington. They died. He learned instead to use a palette of native plants—sea kale and teasels, viper's bugloss, gorse, wild peas, and sea holly—all hardy, somewhat vegetal forms. For high points of color, Jarman turned to the foxgloves that are both wild and contain toxins, and then the blood-red flower that thrives on traumatized soil, the poppy. The result is a garden that is neither rural nor urban, neither simply bucolic nor entirely post-industrial. It grows out of gravel and is decorated with flotsam and jetsam, driftwood, and fishing floats. It speaks of both paradise and ruin.

But Jarman was an artificer, too. His garden combines deference to both the tenet of genius loci—the spirit of place—with the arts of displacement. Jarman followed the landscaper's dictum of "right plants right place," but he also imported large quantities of compost, which he buried under the shingle. How distant was he, really, from Oscar Wilde, who wore a green carnation in defiance of Nature with a capital "N?" Wilde's green carnation was a campy defense of desires that many considered unnatural, and of how the queer is often seen as a "hothouse flower." The queer subject is stigmatized as rootless, aberrant, contrived, or not part of the reproductive tree of life and has often been figured as a terrible deformity away from the natural world. But what we mean by nature, what we recognize as natural, what kinds of growth we nurture, and what we suppress are all constructs and change with each generation and, indeed, across a single generation. Young Jarman embedded himself in landscapes of pulsating life, like the "pre-Lapserian" cruising grounds of Hampstead Heath, rich with anonymous pleasures, and then removed himself to beautify and then die in a post-apocalyptic place redolent of tempests, drowned sailors, and bachelor fishermen. Jarman was an outsider, a transplant to Dungeness, and he didn't take refuge in the dogmas of the autochthonous. "Why shouldn't I," he wrote, "invite people into another garden rather than walk in theirs?"[46]

Jarman's garden, along with his writings about gardening and scenery and belonging, joyfully instigate what Jill H. Casid has called "Landscape Trouble." This title, which she gives to her 2008 contribution to a roundtable on landscape theory, tips the wink to a classic work of queer theory. Judith Butler's groundbreaking book *Gender Trouble* was published almost 30 years earlier in 1990. Casid's piece is about colonialism and landscape and the thorny question of how disindigenation has been a powerful tool of colonialism through deforestation, the transplantation of plantation crops, or the "graft[ing of] one idea of island paradise onto another."[47] This focus on how empires are not only built but also planted and transplanted or "inhumed," to use the term Casid coins, resonates against the way that queer theory and culture have challenged our definitions of nature and the natural.

Casid's earlier work, in *Landscape and Colonialism* (2005), brings postcolonial and queer theory together, examining the trope of "nomadic gardens of queer longing" in Shani Mootoo's 1996 novel *Cereus Blooms at Night*.[48] As Audre Lorde challenged us to ask what tools we needed to dismantle the master's house, this novel asks what it takes to uproot, or overgrow the master's plantation. "To plant," Casid reminds us, "was to make colonies."[49] Sowing seed was one way of staking a claim on the land and metaphorizing that land as an inseminated woman sought to naturalize both imperial agrarian practices and heterosexual reproduction. Casid reminds us that the idiom of "husbandry" yokes together the possession of land with the patriarchal possession of women. In imperial agrarian discourse, landscape is allowed—forced—to be feminine, but it is never allowed to be queer. But the plot of Mootoo's novel—both its narrative and its place/location—"sustains seemingly impossible relations of desires" between a "cast of transgendered, transhuman, queer and ethnic hybrids."[50] The story shows us that plants, allowed to run wild, can have the power to not only over-run the master's garden but also dismantle the master's house; the titular cereus takes hold of the walls of a sexually abusive father's house and pulls them down around his corpse. *Cereus Blooms* is, Casid concludes, a story in which "'nature' takes revenge against the regime of the 'natural.'"[51] The garden can rise up against the patriarchal imperial gardener. Through hybridization, relocation, or simply running rampant, the plantings of the European landscape garden, or the plantation, can become sex rebels and decolonial activists.

This recognition that grown and built environments should be considered to have agency, and be made up of "vibrant matter," to use the term coined by Jane Bennett, is central to the "non-human turn" that has recently brought together queer and environmental studies. In 2008, Noreen Giffney and Myra Hird published *Queering the Non/Human*, an essay collection that proposes that the human is neither central to nor uniquely sentient in the world.[52] The non-human is as thinking, as agent, and as desiring as we have imagined ourselves to be; Giffney and Hird's essays give a collective shove to the anthrocentrism, anthronormativity, and anthropomorphism. As Michael O'Rourke writes in the preface: "Our transimmanence, or allness, a being-with towards others, *all* others, brings about new modes of sociability."[53] Two years later, *Queer Ecologies: Sex, Nature, Politics, Desire*, edited by Catriona Mortimer-Sandilands and Bruce Erickson, similarly explored how "sexualities and environments meet and inform one another."[54] Their essays challenge us to hear how the spatial and the sexual are inextricable when we talk about the "orientation" of our desires and the "environments" that might stimulate or inhibit those desires. "These spatial-sexual processes," they write, "have also affected the spaces of *nature,* not only in formal and designated natures but also across socionatural environments, more broadly."[55] One contributor, Gordon Brent Ingram, seeks to address the "enigmatic gap"[56] of sexuality in the field of landscape ecology, pointing out that its critical vocabulary of "patches," "edges," "ecotones," "flow," and "matrices" are perfectly suited for describing how marginalized sexual subjects find spaces in which to gather, connect, live, and love. In 2013, Nicole Seymour's *Strange Natures: Futurity, Empathy, and the Queer Ecological Imagination* joined this new body of work on "queer ecology," tracing an evolution in which "natural" was "something of a dirty word in queer theory"[57] with an understanding that nature may teach us a thing or two about being strange. As Tim Morton writes in a 2010 guest column in *PMLA*, "All life-forms, along with the environments they compose and inhabit, defy boundaries between inside and outside at every level. When we examine the environment, it shimmers, and figures emerge in 'strange distortion.'"[58] All these theorists of queer ecology agree that everything we are and everything around us exists in a condition of such interspecies intimacy that it doesn't even make sense to think of insides and outsides, centers and peripheries, foreground and background. In other words, we are our landscape.

With this understanding that landscape is, to use Donna Haraway's term, a "natureculture"[59] that is not a backdrop to human life and desire but instead exists in dynamic interrelation with it, let's return now to Stein's definition of landscape with which I opened this chapter: "A landscape is what when they that is I/See and look." Stein's assertion that "A landscape is. . . is I" sounds very like an assertion made by trans theorist Susan Stryker just two years ago: "Because I am there, and because I am trans, this is a transecology."[60] Stryker makes this grounding claim about the ground in the preface to a 2020 essay collection called *Transecologies*. Theorizing bodies-in-place and place-in-bodies is an alternative to both biological essentialism and social constructionism. Nicole Seymour's essay in the collection argues that nature is a site and paradigm of transitioning as central to all life forms. Seymour calls this "organic transgenderism" and proffers the observation that gender transition is "akin to the life-cycle changes of plants and animals."[61] She cites poet Oliver Baez Bendorf, "If you've ever doubted that a body can transform/completely, take the highway north from town. . . . The land where I was/born was born an ocean, and that ocean born of ice."[62] Together, Seymour and Baez turn to landscape as a natureculture (oceans and highways) that manifests how transformation is crucial to all forms of life.

Lucas Crawford begins his 2015 book *Transgender Architectonics* with a premise similar to Stryker's: "Transgender space in general may be defined. . . by those spaces that we visit and must navigate on a daily basis."[63] He goes on, however, to explore the importance of "trans-imaginative worlds," showing that trans subjects don't just exist in or fit into extant places and spaces but can instead be architects of the "acts and collaborations that happen across bodies, buildings, and milieus,"[64] and hermeneuts of accidental or constructed "otherworldly landscapes."[65] Crawford agrees with Seymour: theorizing *trans* and *architecture* together "draws out the always-already trans quality of materiality" and leads us toward an understanding of the "ubiquity of constant transformation for all."[66] This metaphysical theorization of all life and all spaces as universally "always-already trans" does not, of course, gloss over the perils that queer and trans subjects all too often experience when navigating both gender politics and landscapes, nor the frequent ways that the logistics of finding medical care and social resources force trans people to make geographic relocations. In an earlier article, Crawford points out that "the experience of gender modification seemingly demands metaphors of sovereign territoriality as well as literal movement from place to place by those who practice it."[67] Crawford's linking of trans experience with a districting/redistricting trope has a foundation in the much earlier work of Jay Prosser, who in 1998 observed that "metaphoric territorializing of gender and literal territorializations of physical space have often gone hand in hand."[68] This territorialization often takes the form of colonization; Native Studies scholars such as Scott Lauria Morgensen have detailed how "racialized heteropatriarchal control" has been foundational to white settler colonialism, and the sexual policing of Indigenous bodies has been a way into and through stolen land.[69] What if, however, in each of these scenarios, land rises up to meet us? What do I mean by this? Crawford gestures toward one answer when he suggests that "each bodily transition (from gender to gender or place to place) may be a matter of spatial ethics as much as sexual ones, of orientation to place as much to the body, of being moved in certain ways as much as moving."[70] When Crawford hypothesizes that place can "move" us he is allowing the landscape traversed by the trans subject to be affectively agent. It might prompt us to find a land- or eco-centric emphasis within Susan Stryker's assertion; when she affirms, "Because I am there, and because I am trans, this is a transecology," it is possible that the "thereness," which Stryker makes the primary clause of the sentence, is as formative of her trans self as the other way round.

And so we can return to what Ann Whiston Spirn told us: "Landscape moves and shapes each one of us."[71] We must take the capaciousness of her phrase "each one of us" seriously, understanding it to include queer subjects. "Include" is not in the sense that a gate has been opened by a keeper who has decided to let us into an enclosure, but in the sense of *being incorporated into*. In Gertrude Stein's tautological sense of "A landscape is. . . is I," the queer subject emerges in and out of landscapes of refuge and paradise, peril and pleasure, knowing all the while that we exist within each other's embrace.

NOTES

1 Gertrude Stein, *Stanzas in Meditation: The Corrected Edition,* eds. Susannah Hollister and Emily Setina (New Haven and London: Yale University Press, 2012, 1932), 213. *Stanzas* was composed in 1932 and first published in 1956.

2 Eugene W. Holland, Daniel W. Smith, and Charles J. Stivale, eds., *Gilles Deleuze: Image and Text* (London and New York: Continuum, 2009), 52.

3 Gertrude Stein, *Lectures in America* (New York: Random House, 1935), 225. She redoubles the sense that landscape is sovereign unto itself, subject to no jurisdiction when she writes: "It was a landscape. And it belonged to no country," 176.

4 This term is Gilles Deleuze's, applied to Stein by Isabelle Alfandary, "Becoming American, Becoming Agrammatical: Reading Stein with Deleuze," in *Gertrude Stein in Europe: Reconfigurations Across Media, Disciplines, and Traditions*, eds. Sarah Posman and Laura Luis Schultz (London: Bloomsbury, 2015), 129.

5 Gertrude Stein, *Narration: Four Lectures* (Chicago, IL: University of Chicago Press, 1935), 48.

6 They first rented the Bilignin house in 1929, and often spent six months of the year there. Edward Burns, ed., *The Letters of Gertrude Stein and Carl Van Vechten, 1913–1946* (New York: Columbia University Press, 2013), 674.

7 Gertrude Stein, *Last Operas and Plays* (Baltimore, MD and London: Johns Hopkins University Press, 1935), XLVI.

8 Gertrude Stein, *Lectures in America* (New York: Random House, 1935), 264.

9 Kenneth R. Olwig, *The Meanings of Landscape: Essays on Place, Space, Environment and Justice* (Abingdon: Routledge, 2019), 4–9.

10 Rachel Ziady DeLue and James Elkins eds., *Landscape Theory* (New York and London: Routledge, 2008), 92.

11 Sara Ahmed, *Queer Phenomenology: Orientations, Objects, Others* (Durham, NC: Duke University Press, 2006), 1. Using a phenomenological approach, focusing on lived and affective experiences, Ahmed explores how spaces change both our individual bodies and the collective social bodies to which we belong: "spaces 'impress' on the body," shaping and reshaping the body and the "skin of the social," 9.

12 Eve Kosofsky Sedgwick, *Tendencies* (Durham, NC: Duke University Press, 1993). The word "comes from the Indo-European root -twerkw, which also yields the German *quer* (transverse), Latin *torquere* (to twist), English *athwart*," xii.

13 Michael Foucault, "Friendship as a Way of Life," in *Michel Foucault: Ethics, Subjectivity and Truth,* ed. Paul Rabinow (New York: The New Press, 1994), 135–140.

14 Michael Foucault, "Friendship as a Way of Life," in *Michel Foucault: Ethics, Subjectivity and Truth,* ed. Paul Rabinow (New York: The New Press, 1994), 138, 140.

15 Sara Ahmed, *Queer Phenomenology: Orientations, Objects, Others* (Durham, NC: Duke University Press, 2006), 19–20.

16 Catherine Belsey, *Desire: Love Stories in Western Culture* (Oxford: Wiley-Blackwell, 1994), 6.

17 Scott Herring, *Another Country: Queer Anti-Urbanism* (New York: New York University Press, 2010), 5.

18 Scott Herring, *Another Country: Queer Anti-Urbanism* (New York: New York University Press, 2010), 22.

19 David Bell, "Eroticizing the Rural," in *De-Centring Sexualities: Politics and Representation Beyond the Metropolis*, eds. Richard Phillips, David Shuttleton, and Diane Watt (London and New York: Routledge, 2000), 83–101, 84.

20 Exceptions include: Lisa Moore, "Safe Space, Silo Storage, Outhouse with a View: Lesbian Garden History," in *Queering the Interior*, eds. Andrew Gorman-Murray and Matt Cook (London: Bloomsbury, 2018); Will Fellows, *Farm Boys: Lives of Gay Men from the Rural Midwest* (Madison: University of Wisconsin Press, 1998); Karen Lee Osborne and William J. Spurlin, eds., *Reclaiming the Heartland: Lesbian and Gay Voices from the Midwest* (Minneapolis: University of Minnesota Press, 1996); Michael Warner, "Walden's Erotic Economy," in *Comparative American Identities: Race, Sex and Nationality in the Modern Text*, ed. Hortense Spillers (New York: Routledge, 1991), 157–174; Andrew Gorman-Murray, Barbara Pini, and Lia Bryant, eds., *Sexuality, Rurality, and Geography* (Lanham: Lexington Books, 2013).

21 See Joyce Cheney's 1985 book *Lesbian Land* for documentation of lesbian land movements such as Daughters of Earth, HOWL, Spiral Women's Land Cooperative, and WEB.

22 Lisa Moore, "Safe Space, Silo Storage, Outhouse with a View: Lesbian Garden History," in *Queering the Interior*, eds. Andrew Gorman-Murray and Matt Cook (London: Bloomsbury, 2018), 121–122.

23 See, for example, Scott Lauria Morgensen, *Spaces Between Us: Queer Settler Colonialism and Indigenous Decolonization* (Minneapolis and London: University of Minnesota Press, 2011), 127–160.

24 See Katherine Schweighofer, "A land of One's Own: Whiteness and Indigeneity on Lesbian Land," *Settler Colonial Studies* 8, no. 4 (2018), 489–506.

25 Marc Treib, "Moving the Eye," in *Sites Unseen*, eds. Diane Harris and D. Fairchild Ruggles (Pittsburgh: Pittsburgh University Press, 2007), 61–86, 67.

26 Lisa Moore, *Sister Arts: The Erotics of Lesbian Landscapes* (Minneapolis: University of Minnesota Press, 2011), 16.

27 For a detailed tracing of Forster's use of the concept of the greenwood, see Elizabeth Wood Ellem, "E.M.Forster's Greenwood," *Journal of Modern Literature* 5, no. 1 (February 1976): 89–98.

28 E. M. Forster, *Maurice: A Novel* (New York and London: Norton, 1971), 240.

29 E. M. Forster, *Maurice: A Novel* (New York and London: Norton, 1971), 250.

30 E. M. Forster, *Maurice: A Novel* (New York and London: Norton, 1971), 254.

31 E. M. Forster, *Maurice: A Novel* (New York and London: Norton, 1971), 249. For a history of the Arcadian pastoral that informs Forster's concept of the "greenwood," see Byrne R. S. Fone, "This Other Eden: Arcadia and the Homosexual Imagination," *Journal of Homosexuality* 8, no. 3–4 (Spring/Summer 1983): 13–34.

32 Edward Carpenter, *My Days and Dreams* (London: George Allen & Unwin Ltd, 1921, orig. 1916), 148 and 236.

33 E. M. Forster, *Maurice: A Novel* (New York and London: Norton, 1971), 211.

34 Matt Cook, "Wilde Lives: Derek Jarman and the Queer Eighties," in *Oscar Wilde and Modern Culture: The Making of a Legend*, ed. Joseph Bristow (Athens, OH: Ohio University Press, 2008), 285–304, 294.

35 Edward Carpenter, *My Days and Dreams* (London: George Allen & Unwin Ltd, 1921, orig. 1916), 67–68.

36 Anna Tsing, "The Buck, the Bull, and the Dream of the Stag: Some Unexpected Weeds of the Anthropocene," *Suomen Antropologi: Journal of the Finnish Anthropological Society* 42 (2017): 1, 7.

37 Quoted by Tony Peake, *Derek Jarman: A Biography* (Woodstock and New York: The Overlook Press, 2000), 151. Jarman decried the clearing of areas of undergrowth on the Heath as an attempt to discourage cruising, writing "Nature abhors Heterosoc [. . .] Nature is Queer." Derek Jarman, *At Your Own Risk* (Woodstock, NY: The Overlook Press, 1993), 23.

38 Derek Jarman, *Modern Nature* (Woodstock, NY: The Overlook Press, 1994), 84 and 42.

39 Derek Jarman, *Modern Nature* (Woodstock, NY: The Overlook Press, 1994), 118.

40 Derek Jarman, *Modern Nature* (Woodstock, NY: The Overlook Press, 1994), 67.

41 Derek Jarman, *Kicking the Pricks* (Minneapolis: University of Minnesota Press, 2010. Originally published as *Last of England,* 1987), 239.

42 Tony Peake, *Derek Jarman: A Biography* (Woodstock and New York: The Overlook Press, 2000), 395.

43 Tony Peake, *Derek Jarman: A Biography* (Woodstock and New York: The Overlook Press, 2000), 395.

44 Elizabeth Freeman, *Time Binds: Queer Temporalities, Queer Histories* (Durham, NC and London: Duke University Press, 2010).

45 Derek Jarman, *Modern Nature* (Woodstock, NY: The Overlook Press, 1994), 30.

46 Derek Jarman, *Kicking the Pricks* (Minneapolis: University of Minnesota Press, 2010. Originally published as *Last of England,* 1987), 108.

47 Jill H.Casid, "Landscape Trouble" in *Landscape Theory*, eds. Rachel DeLue and James Elkins (New York: Routledge, 2008), 179–187, 182. For more on the instrumentalization

of plants to the colonial project, imported to express control, naturalize occupation and make the colonized landscape fit the imperial tastes and imagination, see also Elizabeth Hope Chang *Novel Cultivations: Plants in British Literature of the Global Nineteenth Century* (Charlottesville and London: University of Virginia Press, 2019).

48 Jill H.Casid, *Landscape and Colonization* (Minneapolis and London: University of Minnesota Press, 2005), xvi. Casid's phrase is an homage to bell hooks' phrase "diasporic landscapes of longing."

49 Jill H.Casid, *Landscape and Colonization* (Minneapolis and London: University of Minnesota Press, 2005), xvii.

50 Jill H.Casid, *Landscape and Colonization* (Minneapolis and London: University of Minnesota Press, 2005), xxi.

51 Jill H.Casid, *Landscape and Colonization* (Minneapolis and London: University of Minnesota Press, 2005), xx.

52 Ground for this queer ecology movement was broken by Greta Gaard over ten years earlier. See: "Toward a Queer Ecofeminism," *Hypatia* 12, no. 1 (1997): 114–137.

53 Myra J. Hird and Noreen Giffney, *Queering the Non/Human* (Aldershot: Ashgate Publishing, 2008), xviii.

54 Catriona Mortimer-Sandilands and Bruce Erickson, *Queer Ecologies: Sex, Nature, Politics, Desire* (Bloomington: Indiana University Press, 2010), 5.

55 Catriona Mortimer-Sandilands and Bruce Erickson, *Queer Ecologies: Sex, Nature, Politics, Desire* (Bloomington: Indiana University Press, 2010), 17.

56 Gordon Brent Ingram, "'The Place, Promised, That Has Not Yet Been': The Nature of Dislocation and Desire in Adrienne Rich›s Your Native Land," in *Queer Ecologies: Sex, Nature, Politics, Desire*, eds. Catriona Mortimer-Sandilands and Bruce Erickson (Bloomington: Indiana University Press, 2010), 262.

57 Nicole Seymour, *Strange Natures: Futurity, Empathy, and the Queer Ecological Imagination* (University of Illinois Press, 2013), 2.

58 Timothy Morton, "Queer Ecology," *PMLA* 125, no. 2 (March 2010): 274.

59 Donna J. Haraway, *The Companion Species Manifesto: Dogs, People, and Significant Otherness Vol. 1* (Chicago, IL: Prickly Paradigm Press, 2003), xxiii. For astute queer commentary on Haraway's neologism, see David Bell "Queernaturecultures," in *Catriona Mortimer-Sandilands and Bruce Erickson,* eds. *Queer Ecologies: Sex, Nature, Politics, Desire* (Bloomington: Indiana University Press, 2010), 134–146.

60 *Transecology: Transgender Perspectives on Environment and Nature*, ed. Douglas A. Vakoch (London and New York: Routledge, 2020), xvi.

61 Nicole Seymour, "'Good Animals': The Past, Present, and Futures of Trans Ecology," in *Transecology: Transgender Perspectives on Environment and Nature*, ed. Douglas A. Vakoch (London and New York: Routledge, 2021), 190–204, 191.

62 Nicole Seymour "'Good Animals': The Past, Present, and Futures of Trans Ecology," in *Transecology: Transgender Perspectives on Environment and Nature*, ed. Douglas A. Vakoch (London and New York: Routledge, 2021), 194.

63 Lucas Crawford, *Transgender Architectonics: The Shape of Change in Modernist Space* (Farnham: Ashgate, 2015), 1.

64 Lucas Crawford, *Transgender Architectonics: The Shape of Change in Modernist Space* (Farnham: Ashgate, 2015), 2, 14.

65 Lucas Crawford, *Transgender Architectonics: The Shape of Change in Modernist Space* (Farnham: Ashgate, 2015), 162.

66 Lucas Crawford, *Transgender Architectonics: The Shape of Change in Modernist Space* (Farnham: Ashgate, 2015), 14.

67 Lucas Crawford, *Transgender Architectonics: The Shape of Change in Modernist Space* (Farnham: Ashgate, 2015), 129.

68   Jay Prosser, *Second Skins: The Body Narratives of Transsexuality* (New York: Columbia University Press, 1998), 171.

69   Scott Lauria Morgensen, *Spaces Between Us: Queer Settler Colonialism and Indigenous Decolonization* (Minneapolis and London: University of Minnesota Press, 2011), xii.

70   Lucas Crawford, *Transgender Architectonics: The Shape of Change in Modernist Space* (Farnham: Ashgate, 2015), 129.

71   Rachel Ziady DeLue and James Elkins, eds., *Landscape Theory* (New York and London: Routledge, 2008), 93.

Chapter 13: Is Landscape Collaborative?

Tarna Klitzner

The discourse of collaboration encourages the exchange of ideas, the sharing of values, and the building of trust. The premise is that through shared conversations in the teaching and the practice of landscape architecture, meaningful, layered, and valued landscapes will evolve. Embedded in this encouragement for conversation and collaborative means of working is the recognition that individuals possess knowledge and values that are limited by our experiences and exposures. In the transition from being in the space of the individual to being in the space of the public, the need arises to understand and engage with various interpretations of the public realm and its potential values and opportunities beyond our immediate understanding.

The premise for a successful collaborative process is that the role players "need to be open, accountable, inclusive, transparent, and legitimate."[1] This requires the desire to engage with opinions and ideas that are diverse in a manner that embraces and is inclusive. The legitimacy as a consultant requires a landscape architect to be knowledgeable in the field of landscape architecture and open-minded with respect to our roles and potential impacts on projects.

Ultimately, the nine lessons learned from Favretto's 2021 analysis, "Collaboration and Multi-Stakeholder Engagement in Landscape Governance and Management in Africa: Lessons from Practice," resonate with my experience working within landscapes of collaboration.[2] The analysis comprises case studies that are referred to within my proposed four "Categories of Engagement" that guide the lens through which collaboration is discussed in this essay.

CONSTRUCTING STRATEGIES

The decision to conceptualize, design, and construct interventions through the collaborative lens requires strategies that enable a process that is open, transparent, and inclusive of a wide range of participants. Therefore, it is essential that this process is embraced and actively strategized to ensure that a resilient construct is established in which the participants can further grow the project resolve the runt.

DOI: 10.4324/9781003148142-14

GATHERING CONTEXTUALITIES

This is the "Landing and Grounding" stage of the project in which the contexts are understood. I have found this is a reiterative process—as one engages with various participants, spaces, and technical information, more layers of knowledge are revealed.[3] This requires a willingness to build on what already exists and acknowledge and embrace the role of histories and contexts (social, environmental, and economic). This is also the stage where various participants convey their visions and concerns. My experience as one who lives outside their community has been that this co-design process is an invaluable means for the landscape architect to gain context and knowledge as participants have a specificity of communal insight that I cannot have.

SPECIFIC INTERPRETATIONS

Specific Interpretations fall squarely in the terrain of the landscape architect. This is where the measurable and immeasurable, as revealed through the Gathering of Contextualities, are interpreted and imagined into opportunities that can be developed within the collaboration process, leading to projects that delightfully embrace the communities' visions and hopes.[4] Success requires a willingness and openness to listen to the various informants and a reimagining of the environment in a manner that adds opportunities for new experiences.

STRATEGIC CONSTRUCTIONS

Strategic Constructions lead to a proposal and physical construction strategy that enables communities to develop agency and capacity through the making and maintenance of the projects. This involves the inclusion within design opportunities for community engagement, knowledge transfer, and capacity building.

The collaborative processes I have worked within have varied from individual relationship-based collaborations to community collaborative constructs. They all engage in varying degrees with the four Categories of Engagement listed above to enable the processes.

INDIVIDUAL RELATIONSHIP-BASED COLLABORATION

In retrospect, individual relationship-based collaborations reveal two forms of engagement: "opportunistic" and "embedded." Opportunistic relationship-based collaboration has occurred on projects where individuals within the project recognize the potential advantages of cooperation and engage collaboratively without the broader support of a collaborative project construct.

The embedded relationship-based collaboration is specific to consultant relationships, where trust, respect, and methodology have become part of the practice's relationship over years of working together. This typology has evolved from an initial opportunistic relationship-based collaboration into a formalized, understood way of working together.

Opportunistic collaborations have required landscape architects to identify the opportunities inherent in a project that could enable change and benefit the project's environment. This process usually requires working closely with a member(s) of the Professional/Client/User team and is dependent on openness to engagement and change. An example of an opportunistic collaboration is the Mitchells Plain Hospital Project, where we were able to address the city's water scarcity issues through the promotion of the replacement of conventional piped stormwater systems with sustainable urban drainage systems.[5] Over and above the spatial, environmental, and economic values that are possible with these collaborations, there is value in using these opportunities as pilot projects that create a precedent for future developments.

An example of an embedded relationship-based collaboration is TKLA's professional relationship with CCNIA, an architectural firm based in Cape Town. We have worked together within various consultancy formats since 1995.[6] The collaboration depends on co-design and co-production from the inception of the projects. The trust we have developed over years of working together enables us to listen and explore potentially contradictory scenarios and look for opportunities within them. This has resulted in the siting and positioning of buildings in ways that simultaneously enable the client's needs and enhance landscape relationships. The typology of interventions includes affordable housing propositions, community libraries, educational infrastructure, and residential and commercial landscapes. Working together, we have developed a synergy between the built and landscape typologies that enable positive relationships between the buildings and the surrounding environment.

The development of form and locating of the intervention sits within both practices. This "Specific Interpretation" is a reiterative process that includes drawings, models, sketches, and written thoughts woven together within a continuous discourse. The result is a reciprocal interchange between the built and landscape forms, with building typologies that have an intimate relationship with their environment and landscapes that are held and facilitated by the location of the built form.

Both the individual relationship-based collaborations, opportunistic and embedded, are methodologies that can occur within the community collaborative construct, which is consciously structured to embrace and facilitate collaboration on all levels and within all spheres of the project.

COMMUNITY-BASED COLLABORATION

The Harare Urban Park is an example of how a community collaborative construct facilitated the implementation of landscape design strategies (Figure 13.1).

CONSTRUCTING STRATEGIES

The Violence Prevention Through Urban Upgrading (VPUU) model is a non-profit company funded by the German Development Bank. It was tasked with

Figure 13.1 Design
exploration, Harare Park,
2007

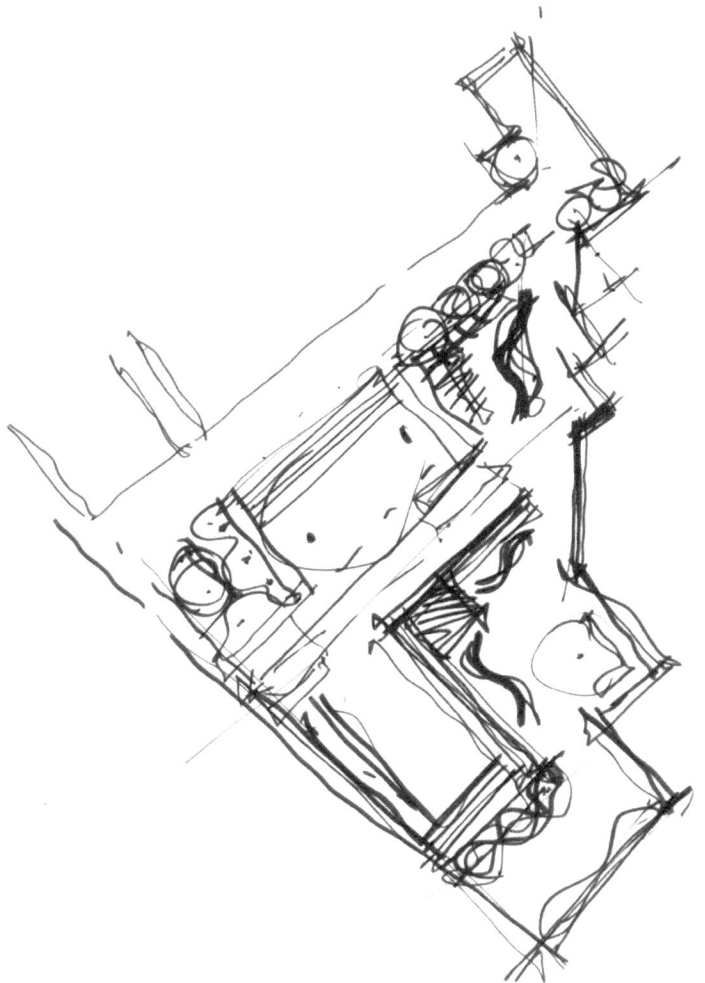

developing a strategy for crime reduction through community participation and
urban upgrading. Their strategy is a bottom-up approach rooted within the communities they are working within. The model they have developed is based on
five spheres of intervention, all of which are founded on four foundational concepts: prevention, cohesion, protection, and research.[7] Through VPUU's baseline
research of Metropolitan Cape Town, they identified the under-resourced town-
ships in Cape Town with the highest crime rates. Khayelitsha was identified as
the township where the violence prevention strategy was initiated.

    The project within Khayelitsha is a partnership between the residents of
Khayelitsha, the COCT (City of Cape Town), and the German Development Bank.
Their primary concern was to assist in crime reduction through the hypothesis
that reclaiming the public domain and positively occupying dangerously labeled
spaces will reduce opportunities for crime.

GATHERING CONTEXTUALITIES

Door-to-door surveys identified the most desired yet perceived dangerous route for pedestrians moving through the community of Harare from the informal settlement in the south to the railway station in the north. This became both the space of intervention and the space of collaboration that landscape architects entered.

Within South Africa, urbanization is situated within a legacy of apartheid planning, which marginalized communities according to race and placed the poorest communities far from the urban center.[8] Khayelitsha is situated within a seasonal wetland coastal plain exposed to a high-water table in winter and salt-laden winds in summer. Engineered during apartheid to house Black urban communities, the Khayelitsha Township was planned with infrastructure and natural systems used as a divisive means of separating and controlling communities. The endemic undulating dune system was bulldozed to create two parallel drainage systems accommodating the one in a 100-year rainfall event. The township is further divided by its road and rail systems and separated from the surrounding communities with undeveloped open spaces used as buffer zones.

The resultant sense of isolation and separation plays down from the scale of the city to the street, where public gathering and socializing spaces were not encouraged or developed under the Apartheid regime (Figures 13.2 and 13.3).

Through interactions with community groups and residents, a route was identified that linked Harare's east and west sections over the 1:100 flood corridor. The chosen route weaves through formal and informal housing fabric. We commenced with an area along this linear system, which had been identified in the community meetings as a potential children's park, sited within the stormwater system, crisscrossed by pathways, and a pedestrian route for scholars and workers accessing the railway station. With little surveillance of the area from the surrounding homes, it had become a dumping ground and an undesirable environment.

When entering this design process within an environment unfamiliar to our experience and reference system, our inquiry as designers is how we access and hear the communities' needs and aspirations to design built interventions that positively contribute to the community.

The VPUU strategy served as a construct we could work within, giving us access to the community and making us accessible to them. This methodology proved beneficial as a landscape architect working within a community with a language I do not speak or understand. Following this approach enabled the sharing of stories and thoughts I would otherwise not have had access to.

SPECIFIC INTERPRETATIONS

The tool used for discussion and design development was a physical model (Figure 13.4). This proved to be an engaging method for the exploration of ideas, as it was possible to locate ourselves within the site, with the three-dimensional

Figure 13.2 Site plan,
Harare Park, 2007

**KHAYELITSHA VIOLENCE
PREVENTION THROUGH URBAN
UPGRADE** (Harare)

DESIGN PROPOSAL(PRECINT 3)

PLAN 1: 500

Drawing no: KH-DP-100 Rev B
Scale 1:100   13 APRIL 2007

KALA - KLITZNER ANDERSON LANDSCAPE ARCHITECTS
PO Box 2607 Clareinch 7740 Tel. 021 671 3987 Fax. 021 671 3989
E-mail. info@mala.co.za

visual representation forming the basis for interactive discussion with community representatives. Structuring elements that informed the design and making of the urban park were as follows (Figure 13.5):

1   The construction of visually clear, unobstructed routes enables pedestrians to choose to enter the park with clear visuals of who else is

Figure 13.3 Sketch plan,
Harare Park, 2007

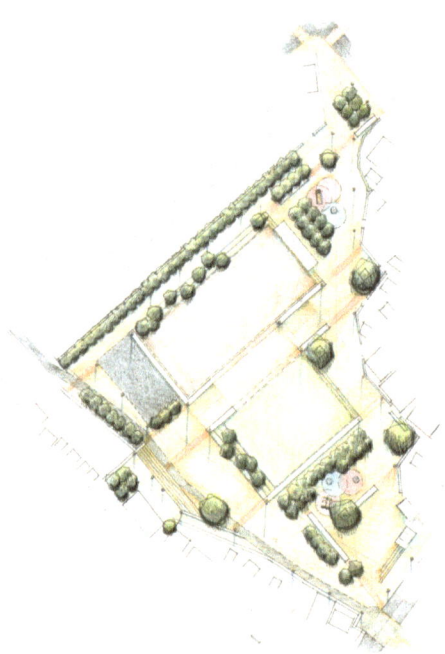

KHAYELITSHA VIOLENCE
PREVENTION THROUGH URBAN
UPGRADE (Harare)

DESIGN PROPOSAL(PRECINT 3)

PLAN 1: 500

Drawing no. KH-DP-100 Rev B
Scale 1:100   13 APRIL 2007

KALA - KLITZNER ANDERSON LANDSCAPE ARCHITECTS
PO Box 2807 Claremont 7740 Tel: 021.671.3987 Fax: 021.671.3989
E-mail: info@kala.co.za

walking on the pathway, allowing them to determine whether they feel safe entering the park.

Initially, trees were not desired as they were perceived as potential hiding places. Fortunately, in one of the workshops, a community member reminisced about the tree groves she grew up with in the rural area of the Eastern Cape. She particularly enjoyed the sound of the birds in the trees and felt that there was a possibility of recreating this

Figure 13.4 Process
model, Harare Park, 2007

Figure 13.5 Diagrams,
Harare Park, 2007. (a)
Multifunctional detention
and play recessed spaces.
(b) Primary circulation
routes. (c) Private and
public boundary seating.
(d) Potential active edge.
(e) Location for children's
playground

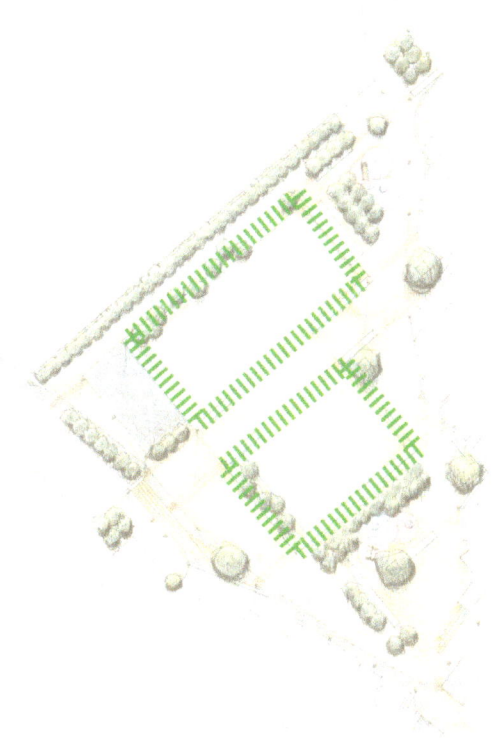

Figure 13.5 (Continued)

Figure 13.5 (Continued)

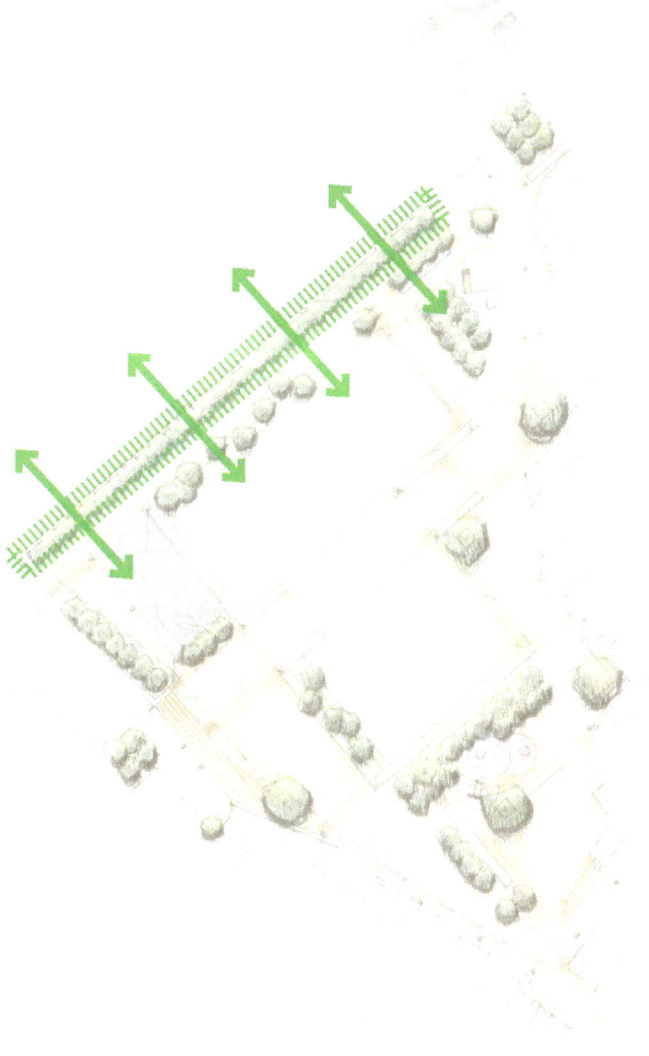

environment in the urban context. This helped ameliorate the fear of trees and informed their placement as lines and groups that are set back from the primary routes to ensure unobstructed visibility along the pathways.

2 The defining of the edge/boundary between the public and private realms: the construction of a low boundary wall defines a clear, recognizable edge between the private and public realms, preventing intrusion into the one and unofficial occupation of the other.

Figure 13.5 (Continued)

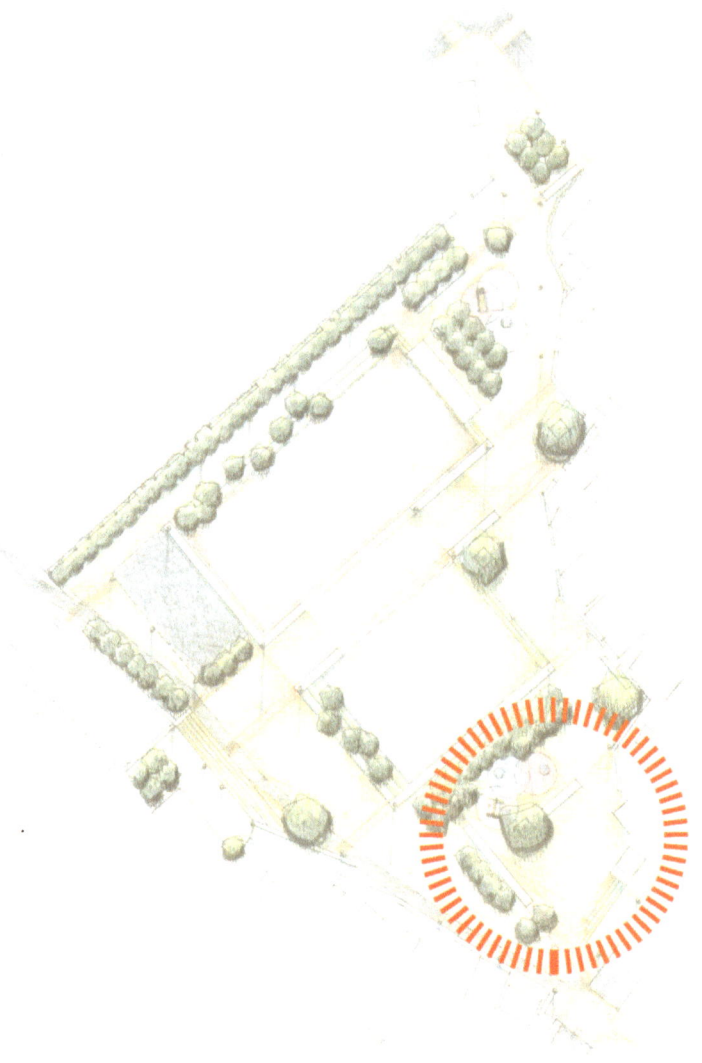

3   The accommodation of stormwater detention areas as dual use play areas gave us the opportunity to construct seating steps, plant rows and groups of trees for shade, and thick walls for trading and seating.

4   The facilitation of passive surveillance as crime prevention is informed by Jane Jacob's theories of positively occupying the public realms.[9] These theories facilitate safety by encouraging and creating opportunities for access from the square to the private properties, with the hope that neighboring homes will extend up to the edge and facilitate communal surveillance of the square.

The location also facilitates Jacob's theories with public buildings built within public spaces. This is a strategy developed by VPUU to ensure that there are eyes on the street. The intention is that these buildings (Active Boxes) are a minimum of three floors high so that they are visible from and have visibility of the public routes. They can differ programmatically between projects; however, they all house community spaces and a flatlet that ensures a 24-hour presence in public spaces.

5   To create safe places for children, unconventional play equipment, created from salvaged alien vegetation, was workshopped with the community crèche principles, who were very positive about the explorative opportunities of the play structures. This resulted in revealing discussions, especially around the location of play areas and trees.

We had initially positioned play areas at both entrances/exits, as informed by Jan Gehl's[10] facial recognition distances, to ensure that children are constantly observed and recognized when at play. During workshops with the community, it was decided that only the play equipment close to the Active Box ensured consistent surveillance of the children.

6   High illumination at night is critical for safety. This is achieved with closely spaced pedestrian-scaled post-top lights, which facilitate well-lit areas for more secure movement. High mast spotlights on the play courts facilitate extended park use in the evening (Figures 13.6–13.8).

The design strategy evolved as a layering of the key informants anchored with the playing court and active box to the south and opening to the north into the pedestrian route linking to the station (Figures 13.9 and 13.10).

Figure 13.6 View across recreational spaces, Harare Park

Figure 13.7 Dense lighting layout, Harare Park

Figure 13.8 Street furniture palette, Harare Park

STRATEGIC CONSTRUCTION

Many voices and facilitators informed and acted upon the making of Harare Park's landscape (Figure 13.11). Through the project's access to the Social Development Fund housed in the COCT, which finances small community-based projects that occurred concurrently with the park's infrastructure contract,

Figure 13.9 Primary
pedestrian route looking
north, Harare Park

Figure 13.10 Primary
pedestrian route looking
south, Harare Park

community participation was enabled in their own urban environment. The ethos
of embedding skills within the community and the application of sustainability
principles within the projects, in combination with the Harare community,
expressed the need for work opportunities informed by the aesthetic, design,
and detailing of the hard and soft landscaping. A local stone mason trained community members to work with the calcrete stone (found under the Cape Flats
sands), which was used to clad the walls and seats. Artists within the community
of Harare were invited to participate in a mosaic training facilitated by Lovell
Freedman and Jane Solomon. The community's artists developed the themes,
designs, and colors for the mosaic. In addition, children from the adjacent
creches made handprints on tiles, which were fired and included in the mosaic

Figure 13.11 Mosaic mural in process, Harare Park

mural. The stone and mosaic work contribute to the character and ambiance of the spaces. This process enabled the production of beautiful, inspiring artworks and cemented bonds between the participants through their collective understanding of the common issues they share (Figure 13.12).

The Harare Park Project was completed in 2010. However, in physical terms, large areas of the urban infrastructure are currently undermined due to factors outside the control of the Khayelitsha Township and the community of Harare. The park supports a busy pedestrian route, with small businesses activating the edges and the formalization of the surrounding housing infrastructure. The endurance of the play areas, mosaic artworks, boundary seat walls, and active pedestrian routes manifest the community's valuing of the intervention. The site's transformation from a space no one wanted to own to a space that fulfills the need within the communities as a safe route and active community space is a manifestation of the appropriateness of the project, which we attribute to the investment in the collaborative strategy applied to the various development stages.

I acknowledge that many landscape architectural projects are not constructed within a collaborative framework. However, it seems to me that even from the smallest opportunistic collaborative engagement, the value gained through sharing knowledge and experiences presents opportunities for interventions that privilege the unique characteristics of place and resonate with Norbert Schultz's "Genius Loci."[11]

Figure 13.12 Site south of Harare Park pre-development

Through the four categories of engagement—constructive strategies, gathering contextualities, specific interpretations, and strategic constructions—I have found an enabling and engaging way of making landscapes. These four stages reveal opportunities for collaboration, which are not immediately apparent at the commencement of the design process. They facilitate the making of interventions that have the potential to resonate with the users of the spaces, which, in my experience, has resulted in interventions that are positively utilized and sustained.

Is Landscape Collaborative? Certainly, my experience has revealed that for landscapes to be relevant and embraced, the process of making these landscape interventions should endeavor to be collaborative.

NOTES

1 Nicola Favretto et al., "Editorial for Special Issue: 'Collaboration and Multi-Stakeholder Engagement in Landscape Governance and Management in Africa: Lessons from Practice,'" *Land* 10 (2021): 285.

2 Nicola Favretto et al., "Editorial for Special Issue: 'Collaboration and Multi-Stakeholder Engagement in Landscape Governance and Management in Africa: Lessons from Practice,'" *Land* 10 (2021): 285.

3 Christophe Girot, "Four Trace Concepts in Landscape Architecture," in *Recovering Landscape: Essays in Contemporary Landscape Architecture*, ed. James Corner (1999). Princeton Architectural Press, ISBN, 1568981791, 9781568981796.

4 James Corner, "Eidetic Operations and New Landscapes," in *Recovering Landscape: Essays in Contemporary Landscape Architecture*, ed. James Corner (1999).

5 Tarna Klitzner, "A Landscape of Healing on the Cape Flats," *African Journal of Landscape Architecture* 2 (2021). Article 07.

6 Charlotte Chamberlain and Nicole Irving, CNIA; www.ccnia.co.za (accessed May 28, 2024).

7 M. G. Krause, C. Shay, D. Cooke, J. Smith, E. Taani, and I. Lange et al., "*Violence Prevention through Urban Upgrading*, VPUU" (May 2015).

8 P. M. L. Anderson and P. J. O'Farrell, An Ecological View of the History of Cape Town, *Ecology and Society* 17, no. 3 (2012).

9 Jane Jacobs, *The Death and Life of Great American Cities* (New York: Random House, 1961).

10 Jan Gehl, *Life Between Buildings Using Public Space*, trans. Jo Koch Van Nostran Reinhold (New York: Van Nostrand Reinhold, 1987).

11 Christian Norberg-Schultz, *Genius Loci: Towards a Phenomenology of Architecture* (London: Academy Editions, 1980).

Chapter 14: Is Landscape Reconciliation?

Ujijji Davis Williams

> By our unpaid labor and suffering, we have earned the right to the soil
> many times over and over, and now we are determined to have it.
>
> —Anonymous, 1861[1]

In the context of land and landscapes, the United States is experiencing a continuance of the Indigenous "Land-Back" movement. This extension of the Indigenous sovereignty movement is refueling conversations around Black American community-building through a lens of self-determination, self-image, and uncovered narrative. This reemerging debate centers around the agency of the American landscape, its ownership and authorship, and equitable access and use. It also begins to define a form of reparations for historic grievances, specifically for Black people in the United States, in an ongoing national debate that considers both financial and structural shifts to respond to significant—generations deep—injustice.

As the United States continues to grapple with its history of oppression and the current misgivings of equality, the impact of racism and its stipulations on land, land use, and environmental response seem deeper rooted than historically appreciated. In fact, these conditions have controlled the very fabric and culture of the American relationship to land, which often excludes or inhibits Black people and other people of color from establishing or maintaining spaces and places of their own image, story, culture, and creation. As we look toward the historical, current, and planned responses to climate change and significant climate-related disasters, Black people and other people of color remain the most vulnerable to damage and the most disadvantaged to recovery—much of which is linked to historic legacies that ordained these groups to live in paths of destruction with limited institutional support.

As designers and land advocates who are committed to the principles of justice, equity, diversity, and inclusion within landscape architecture and design, we must evaluate the way we think about and apply our design processes, especially to the landscape. This moves beyond notions of professional representation into a full internal evaluation of how our design approach and projects reflect justice in the built and natural environment for all peoples. In the context of the American landscape, reparative justice is a necessary additive to design

DOI: 10.4324/9781003148142-15

pedagogy, practice, and purpose, especially given the rising interest in acknowl-edging, memorializing, and reclaiming sites related to Black histories and futures.

Now is the time to redefine our commitment to the American landscape by challenging our societal disposition toward institutions, structures, laws, and responses that disproportionately impact Black people and people of color within both the natural and built environments. Landscape architects can and should be at the forefront of this shift in dialogue and action in a concerted effort to restore the quality and value of the damaged environments that have dispropor-tionally impacted the health, safety, and well-being of marginalized groups in the United States. Moreover, in that same effort, landscape architects should lever-age their design skill sets to elevate important, previously erased narratives that highlight the full truth of events and histories of all people within the American landscape.

THE LANGUAGE OF THE LANDSCAPE

The landscape lends itself to be broad, open-ended, and thus inclusive as a con-ceptual space. However, the landscape also comprises sites with physical bound-aries, specific cultural indicators, and political limits and overlays[2] that impact its condition, use, quality, and value. The interchange of "landscape" and "site" is "an active and strategic agency of culture,"[3] which defines our societal value of places and spaces.

The landscape is the first language, or the first medium, used and manipu-lated to define shapes, forms, patterns, and functions that directly relate to and are influenced by our societal values.[4] The landscape itself is the direct link between people and place, and its manipulation and interpretation can evoke powerful emotions, responses, and representations.[5] Within these constructs, the landscape remains abstract and equally malleable. Yet, with the overlay of the United States race-based and class-driven political, institutional, and societal structure, the landscape is indeed finite and limiting and, in many ways, con-structed within a singular image or a singular set of values. This interpretation exposes conflicting binaries within the American landscape, including power versus powerlessness or a sense of permanence versus temporality.

In his 1963 speech, "Message to the Grassroots," Malcolm X professed, "Land is the basis of all independence. Land is the basis of freedom, justice, and equality,"[6] a statement that remains true to this day. Land ownership (and control) continues to be a vehicle for independence and wealth, and yet as physi-cally tangible as land is, the prospect of landownership has been incredibly elusive for Black people and people of color—who have systematically suc-cumbed to political tools, violence, and other means of land usurpation. Black people and people of color have an enduring relationship with the American landscape, one that is rooted in toil and perpetual landlessness. With that legacy, the language of the American landscape is often missing the vernacular of Black people and other people of color. This omission strengthens the boundaries of otherness, enforces both longstanding and ephemeral political lines, and limits the variety of cultural indicators that could support the notion of a truly diverse and rich society.

In the growing concern of climate change, landlessness is further reinforced as many Black people and people of color live in the paths of natural disasters, a remnant from historic planning, zoning, and other discriminatory tools that maintained racial and class-based separations. With the disproportionate impacts of hurricane damage, flooding, fires, and other environmental surges,[7] in many ways, Black people and people of color have become—or have continued to be—victimized by the country's racialized structure that permeates into the American landscape. This structure maintains social stratification and strengthens the vulnerabilities of these marginalized groups. As a language, the landscape lends itself to being either the tool that continues these disparities—or the tool that challenges them.

The landscape is a physical and emotional construct that has been used as a conduit to reflect the values of our society. Today, those values are changing rapidly. A multifaceted reckoning and reevaluation are needed to repair the disparaging value system of the past and establish one that equitably prepares for the future. Landscape reconciliation, as a concept and strategic base, offers an opportunity to respond to the past and prepare for the future through the landscape. The time is upon all of us to shift away from what has worked for some to understand how we might lift into a new tenet in practice that works for many more people.

LANDSCAPE RECONCILIATION

Ujijji Davis Williams, *Split and Buried*, Landscape Reconciliation Series, 1 of 3, 6″ × 9″, Collage on Paper, 2021 (Figure 14.1).

Ujijji Davis Williams, *Growth*, Landscape Reconciliation Series, 2 of 3, 6″ × 9″, Collage on Paper, 2021 (Figure 14.2).

Ujijji Davis Williams, *Breach,* Landscape Reconciliation Series, 3 of 3, 6″ × 9″, Collage on Paper, 2021 (Figure 14.3).

Landscape reconciliation is a school of thought that begins to marry the effort of reparative justice for Black people and other people of color through critical efforts around landscape stewardship, land justice, and responses to climate change. It pairs the necessity of rectifying historic grievances related to the racialized context of land with the *futuring* of land-based interventions that support environmentally just efforts and proposals that reflect a blend of languages within the landscape through various means of cultural or vernacular interpretation. In short, landscape reconciliation is (or can be) a land-based approach to reparations—an intentional effort to return land to Black people and other people of color who have been victimized by their disenfranchisement to the land through ownership, authorship and other environmentally resilient means of placemaking and placekeeping. The term also begins to define what

Figure 14.1 Landscape reconciliation series, 1 of 3 "Split and Buried"

other roles landscape architects—as premier landscape stewards—can play in repairing these social and environmental relationships in a way that challenges and dismantles the racialized attribution of American landscapes and the responses (or lack thereof) to its ongoing depletion.

Reconciliation, by definition, focuses on relationships. Reconciliation offers an opportunity to repair bonds, settle and resolve issues, and "bring into harmony" by establishing a closeness that may not have been present previously.[8] The definition also implies previous wrongdoing or established distance between groups that need to be rectified. According to the 2015 report of Canada's Truth and Reconciliation Commission—for reconciliation to occur, there must be an "awareness of the past, acknowledgment of the inflicted harm, atonement for the impacts, and action to change behavior."[9] In the context of racism and the American landscape, landscape reconciliation begins to specify which relationships require resolve and how.

Landscape reconciliation is integral to establishing a working framework and is defined in the following:

1   The act of restoring assets and peoples that have been forcibly removed from previously occupied or owned lands through public and private racist policies and practices, ranging from local to federal authority

Figure 14.2 Landscape reconciliation series, 2 of 3 "Growth"

2   The act of restoring the quality and values of lands intentionally or inadvertently polluted or exploited, prioritizing lands that have disproportionately impacted the health and safety of Black people and other people of color

3   The redesign or inclusion of landscapes that elevate the narratives of landscape histories intentionally or inadvertently omitted and erased

Landscape reconciliation is a multifaceted concept, aligning design and activism as tools that use the landscape as a medium. It refers to the "land back" movement and the demand for compensation for those damages that also used the landscape as a medium. A major theme in the history of the United States is the movement and removal of peoples from their heritage lands—beginning with First Nations forcibly and violently transported to arid reservations,[10] continued with the redlining and blockbusting of predominately Black neighborhoods in major metropolitan areas,[11] and recently demonstrated through dramatic demographic shifts due to gentrification and urban land speculation.[12] This cultural pattern of the forced migration of disenfranchised people within the American landscape made way for the prosperity of whites, the efficiency of infrastructure, and the siloing of people away from each other. This pattern, in essence, weaponized land against those who could not exercise power under the law that gov-

Figure 14.3 Landscape
reconciliation series, 3 of 3
"Breach"

erned it. Landscape reconciliation considers the "land back" movement and a systematic end to forced migration as a primary goal, whereas short- and long-term design outcomes should mitigate—if not combat—the impacts of physical, cultural, and economic displacement.

Landscape reconciliation also responds to the impacts of environmental injustice and violence. "Sacrifice zones,"—a term coined by Steve Lerner—refers to low-income and racialized communities "shouldering more than their fair share of environmental harms related to pollution, contamination, toxic waste, and heavy industry."[13] These communities also experience little to no change in their local environmental condition as many responses are executed at a deliberate speed.[14] It appears that our country's institutionalization of racism—through structural oppression, law and policy, and negligence—permeates into our lack of response toward environmental justice and land-based environmental challenges that specifically and disproportionately impact Black people and people of color. This relationship results in the form of environmental violence through the degradation of our native habitats, exacerbated health impacts, depleted resources, land devaluation, and the erasure of lands with important historical contexts and stories. And so, as we look toward our collective responses, it appears that our duty to respond to climate change can be a form of reparations for Black people and other people of color in the United States. Landscape reconciliation focuses on this act of restoration through the landscape to remedy the past environmental-based grievances that were not wholly addressed while simultaneously preparing for a more resilient future.

The 2020 Green New Deal is the first political positioning of its kind that openly acknowledges environmental racism in the United States and the federal government's historic lack of action in environmental stewardship, sustainable practice, and justice, especially for Black people and other diverse communities. The Green New Deal calls on "the federal government to. . . end all forms of oppression"—much of which is driven by land policies and environmental impact.[15] The 2020 Green New Deal proposes a federal investment in policies and projects that will impact land use through building design and construction, systems infrastructure, travel, and food production.[16] While there is no specific law, and the bipartisan government has yet to agree on the proposal's values, the Green New Deal offers a lens to where the United States priorities should be: to effectively respond to climate change, as well as address the growing disparities of environmental injustice and economic participation, especially by race and income.[17] Landscape reconciliation is an effort that must be scaled to achieve a positive impact, mirroring the ways that the historic damage was scaled.

Lastly, landscape reconciliation refers to the elevation of landscape narratives through design or providing the presence of diverse histories within the American landscape. This component of the definition may be the most tangible element as this theme is part of a current conversation within placemaking and placekeeping, especially for sites reflecting Black and Brown heritage, resilience, and suffering. The current theme of belonging within the built environment is an intrinsic value that is resurfacing within the design process, a theme that considers what spaces say, or how spaces feel, across various fronts of people who share different identities, histories, and experiences within the United States. Part of this value is the recovery of lost or erased histories that the American landscape does not reflect, especially for narratives related to the Black contribution and legacy in the United States. While this tenet seems to emerge as a new or transformative idea, it is rooted in core landscape architectural design theory. In his essay, James Corner pontificates that the landscape architecture profession must approach design with a "unity of techne and poiesis"[18]—or a unity of understanding and formation within design. Corner offers that landscape architecture must explore the dimensions of "revelatory knowledge" and "creative symbolic representation"[19] and that these components must refer to each other consistently through the design process and then in the design execution. The outcome allows the practice to reconcile a cluster of binaries, including the historical with the contemporary, the eternal with the moment, and the universal with the specific.[20]

In this construct, and within its definition, landscape reconciliation joins advocacy, activism, and design. It introduces land, its use, value, and condition into the conversation around reparative justice for those disproportionately impacted by land-based strategies. This expands the debate around reparations from solely a monetary response but also an institutional and physical one.

Design has the capacity—and should have the audacity—to tackle these challenges and centralize landscape reconciliation as a core tenet in practice.

THE VALUE OF OWNERSHIP

Landscape reconciliation has an opportunity to uncover and understand the intersection of racism and the implications that "racism culture" has on land or the responses to land-based challenges. As a framework, landscape reconciliation looks to uncover the value of the past to define pathways toward a more inclusive—and less racialized future.

The first theme within the definition framework focuses on the act of restoring people and assets to the land that was unjustly taken by either an act of intimidation or legal overhaul. Given the history previously described, what does it mean to have "ownership" of land? How can we truly empower people through the land? Is authorship of the landscape more powerful than ownership of the land asset itself?

While land ownership has been a tool for wealth and economic freedom, Black people and other people of color have been systematically excluded from long-term land ownership through a myriad of means. This is a recurring theme, not just historically but also within the contemporary examples of land usurpation or land loss. For a city like Detroit, this challenge is still very much present. The history of land here in Detroit is checkered, but today, it reflects a long history and presence of loss: loss of homes, neighbors, people, and sense of place. In 2020, the Detroit Free Press reported that the City of Detroit "overtaxed homeowners by at least $600 million after it failed to accurately bring down property values in the years following the Great Recession [of 2008]."[21]

Between 2010 and 2016, a span of seven years after a debilitating national economic downturn and within the city's declaration of bankruptcy, it was found that Detroit residents (78% Black,[22] $33,965 Median Household Income[23]) were paying nearly twice as much on their property taxes due to the overassessment. City records show that a substantial number of residents have remained in their homes since 2010, indicating that these homeowners endured the "full brunt" of the over-taxation for the entire seven years.[24] Furthermore, a significant portion of overtaxed households are still paying a total of $153 million in back taxes.[25] While many in leadership consider this a terrible oversight, there are limited options and solutions to make these families whole for the money they were required to pay to keep the land and the homes they had.

The city that once boasted the highest Black homeownership and landownership rates in the country also has a strong history of migration of Black people and their removal from the land, considering the origin saga of Black Bottom.[26] Today, Detroit still struggles to activate vacant land and structures outside of its central business districts that were once home to families, churches, and "neighborhood things."[27] Landownership as the medium to own, preserve, and hold the Black experience spatially becomes trickier as the tools for land removal become less overt—even inadvertent—but equally as effective.

Landscape reconciliation places landscape architects in this conversation: land, ownership, taxes, legacy, and tenure. These components contribute to the changing identity of land, its vernacular, and the various histories to respond to or uncover. It becomes a critical attribution to the profession because the American system around the American landscape historically and presently keeps land away from specific groups of people. How can we continue to be land stewards with an ethos of providing access to all peoples when we are aware of—and support—systems that limit access and tenure to the land itself? Our role as landscape architects can expand in a way that requires our society and the people we serve to broaden their scope of impact and inclusion.

Landscape reconciliation should not be confused with the idea of a "right to landscape," a concept often articulated as grounded in "white, propertied citizenship."[28] This notion of a "right to landscape" actively excludes people of color from the conversation and ignores the historical and contemporary contributions to American landscapes. The "right to landscape" is a movement that protests Indigenous sovereignty—and all other nonwhite sovereignty—by centralizing the settler-colonial imperative of erasing and denying Indigenous groups access to land and its control.[29] Thus, the "right to landscape" ethos reinforces norms of propertied citizenship and whiteness[30] and thus threatens the "land-back" values of landscape reconciliation.

PLANNING AS JUST ACTION

Landscape reconciliation establishes another purpose for designers: to incorporate just action as part of their practice. Environmental justice needs to be juxtaposed with social justice in the context of design. It is at this intersection that designers—landscape architects especially—can influence planning justice within the built environment. This is urgent because there is a consistent overlap of poor environmental conditions where Black people and other people of color live due to increasingly profound impacts from pollution, site exploitation, and climate change. If not, we ourselves perpetuate the same violence on the land and the people disproportionately impacted by environmental injustice. As a framework, landscape reconciliation requires layered actions that can help reverse some of these long-term intergenerational damages.

The second theme within the definition framework focuses on the act of restoring the quality and values of lands intentionally or inadvertently polluted or exploited, prioritizing those that have disproportionately impacted Black people and other people of color. What power do designers have to address these types of sites? What power do they have to address the policies that control these sites and safeguard environmentally resilient and healthy futures for Black people and people of color?

Gordon Plaza in New Orleans offers an interesting look at the intersection of design and political action through the lens of environmental justice, resiliency, and futureproofing. For 50 years, from 1909 to 1958, the City of New Orleans used this site to dispose of medical, municipal, and industrial waste.[31] By the 1980s, the City of New Orleans had turned the site into a subdivision for single-family residential housing. At the time, the residents were allegedly not told that the site was previously a landfill, and the school board that built the elementary school in the neighborhood was allegedly never informed of any soil contaminants.[32] Over the course of their tenure, hundreds of residents contracted—and died from—a myriad of terminal diseases, including various cancers, skin conditions, respiratory issues, and other unknown health conditions.[33] When metal drums began literally rising from the earth onto residents' front lawns, the Environmental Protection Agency (EPA) began testing the neighborhood for contaminants.[34] By the early 1990s, Gordon Plaza was placed on the EPA's priority list and thus was declared a Superfund site in 1994.[35] The EPA, Gordon Plaza residents, and other stakeholders led a class action lawsuit to bring justice to the families for the environmental harms of the landfill, but many of the residents received minimal compensation.[36] After the lawsuit, the school and the public housing within Gordon Plaza were closed and evacuated, but the other Black working-class families that invested in homeownership—and land ownership—did not have the means to move or sell their homes for profit.

Landscape architect, professor, and New Orleans resident Diane Jones Allen knew the history of the site and leveraged her position as a landscape architect to mobilize and organize around significant change. She led a field session during the 2016 ASLA National Conference[37] held in New Orleans to expose this injustice with other landscape architects, looking to learn more about the intersections of urban planning, racism, and environmental apathy. This prompted several of her colleagues to write letters to the City of New Orleans to prevent the development and construction of future residential subdivisions on landfills through the city code and develop other local guidelines for site remediation, capping, and site quarantining.[38] In February 2017, Allen and her supporters were successful in their efforts when the City of New Orleans approved the city code amendments, and later in September 2017, when the New Orleans City Council approved the relocation of the residents off the Superfund site.[39] Presently, Diane Jones Allen, founder and principal of DesignJones LLC, is leading the planning and design for a new residential community for the Gordon Plaza residents that relocates families to city-owned land in the Lower Ninth Ward and away from the Superfund site.[40] This will allow these families to stay in New Orleans, to stay in a community together, and alleviate the environmental harms they've been living with since the 1980s. While all of the harm cannot be undone, establishing and protecting an environmentally just future can offset future trauma and risk to the next generation, especially for those already disproportionately impacted.

Considering the range of impact, built environment professionals have a very interesting opportunity here. While we can physically manipulate the landscape, we can also inform policies that dictate how land can and should be manipulated. This task requires us to look at all our landscape systems as one, especially the social and ecological systems. If we can respond to the social impacts related to the environment that disproportionately impacts Black people and people of color negatively, perhaps there is an opportunity to effect change that disproportionately impacts Black people and people of color positively. This uneven response helps to respond to the historic uneven treatment, promoting a true improvement in quality of life for those who have long endured the opposite.

Landscape reconciliation challenges landscape architects to look beyond the surface and learn from the various systems—both historical and contemporary—that impact our relationship and authority over the American landscape. The framework pushes a quality of stewardship beyond immediate care toward one that futures with communities and agencies looking to repair these damages.

REVERSING ERASURE

As landscape architects, our mediums include time and narrative, two planes that can land statically and dynamically within a physical context. Landscape architecture has the ability to weave times of before in a way that makes the past accessible and establishes a clear pathway to the future—toward what's to come ahead of us. This includes—if not requires—a historical understanding of our landscape systems, ecology, land formation, and other critical biological and physiological shifts. This understanding almost inherently includes human intervention, from which social and societal separatism have impacted the value, history, and perception of the land itself.

In this investigation, the process and cycle of landscape erasure are becoming increasingly prevalent. Landscape reconciliation centers the landscape as a catalyst and medium to honor the stories, narratives, and grievances endured by marginalized people whose presence was intentionally eliminated from the experience of the American landscape. The third and final theme within the definition framework focuses on the act of redesigning or including landscapes that elevate the narratives of those landscape histories intentionally or inadvertently omitted and erased. This typically describes undoing the process of "forgetting" and allowing the American landscape to reflect the full truths of its legacy. This requires unearthing the land-based histories of Black people and other people of color.

One piece of history that consistently haunts the American psyche is the gruesome murder of Emmett Till in 1955. Often, when there is an unjust killing of a young black boy in the United States, particularly at the hands of police, local outlets stem the pattern of injustice to this specific incident. Till, a young black boy of 14, was lynched in Mississippi in 1955 after being accused of whistling to a white woman at a small-town convenience store.[41] Several nights after the alleged incident, two adult white men abducted Till from his family's house, tortured and beat him before sinking his body in the Tallahatchie River.[42] Till's body was found a few days later and was returned to Chicago, where he was born.[43] The men accused of the kidnapping and murder were acquitted by an all-white jury, although they later confessed to their crimes proudly in a magazine interview.[44]

One significant theme of Emmett Till's murder is the landscape itself. Upon abduction, Till was taken to several different places throughout the county, a tour of abuse and death. From the time Till was killed to the present, the landscape of Sumner County does not reflect any history of what happened to Till or any other lynching victims. It appears as though the landscape itself also is complicit with what happened. The Emmett Till Memory Project focuses on the preservation of Emmett Till's story through the site, encouraging visitors to understand the severity of this historic incident by traversing the same landscape.[45] There are nine signs scattered around the Mississippi Delta that mark some of the sites of trauma related to the Till case. However, vandals regularly deface the Graball Landing sign, which was Till's body recovery site. The sign has been spray-painted in support of the Ku Klux Klan, upended and thrown into the river, disfigured with acid, and riddled with bullets. These violent acts point toward a violent culture against Black people and people of color, as well as a broad acceptance of that violent culture that has persisted to the present day. Emmett Till's story remains horrifically and ironically relevant to the national conversation about the unlawful killing of Black Americans. The landscape plays a significant role in what happened to Emmett Till. It can also be a canvas that underscores his death's significance as a precursor to racial reconciliation in the South.[46]

In pursuit of this, challenges in what to elevate become of concern. Is site interpretation landscape reconciliation? If we put elements in the landscape that people can read and learn from, does that reconcile the history of what happened in that place? Does it give people—Black people and people of color—a return to access to the land for healing? Does it offer an additional identity within the landscape?

Landscape reconciliation is not meant to be exclusive but part of a more extensive process of acknowledging power and its influence on the American landscape. There is great power in being able to erase, omit, rewrite, or whitewash the contributions and struggles of Black people and people of color, as it might relate to a physical place. Artist and professor Walter Hood defines these conditions as "hybrid landscapes"—landscapes that are already transformed by history and time but present specific themes or objects that can be accentuated within the design process.[47] For Hood, the design practice should identify and elevate these key hybridizing components that can evoke a shared or collective identity and illuminate a new experience.[48] In this effort, there is a quality of accepting the "cultural baggage" of a landscape that challenges the "ostensible past" as we know it.[49] This approach offers landscape architects and designers an opportunity to exhume histories and narratives that were intentionally erased or forgotten—can you bring something forward so that people have to participate with it in the future?[50]

This is part of the healing that landscape reconciliation centers around—understanding why one narrative was elevated over another, who was given the power to erase the other one's narrative, and where it stands today in the American landscape.

AND NOW

Landscape reconciliation is a multifaceted concept, aligning design and activism as tools that use the landscape as a medium. It refers to and builds from the Indigenous "Land Back" movement, demanding the restoration of land that had been taken through legal or discriminatory practice. It responds to the impacts of environmental injustice and violence, looking to rectify disproportionate negative land-based impacts. It also encourages the elevation of landscape narratives through design or providing the presence of diverse histories within the American landscape. Each component is critical in the current conversation of placemaking and placekeeping of the heritage lands of Black and Brown people in the United States.

Designers, especially landscape architects, play a significant role in facilitating this healing by leading the charge toward reconciliation for a healthy, inclusive, and just future. Our environments—the ecological, the built, and the social—require a repair that brings broken systems back together, bridges important connections, and makes those who have lost whole. If we are to be committed to the principles of justice, equity, diversity, and inclusion within architecture and design, it must reflect in the impacts of our work. Now is the time to make new commitments to the American landscape and all its people who have earned their right to the land. We need to consider how the past, present, and future link together within a physical context as we define a free future that encompasses the qualities of equity, equality, and the other components that we believe are necessary to create a sustainable future that includes everyone.

"The end is reconciliation; the end is redemption; the end is the creation of the Beloved Community. . . . It is this type of understanding goodwill that will transform the deep gloom of the old age into the exuberant gladness of the new age,"[51]—*The Beloved Community*, Dr. Martin Luther King Jr. (1956)

NOTES

1   As printed in, Ta-Nehisi Coates, "The Case for Reparations," in *The Guardian* (2014).
2   Michelle Laboy, "Landscape as a Conceptual Space for Architecture: Shifting Theories and Critical Practices," *The Plan Journal* (2016): 77–96.

3 Michelle Laboy, "Landscape as a Conceptual Space for Architecture: Shifting Theories and Critical Practices," *The Plan Journal* (2016): 77–96.

4 Ann Whiston Spirn, "The Language of Landscape," in *Theory in Landscape Architecture: A Reader*, ed. Simon Swaffield (Philadelphia: University of Pennsylvania Press, 2002), 125–130.

5 Ann Whiston Spirn, "The Language of Landscape," in *Theory in Landscape Architecture: A Reader*, ed. Simon Swaffield (Philadelphia: University of Pennsylvania Press, 2002), 125–130.

6 X. Malcolm, "Message to the Grassroots," Speech, Northern Negro Grass Roots Leadership Conference, Detroit, Michigan, November 10, 1963.

7 Ayana Byrd, "STUDY: After Natural Disasters, Whites Accumulate Wealth While People of Color Lose It," Colorlines.com, August 22, 2018, https://www.colorlines.com/articles/study-after-natural-disasters-whites-accumulate-wealth-while-people-color-lose-it#:~:text=STUDY%3A%20After%20Natural%20Disasters%2C%20Whites%20Accumulate%20Wealth%20While,be%20hit%20by%20natural%20disasters%20such%20as%20floods (accessed May 28, 2024).

8 "Reconciliation," in *American Heritage Dictionary*, 4th edition.

9 Joseph J. Z. Weiss, "Challenging Reconciliation: Indeterminacy, Disagreement, and Canada's Indian Residential Schools' Truth and Reconciliation Commission," *International Journal of Canadian Studies* 51 (2015): 27–56.

10 Doug Kiel, "American Expansion Turns to Official Indian Removal," National Parks Service, https://www.nps.gov/articles/american-expansion-turns-to-indian-removal.htm (accessed August 26, 2021).

11 David M. Dworkin, "The history of Redlining," National Housing Conference, April 1, 2019, https://nhc.org/the-history-of-redlining/ (accessed May 28, 2024).

12 "Understanding Gentrification and Displacement," The Uprooted Project, https://sites.utexas.edu/gentrificationproject/understanding-gentrification-and-displacement/ (accessed August 26, 2021).

13 Steve Lerner, *Sacrifice Zones* (Cambridge: The MIT Press, 2010).

14 Steve Lerner, *Sacrifice Zones* (Cambridge: The MIT Press, 2010).

15 Lisa Friedman, "What Is the Green New Deal? A Climate Proposal, Explained" *New York Times*, February 21, 2019, https://www.nytimes.com/2019/02/21/climate/green-new-deal-questions-answers.html (accessed May 28, 2024).

16 Lisa Friedman, "What Is the Green New Deal? A Climate Proposal, Explained" *New York Times*, February 21, 2019, https://www.nytimes.com/2019/02/21/climate/green-new-deal-questions-answers.html (accessed May 28, 2024).

17 Lisa Friedman, "What Is the Green New Deal? A Climate Proposal, Explained" *New York Times*, February 21, 2019, https://www.nytimes.com/2019/02/21/climate/green-new-deal-questions-answers.html (accessed May 28, 2024).

18 James Corner, "Origins in Theory," in *Theory in Landscape Architecture: A Reader*, ed. S. Swaffield (Philadelphia: University of Pennsylvania Press, 2002), 19–20.

19 James Corner, "Origins in Theory," in *Theory in Landscape Architecture: A Reader*, ed. S. Swaffield (Philadelphia: University of Pennsylvania Press, 2002), 19–20.

20 James Corner, "Origins in Theory," in *Theory in Landscape Architecture: A Reader*, ed. S. Swaffield (Philadelphia: University of Pennsylvania Press, 2002), 19–20.

21 Christine MacDonald, "Detroit Homeowners Overtaxed $600 Million," *The Detroit News*, January 9, 2020, https://www.detroitnews.com/story/news/local/detroit-city/housing/2020/01/09/detroit-homeowners-overtaxed-600-million/2698518001/ (accessed May 28, 2024).

22 "QuickFacts: Detroit city, Michigan; Michigan," United States Census Bureau, 2021, https://www.census.gov/quickfacts/fact/table/detroitcitymichigan,MI/PST045219 (accessed August 27, 2021).

23 "Detroit, MI," Data USA, 2021, https://datausa.io/profile/geo/detroit-mi/ (accessed May 28, 2024).

24 Christine MacDonald, "Detroit Homeowners Overtaxed $600 Million," *The Detroit News*, January 9, 2020, https://www.detroitnews.com/story/news/local/detroit-city/housing/2020/01/09/detroit-homeowners-overtaxed-600-million/2698518001/ (accessed May 28, 2024).

25 Christine MacDonald, "Detroit Homeowners Overtaxed $600 Million," *The Detroit News*, January 9, 2020, https://www.detroitnews.com/story/news/local/detroit-city/housing/2020/01/09/detroit-homeowners-overtaxed-600-million/2698518001/ (accessed May 28, 2024).

26 Ujijji Davis, "The Bottom: The Emergence and Erasure of Black American Urban Landscapes," *The Avery Review* 34 (2018); https://averyreview.com/media/pages/issues/34/the-bottom/2eb52833ea-1566362559/davis-the-bottom.pdf (accessed May 28, 2024).

27 Ujijji Davis, "The Bottom: The Emergence and Erasure of Black American Urban Landscapes," *The Avery Review* 34 (2018); https://averyreview.com/media/pages/issues/34/the-bottom/2eb52833ea-1566362559/davis-the-bottom.pdf (accessed May 28, 2024).

28 Dayna Nadine Scott and Adrian A. Smith, "'Sacrifice Zones' in the Green Energy Economy: Toward an Environmental Justice Framework," *McGill Law Journal / Revue de droit de McGill* 62, no 3 (2017): 861–898, https://doi.org/10.7202/1042776ar (accessed May 28, 2024).

29 Dayna Nadine Scott and Adrian A. Smith, "'Sacrifice Zones' in the Green Energy Economy: Toward an Environmental Justice Framework," *McGill Law Journal / Revue de droit de McGill* 62, no 3 (2017): 861–898, https://doi.org/10.7202/1042776ar (accessed May 28, 2024).

30 Dayna Nadine Scott and Adrian A. Smith, "'Sacrifice Zones' in the Green Energy Economy: Toward an Environmental Justice Framework," *McGill Law Journal / Revue de droit de McGill* 62, no 3 (2017): 861–898, https://doi.org/10.7202/1042776ar (accessed May 28, 2024).

31 Lauren Zanolli, "'We're Just Waiting to Die': The Black Residents Living on Top of a Toxic Landfill Site," *The Guardian*, December 11, 2019.

32 Diane Jones Allen, "Black in Design Session 3: Mobilizing and Organizing," filmed October 6–8, 2017 at Harvard University, Cambridge, MA, video, https://www.youtube.com/watch?v=gY0YFHomBr8&list=PLqxr4aBubkPaZrylp5V72VlSc5s2Jv41u&index=7&ab_channel=HarvardGSD (accessed May 28, 2024).

33 Diane Jones Allen, "Black in Design Session 3: Mobilizing and Organizing," filmed October 6–8, 2017 at Harvard University, Cambridge, MA, video, https://www.youtube.com/watch?v=gY0YFHomBr8&list=PLqxr4aBubkPaZrylp5V72VlSc5s2Jv41u&index=7&ab_channel=HarvardGSD (accessed May 28, 2024).

34 Lauren Zanolli, "'We're Just Waiting to Die': The Black Residents Living on Top of a Toxic Landfill Site," *The Guardian*, December 11, 2019.

35 Lauren Zanolli, "'We're Just Waiting to Die': The Black Residents Living on Top of a Toxic Landfill Site," *The Guardian*, December 11, 2019.

36 Diane Jones Allen, "Black in Design Session 3: Mobilizing and Organizing," filmed October 6–8, 2017 at Harvard University, Cambridge, MA, video, https://www.youtube.com/watch?v=gY0YFHomBr8&list=PLqxr4aBubkPaZrylp5V72VlSc5s2Jv41u&index=7&ab_channel=HarvardGSD (accessed May 28, 2024).

37 Diane Jones Allen, "Black in Design Session 3: Mobilizing and Organizing," filmed October 6–8, 2017 at Harvard University, Cambridge, MA, video, https://www.youtube.com/watch?v=gY0YFHomBr8&list=PLqxr4aBubkPaZrylp5V72VlSc5s2Jv41u&index=7&ab_channel=HarvardGSD (accessed May 28, 2024).

38 Diane Jones Allen, "Black in Design Session 3: Mobilizing and Organizing," filmed October 6–8, 2017 at Harvard University, Cambridge, MA, video, https://www.youtube.

com/watch?v=gY0YFHomBr8&list=PLqxr4aBubkPaZrylp5V72VlSc5s2Jv41u&index=7&
ab_channel=HarvardGSD (accessed May 28, 2024).

39 Diane Jones Allen, "Black in Design Session 3: Mobilizing and Organizing," filmed
October 6–8, 2017 at Harvard University, Cambridge, MA, video, https://www.
youtube.com/watch?v=gY0YFHomBr8&list=PLqxr4aBubkPaZrylp5V72VlSc5s2Jv41u&in
dex=7&ab_channel=HarvardGSD (accessed May 28, 2024).

40 Diane Jones Allen, "Black in Design Session 3: Mobilizing and Organizing," filmed
October 6–8, 2017 at Harvard University, Cambridge, MA, video, https://www.
youtube.com/watch?v=gY0YFHomBr8&list=PLqxr4aBubkPaZrylp5V72VlSc5s2Jv41u&in
dex=7&ab_channel=HarvardGSD (accessed May 28, 2024).

41 Dave Tell, *Remembering Emmett Till* (Chicago, IL: University of Chicago Press, 2019),
chapters 2–3.

42 Dave Tell, *Remembering Emmett Till* (Chicago, IL: University of Chicago Press, 2019),
chapters 2–3.

43 Dave Tell, *Remembering Emmett Till* (Chicago, IL: University of Chicago Press, 2019),
chapters 2–3.

44 Dave Tell, *Remembering Emmett Till* (Chicago, IL: University of Chicago Press, 2019),
chapters 2–3.

45 "The Emmett Till Memory Project," https://tillapp.emmett-till.org/.

46 "The Emmett Till Memory Project," https://tillapp.emmett-till.org/.

47 Walter Hood, "When Memory is not Enough," filmed February 23, 2021, at Harvard
University, Cambridge, MA, video, https://www.gsd.harvard.edu/event/walter-hood/
(accessed May 28, 2024).

48 Walter Hood, "When Memory is not Enough," filmed February 23, 2021, at Harvard
University, Cambridge, MA, video, https://www.gsd.harvard.edu/event/walter-hood/
(accessed May 28, 2024).

49 Walter Hood, "When Memory is not Enough," filmed February 23, 2021, at Harvard
University, Cambridge, MA, video, https://www.gsd.harvard.edu/event/walter-hood/
(accessed May 28, 2024).

50 Walter Hood, "When Memory is not Enough," filmed February 23, 2021, at Harvard
University, Cambridge, MA, video, https://www.gsd.harvard.edu/event/walter-hood/
(accessed May 28, 2024).

51 Martin Luther King, Jr., "The Beloved Community," video, https://www.bing.com/
videos/search?q=the+beloved+community+mlk+&view=detail&mid=AC1DE8C78625A
E3527C9AC1DE8C78625AE3527C9&FORM=VIRE (accessed May 28, 2024).

Chapter 15: Is Landscape Insurgency?

António Tomás

The famous battle for Cuito Cuanavale marks the highest point of the military confrontation in which an Angolan army, reinforced by military contingents from Cuba and Soviet advisers, clashed with the regular forces of UNITA, backed by military units from South Africa. The question that has puzzled countless analysts is why Cuito Cuanavale? Why so much destruction in a village so far away and lacking in any strategic importance? In this chapter, I argue that to engage with this question, one needs to situate Cuito Cuanavale in its proper historical and geographical context and how it was created through the tension between the colonial empire and African polities. The point here is not to trace a direct correspondence between war and landscape but to reflect on how specific landscape configurations beget or allow for forms of insurgent violence.

Terras do fim do mundo, or "Lands at the end of the world," is the name given to a long stretch of territory in the southeastern corner of Angola where the borderlands of Angola, Namibia, and Zambia converge. In the late nineteenth and early twentieth centuries, these immense territories were claimed and then divided by the British, the Portuguese, and the Germans. However, colonial efforts to take control didn't begin until the early 1900s. The Portuguese-dominated territory, Colónia Penal do Moxico, where the administration of Moxico commenced, was founded in 1894.[1] Even so, the Portuguese didn't build their southernmost settlement, Moçamedes, until the 1940s. Once this was established, they attempted to subdue the rest of southern Angola.

Central to these processes of occupation is the status of marginality, both literally and figuratively. They have been marginalized from centralized decision-making that occurs in Luanda, and they are also living at the margins of three vast territories. It is important to understand how these territories and their corresponding African polities managed to preserve their independence and political autonomy despite this marginality. Yet, their location, at the node of imperialist desires and, consequently, border formations, also meant that these lands have witnessed extreme forms of violence in both colonial and postcolonial times. The instances of violence that occurred in these places have not been accidental. Violence has historically occurred there because these territories sit at the fringes of empires and nation-states. This point is reminiscent of how geographers have paid attention to the specificity of certain territories, for instance, the

DOI: 10.4324/9781003148142-16

emergence of the practices deemed as "maroonage."[2] In places like the Qui-ilombo of Zumbi dos Palmares, Brazil (and many other places in the United States and Latin America), certain forms of geographical configurations have provided havens for communities of fugitive slaves or recently freed slaves.

With this context established, I will be calling those vast and immense territories that extend across the border between Angola and Namibia "landscapes of war." In doing so, I will provide a reflection on how certain kinds of historical and geographical configurations may induce or allow for forms of insurgency. The main concern is the relationship between war and landscape or war and nature. Even though these issues fall mostly under the scope of military studies, I will be gauging the extent to which these notions are relevant to the social sciences. For instance, more frequently than not, scholarly literature that deals with forms of resistance, such as the Mau Mau in colonial Kenya, disregards how such insurgencies are anchored and rely on geographical conditions. These conditions are often so integral to these insurgencies that without paying close attention to landscape determinants, one would hardly explain their eruption. On these entanglements, it is worth paying attention to what Charles Parrack mused in *Architectures of Borderlands*:

> The aim of these wars is to purify specific territory. Land and space become intensely political. Each square meter becomes worth of fighting over. The dispossessed are at the mercy of those who want to redraw the maps, and to erase the history of the other by demolishing the marks and inscriptions that have been made on the landscape.[3]

However, my intention here is not to postulate a sort of landscape agency in which spatial configurations take precedence over everything else. Instead, I aim to provide a fine analysis of these spaces and the kinds of actions that take place in them. I am also not trying to argue that landscape *is* war per se; my intent, rather, is to locate violence in the specific historical conditions through which these spaces were given political status, either in the context of the empire or the nation-state. More specifically, I will discuss the extent to which territorial regimes at the border and fringes with their remoteness—between national and imperial order—create the conditions for forms of military and militaristic activity.

In dealing with these forms of landscape, the most important thing to bear in mind is how we attempt to understand these problems. In this chapter, I am interested in the history of conquest and particularly in how these territories and their populations have been integrated into the larger political units of an empire or nation-state. As I will demonstrate, a great part of the violence that occurred in these territories was the result of both the imposition of the imperial order upon them and the tensions between the imperial and burgeoning nation-states.[4]

Contemporary scholarship has shown the extent to which landscape is not tantamount to nature and how landscape is, in many cases, a project of political ecology.[5] To put it in slightly different terms, this scholarship has advanced the understanding that landscape is not nature, per se, but produced in different ways. The literature on how fauna and flora are produced for the sake of preservation does not have to be rehearsed here. Parks and other forms of natural reserves played a central role in the ways in which the regions I am concerned with, and which are the topic of this chapter, came under the purview of colonial administrations. Nature was, in fact, the main factor through which these places came to be controlled and their populations divided and compartmentalized. As Giorgio Miescher has shown, the circulation of cattle was instrumental not only in the delimitation of the border between Angola and Namibia but also in the establishment of Namibia's internal border, which cuts the country into two halves and is called the "Red Line."[6] My argument here is that the status of these territories explains, to a great extent, the forms of militarized violence that happened there.

COLONIAL ENCROACHMENTS

When the Portuguese arrived in what today is known as Angola in 1492, their presence was, for the most part, restricted to the coastal territories. This, however, did not prevent them from entertaining productive commercial relations with the people of the hinterlands. Effective domination didn't take shape until the late 1800s, particularly with the foundation of the settlement of Moçamedes, from which the Portuguese planned to conquer the adjacent hinterlands. On the side of Namibia, things were no different. Northern Namibia was known by Boers settlers from their arrival and the foundation of the Cape Colony in the 1600s, but only later, through Germany, was the territory given international legitimacy. According to Wallace, this late incorporation was due to the desire of the various groups that occupied the region to maintain independence, their capacity to engage in long trade operations, and the paucity of the land they occupied.[7]

The Scramble for Africa changed the nature of these territories as a massive horde of European operators, from soldiers to missionaries, scientists to merchants, gold-diggers and other metal prospectors to agriculturalists, flowed into the continent. However, the European dynamics that preceded the First World War put the last nail in the coffin of the freedom of the peoples who inhabited these areas. Negotiations for the demarcation of the boundary between Angola and Namibia took place around this time. Both sides agreed with the notion that the border should be established using the natural reference point of the Cunene River. However, tracing the border from its mouth, in the Atlantic Ocean, up north in the Angola territory was impossible because the river runs almost parallel to the ocean through the middle of the Angola territory. So, there was a need to trace a line from the Ruacana Falls to Cubango River at the same parallel. However, the precise boundary between Angola and Namibia oscillated whimsically throughout decades, although both parties agreed on the formation of a neutral zone, a triangle of 400 km by 11 km running from south to north (Figure 15.1). This was the zone over which none of the colonial powers would

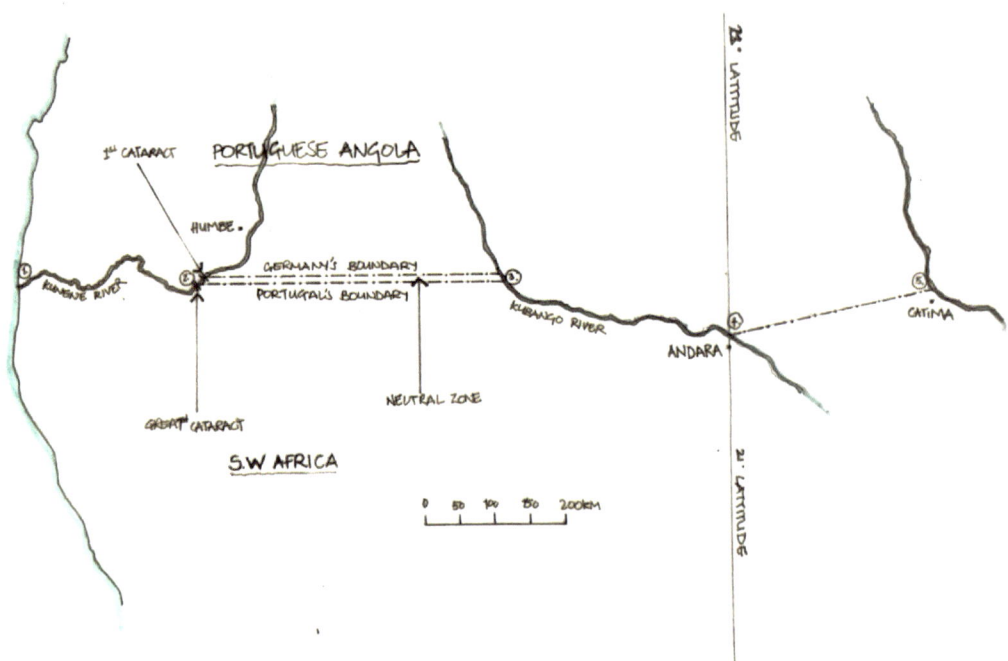

Figure 15.1 Border agreement between Portugal and Germany. Courtesy: P. Bandora, 2023.

effectively control. Or at least, they would not until the Portuguese built fortifications along the neutral zone (Figures 15.2 and 15.3). This provocative act attracted the fury of the Germans and came to be the reason behind the battle of Naulila, in which the Germans inflicted a severe defeat on the Portuguese. To avenge this humiliation, the Portuguese government dispatched a 10,000-strong force led by General Pereira d'Eça to southern Angola. He confronted and defeated the beleaguered German forces in the region and, in the process, founded a few more settlements in southern Angola, such as the one named after him, Pereira d'Eça, in 1915. Subsequently, by confronting and subduing warring African polities, like the Kwanyama, d'Eça also managed to expand colonial domination over those territories.

Nevertheless, the settlements advanced slowly, without event, from the coast to the eastern border of Angola. The distance was significant, as Ricardo Roque explains, "communications between the interior, the capital city, and the coastline were slow and erratic, dependent upon moving large groups of people, supplies, and animals across dense jungle and treacherous rivers."[8] Things only started to pick up with the construction of the Benguela Railway, which, cutting the country practically in half, was thought to provide British-dominated Rhodesia with access to the sea. It was only in 1890 that Portugal sent the party, led by Captain Trigo Texeira, to whom the occupation of Moxico is singlehandedly attributed.[9] A Penal and Agricultural Colony was established in 1894, which was later transformed into the administration of Moxico.[10] Cuando Cubango was only founded in the 1930s. Despite these administrative advances, for the better

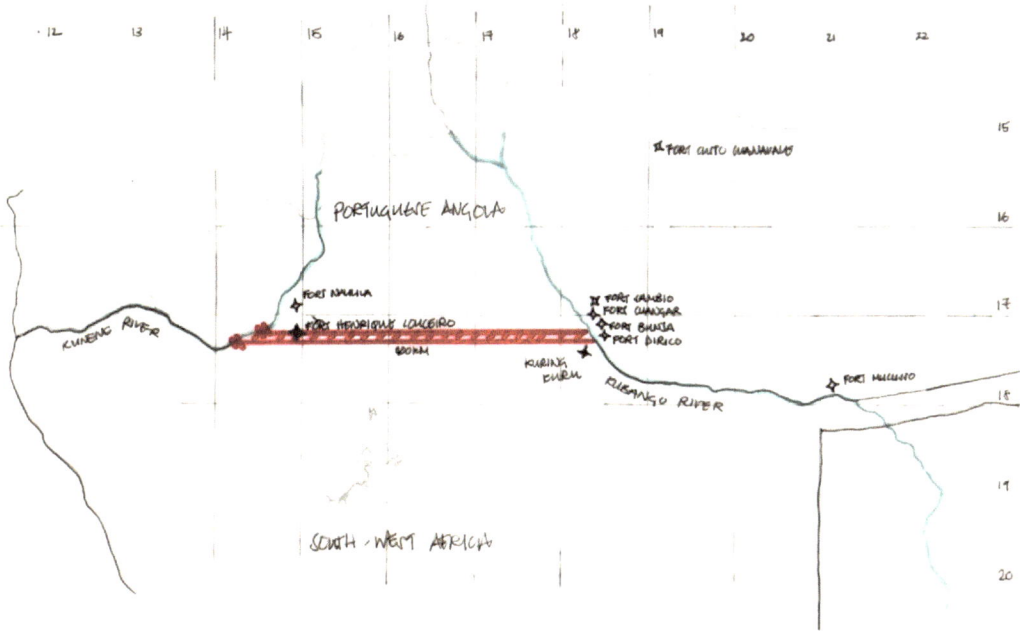

Figure 15.2 Portugal fortifications along the neutral zone. Courtesy: P. Bandora, 2023.

part of the 1900s, these territories south of the railway could not shake off the nickname of *Terras do fim do mundo*. The territory between the railway and the border with Namibia was the site of the military confrontations I will be discussing in the next sections (Figures 15.4 and 15.5).

THE FORMATION OF THE WAR ZONE

After almost four decades with little or almost no conflict, violence erupted in the 1960s. In 1961, nationalists from the country's north started the anti-colonial war to liberate Angola. From then on, both Congo and Zaire supported two different contenders in the struggle for the liberation of Angola: the FNLA and the MPLA. Since the FNLA was the most well-organized formation in the north of the country, the MPLA, to survive, was left with little choice but to move to the east by the border with Zambia. Since John Marcum's book on nationalism in Angola, it has been a staple of interpretations that the anti-colonial war had its ethnic roots and motivations. However, seen from a purely geographic point of view, this argument does not make a great deal of sense. While the FNLA was fighting in the north, in the part of the country where they had their sociological base, the MPLA, and UNITA were fighting in the east, a hugely underpopulated part of the country where they were deprived of their bases (Figure 15.6).

The Civil War in Angola is said to have had a purely ethno-linguistic character in that the three national liberation movements were heavily associated with ethnic groups or the lack thereof.[11] FNLA was linked with the Bakongo, the

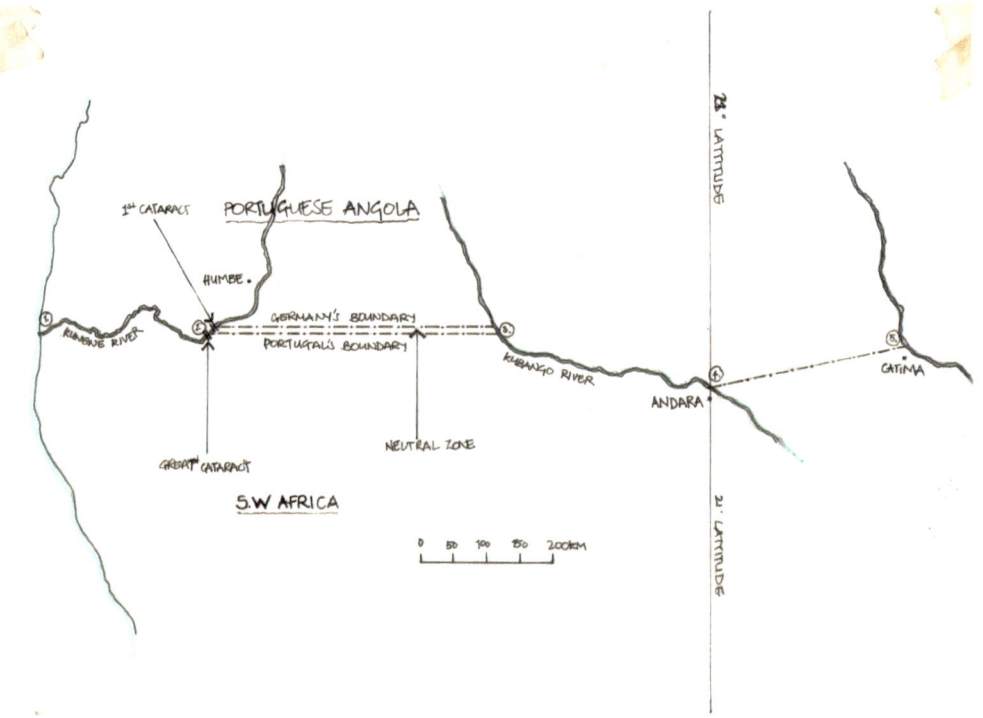

Figure 15.3 Border agreement between Portugal and Germany in the 1890s. Courtesy: Google Earth.

MPLA with the Mbundu, and UNITA with the Ovimbundu. However, a strong reaction to this view has emphasized the more geo-political aspect of the problem by ascribing the conflict to geographical implications, in which FNLA is associated with the north, the MPLA with the center, and UNITA with the center-south. However, there was a geographical component to the war that played out in the ways in which geopolitics was being determined in the region.

If taken in relation to the theory of counterinsurgency as expounded by Mao and many others, none of this makes a great deal of sense. The existence of the guerrilla has always been predicated on the support of the population, and, as Mao famously said, the guerrilla needs the population as the fish needs water.[12] However, even though UNITA was formed by the Ovimbundu and Bailundo component of the Angolan population, it rarely operated in that part of the country, except perhaps in the period after the Carnation Revolution when it was allowed to operate in Huambo. During the Civil War, UNITA occupied a region of the country, Jamba, which was, for the most part, sparsely populated. However, from the border point of view, such a placement made a lot of sense. South African Defence Force had just established their HQ on the Namibian side of the border, as Jannie Geldenhuys expounds:

> Another worrisome aspect was the exclusivity of the group of staff officers in Pretoria who controlled the operation. During October–November 1975, I argued strongly that we should implement the normal command and control structures for conducting the

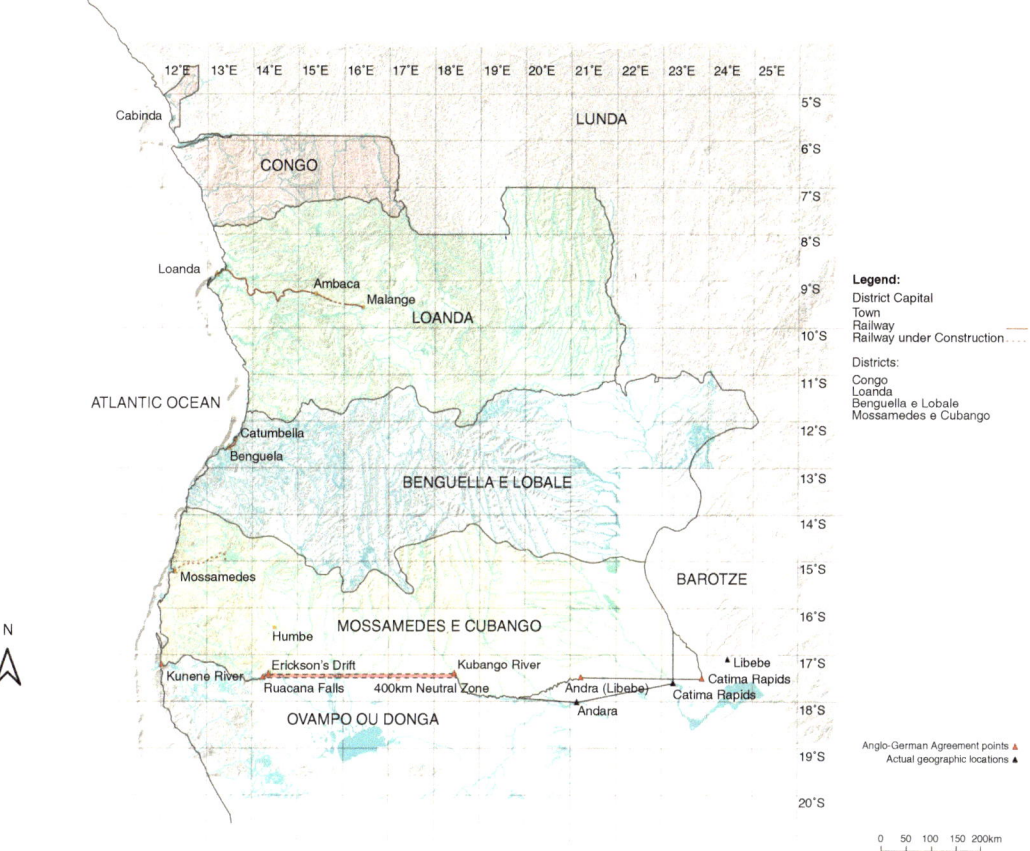

Figure 15.4 Angola border formation in the 1890s. Courtesy: P. Bandora, 2023.

operations, and revert to the more formal procedures. General Malan supported me. This led to the establishment of HQ 101 task force at Rundu, with General André Van Deventer taking over from Dawie Schooman and assuming responsibility for Operation Savannah.[13]

It was not only legitimate political grievances that sustained UNITA's resistance but also the particular configurations of the terrain UNITA's men came to occupy. UNITA constituted itself on the country's border, where it created its headquarters, and it was a few kilometers from Rundu, on the other side of the border, where the SADF had its military bases.

Whereas the MPLA was pushed to the southeastern part of the country because of the difficulty of waging war from either the center or the north of the country, the case of UNITA was rather different. UNITA, in fact, only came into being because of the entanglement between insurgency and nature. Minter has shown the extent to which the first sponsors of UNITA were wood prospectors in the southeast part of Angola who wanted to prevent the MPLA from accessing the south. From their first operations in Kasamba, Vila Texeira de Sousa, UNITA

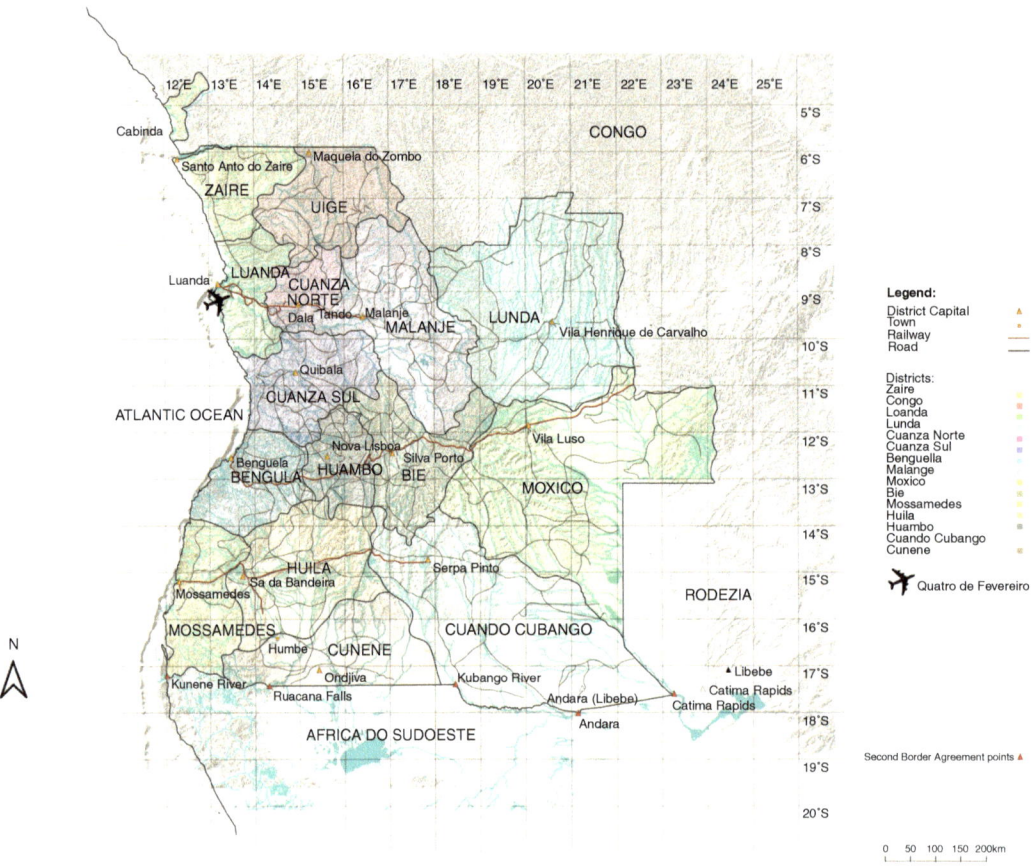

Figure 15.5 Angola border formation in the 1930s. Courtesy: P. Bandora, 2023.

would, for the most part, show a stubborn tendency to operate to the south of the Benguela Railway, in the *Terras do fim do mundo*.[14] This made the work of Portuguese insurgency easier, and in that part of the country, some of the harshest counterinsurgency strategies were implemented, almost wiping away the MPLA military wing (Figure 15.7).

ESCALATION ON THE BORDER

When the anti-colonial struggle started in Angola, South Africa did not concern itself with insurgency in the neighboring country. But the security of the border began to quickly deteriorate with the formation of the South West Africa People's Organization (SWAPO) and its military wing, PLAN, in 1966, which vowed to fight for Namibia's independence. While the Portuguese were fighting a war on the northern side of the border, the South Africans were gearing up for the same in the South. Since Portugal and South Africa both upheld the hegemony of minority settlers' communities in the region, they found ways to collaborate. This effort was named the "Alcora" project and consisted of a highly secretive

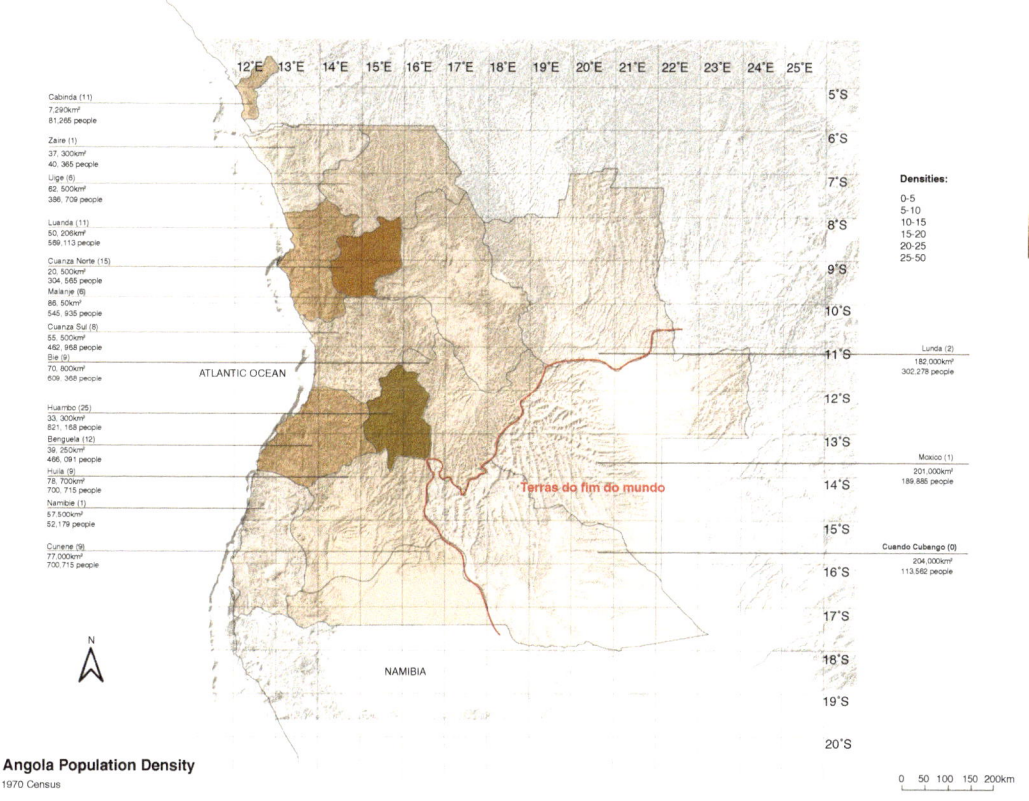

Angola Population Density
1970 Census

Figure 15.6 Population density in Angola during the 1960s. Courtesy: P. Bandora, 2023.

plan formulated by South Africa and Portugal to fight both insurgencies around the borders and in the borderlands between Angola and Namibia.[15] Through the Alcora project, the South African army was permitted to enter Angolan territory whenever it was necessary to pursue Namibian freedom fighters. In exchange, the Portuguese received military, economic, and political aid to fight their own insurgents.

All of this seemed unproblematic until 1975 when the military actions took place under the auspices of two neighboring empires. However, everything changed in 1975 when Angola became independent and, as a nation-state, eager to protect its sovereignty. According to Fred Bridgeland, the first to break the news, contingents of the South African army entered Angolan territory even before the proclamation of independence in 1975. Attempting to take advantage of the chaos that followed the Carnation Revolution and the subsequent erosion of the Portuguese empire, the South African troops moved from the border northbound to link up with the FNLA, reinforced by Zairian troops, whose goal was to prevent the MPLA from proclaiming independence. Cuba came to the MPLA's rescue, and South Africa was forced to leave the Angolan territory in early 1976. In the process—and taking advantage of the fragmentation of the FNLA's army—South Africa recruited hundreds of operatives and their families,

Distances

The Battle of Cuito Cuanavale

Figure 15.7 SADF headquarters established in Angola, Namibia, and South Africa. P. Bandora, 2023.

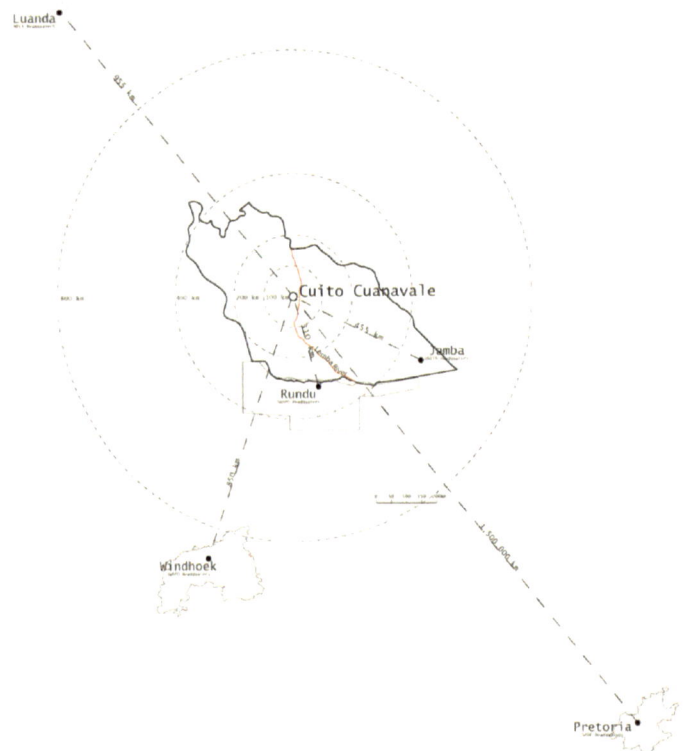

which were then used as the backbone for the formation of one of the most singular and resourceful fighting units on the continent: the 32 Battalion.

The independence of Angola had fundamental consequences for the dominance of South Africa over Namibia. As the SADF Chief of Staff, Jannie Geldenhuys, later mused in his memoirs, "SWAPO was now potentially in a stronger military position than before, because for the first time it had obtained what is more or less a prerequisite for a successful insurgent campaigning, namely the border that provided safe refuge."[16] On the way, the whole line separating Angola from Namibia became the battleground in which issues of independence, democracy, and sovereignty would play out quite dramatically.

Having an enemy on the other side of the border supporting an insurgency in the territory they dominated could only lead South Africa to collaborate with Jonas Savimbi's UNITA. UNITA was expelled from Luanda in July 1975 and, along-side FNLA, proclaimed independence in Huambo. It controlled the city until March 1976, when combined forces of the recently created Angolan army and the Cubans drove them away. UNITA suffered enormously, losing men and equip-ment, and for years, its members roamed the country in a fight for survival. They had been almost obliterated by the time the *Washington Post* journalist Leon Dash found them. He spent months traveling thousands of kilometers with the guerrillas. "North of Kavango River and western Caprivi, in the Cuando Cubango province of Angola, Jonas Savimbi survived, got back on his feet, was able to build up his forces and begin to control the area."[17] So now, along the border, there were not only two organized armies, the Angolan and the South African, fighting each other, but also two insurgent groups fighting for their own objec-tives, namely Angola's UNITA and Namibia's PLAN. Since UNITA and SWAPO had become "arch-enemies" and were fighting in Ovamboland for the right to uphold their safe zones in the fights against Angola and South Africa, the imme-diate result of this was the "termination of infiltration from Angola to Kavango and western Caprivi."[18] Geldenhuys emphatically explains this imbroglio:

> From SWAPO's point of view the new situation also held certain disadvantages, one being that [Sam] Nujoma's previous ally, Jonas Savimbi, was now his greatest enemy. This created a problem for Nujoma since they both operated in Southern Angola. The MPLA also required SWAPO to make manpower available for their campaign against Savimbi. Savimbi, however, was retreating to the remote expanses of the southeast because he was fighting for the very existence of his movement.[19]

In the next few years, the South Africans conducted countless operations to clear the area of insurgent and nationalist activities in what was perhaps one of the longest battlegrounds in military history. Early recruiters of the South African army simply called this area the Cutline.[20] Geldenhuys surmises in this battleground, "the vastness of the combat zone,"[21] 1,400 km of the border between Angola and Namibia, from the head of the Cunene River in northeast Namibia and south-east Angola to the Caprivi region, and stretching even beyond that: "The distance between Bloemfontein in South Africa and Cela in Angola is 2,386 (1,205 nautical miles), and the distance to Cassinga, which two years later became the main object of Operation Reindeer—is 1,928 km (1,038 nautical miles)."[22] For him, the

> operational area literally stretched from the Atlantic Ocean in the west to Mpalela in the east (the point where four countries met–South West [Africa], Botswana, Zambia and

Zimbabwe) and from the Red Line in the south, the foot-and-mouth cattle disease boundary, to an east-west line approximately 45 km north of the border.[23]

On top of this difficulty was also the variety and diversity of the terrain itself:

> The zone included regions abounding with trees, such as Caprivi; while other areas, such as the southern parts of Angola, consisted of bushveld. There were areas with large rivers, such as Kavango, while others were flat and empty, for example Ovamboland, and areas comprising savannah-type grasslands. Yet other regions like Kaokoland consisted of high mountains and deep valleys. And added complication was that the population differed from region to region.[24]

This zone, the combat zone, becomes even more problematic if one comes to terms with how Namibia came into being and how it was managed by South Africa. In fact, very little has changed in how South Africa differentiated between the zones for the natives and for the white settlement, the so-called Police Zone. So the combat zone was, in fact, all the area above the Red Line, "the boundary set up in the north to control foot-and-mouth disease, which was much feared in cattle-owning South West Africa. This meant that a number of northern areas were excluded from our purview, including all the future hot spots of the 'border war': Kaokoland, Ovambo, the Kavango, western and eastern Caprivi and Bush-manland,"[25] which, from 1973, was under the protection of the SADF.[26]

During the first years of Angola's independence, the country was, for the most part, divided along the line of the Benguela Railway. Very little in terms of development took place south of the railway, and this dramatized the name the Portuguese gave to this region: "Lands of the end of the earth." However, in the 1980s, the military situation started to change with the increase in the military presence of the better-equipped Angolan army on the northern side of the border. Cuban troops and Soviet advisers were instrumental in this escalation since they used the most advanced and sophisticated military equipment, par-ticularly Soviet-built aircraft, which quickly started to challenge South African aerial superiority in the region.

It was in this optimistic context, based on the difficulty South Africans were experiencing—particularly since the arrival of the Angolans of sophisticated mili-tary equipment from the Soviet Union, that the Angolan government-initiated preparations for the "Salute October" operation, whose main objective was to wipe out UNITA guerrilla from the south of the country. The operation consisted of concentrating military means in the small locality of Cuito Cuanavale and, from there, crossing the Lomba River and advancing to Mavinga, which was the last fortification before reaching Jamba and was where UNITA had concentrated their forces and built their headquarters (Figure 15.8).

The SADF had little choice but to intervene directly in the conflict, fearing that such a possibility would cause devastating harm to UNITA and provide SWAPO with control of the entire borderland for their activities. UNITA and South African forces managed to destroy one of the Angolan brigades, the 47,

Scar(iver)s in the Landscape
The Battle of Cuito Cuanavale - Operation Moduler

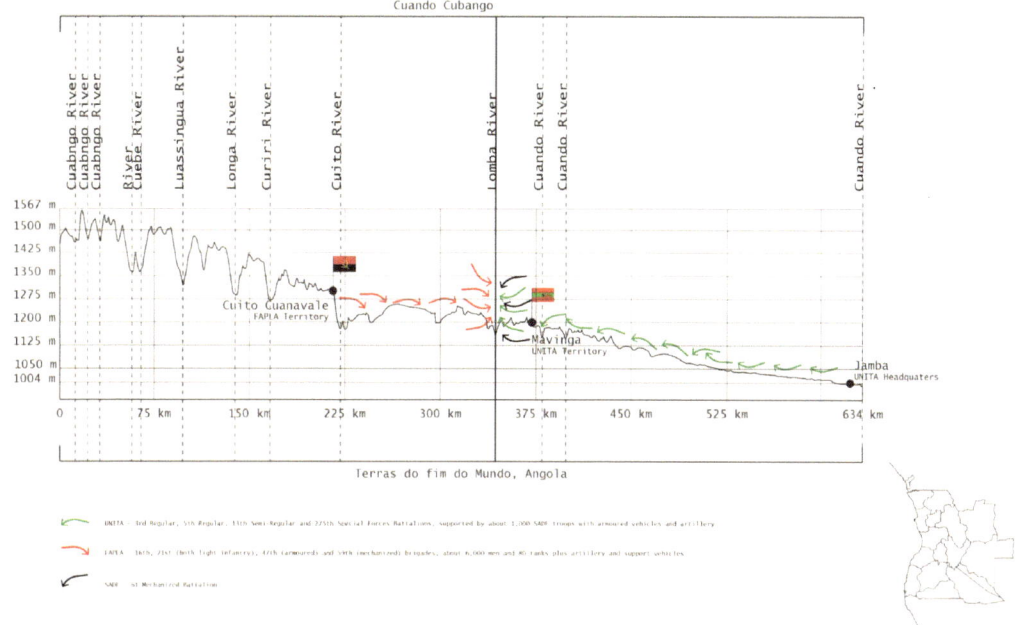

Figure 15.8 Elevation of Cuando Cubango showing MPLA forces crossing Lomba River. Courtesy: P. Bandora, 2023.

and heavily damaged other brigades, including the 16. What came of this is very confusing. South Africans claim that they did not have any desire, or were even equipped, to overtake the locality of Cuito Cuanavale. However, they lost at least three battles in the marshes that gave direct access to Cuito Cuanavale, the Tumpo Triangle, before giving up and moving their units back to the border (Figure 15.9).

THE LAST CHAPTER

The same landscape would also play a crucial role in the last chapter of the war. In 1991, the peace agreement between the Angolan government and the military opposition was signed in Portugal. Both parties agreed to take part in general elections to be convened the next year. Elections took place, and the ruling party won. UNITA refused to accept the results, and Savimbi abandoned the peace negotiations and moved UNITA's general quarters to Huambo, the exact same place where UNITA had proclaimed independence in 1975. Even though UNITA managed to force the government troops to evacuate, it was not long before the government forces reoccupied the city. UNITA then moved to

Figure 15.9 The Trumpo triangle. Courtesy: Google Earth.

Kuito, Bié, remaking the old colonial geography by posting their troops along the line of the Benguela Railway, the line that divides the country into two.

In December 1999, the government sent their troops to take over Huambo, and UNITA had to withdraw. But this time, the conditions were far from similar. Namibia was independent and would not side with the guerrillas, particularly guerrillas who were on the side of Apartheid South Africa when independence was being fought. As such, UNITA's forces were not only deprived of the support they had been given during the civil war, but they were also unable to reach the border, an essential aspect of any civil war. For three of four years, they roamed the *Terras do fim do mundo* until Jonas Savimbi was killed in February 2002.

As such, the Angolan government implemented the same harsh counterinsurgency policies the MPLA faced during the anti-colonial war, which they could not deploy when they fought the guerrilla because of the South African intervention. The landscape played a crucial role in these operations.

If guerrillas live off the population, as Mao would say, the first thing the government needs to do is to isolate the guerrillas from the population. The fact that these territories were not densely populated only helped in this approach. The government could easily transport the population out of these areas, which counterinsurgency manuals call "resettlement." But the government went even further in that it destroyed all the culture—particularly the cassava—the guerrillas could use to survive. So, after losing the possibility of taking the anti-colonial war to the most populated areas, it was no surprise that the MPLA was left with no other choice but to recede to within the country a little above the north of the area being discussed. They were decimated, and the way they were fought was through the destruction of the cultures, particularly the cassava, to the extent that the Portuguese army almost defeated the guerrillas during the times of the anti-colonial war.

The forces of the Angolan government were organized in combat units that clashed with the units of UNITA for years until the so-called presidential column was so depleted in terms of resources, strength, and so on that by February 2002, when Savimbi was killed, he was left with only a few dozen soldiers. That was the end of the civil war in Angola and also the end of the use of this region for subversion and insurgency.

CONCLUSION

In this chapter, I discussed the relationship between landscape and war by focusing on a very particular kind of terrain: the so-called *Terras do fim do mundo*. Whereas there is a great deal of commentary on Cuito Cuanavale and the role this place played in the end of the civil war in Angola, the particularities of the terrain have never been taken into serious consideration. This does not mean that geographical incidents are not taken seriously. Rather, the contrary: landscape, or fieldcraft, has been central to the conduct of the war.

What I have tried to do here is to position this place in its particular historical dimension. These hinterlands were prone to war and conflict because of how they were constituted throughout history, particularly because of how imperial formations, and particularly the tension between the empire and the nation-state, have played out in their constitution. In this way, it was only at the end of South Africa's imperial dreams or its conversion to a traditional nation-state that everything went back to normal.

NOTES

1 Ricardo Roque, "The Razor's Edge: Portuguese Imperial Vulnerability in Colonial Moxico, Angola," *The International Journal of African Historical Studies* 36, no. 1 (2003): 105–124.

2 Robbie Shilliam, "Colonial Architecture or Relatable Hinterlands? Locke, Nandy, Fanon, and the Bandung Spirit," *Constellations* 23, no. 3 (2016): 465.

3 Charles Parrack, "The Territory of Hatred and Compassion," in *Architectures of the Boderland*, ed. Teddy Cruz and Anne Boddington (Chichester, West Sussex: Academy Editions, 1999), 10.

4 Coen makes a very similar argument. There is something about the interest of the empire in the ecological that the nation-state cannot emulate. See, Coen, "Seeing Planetary Change."

5 Bridget L. Guarasci, "The National Park: Reviving Eden in Iraq's Marshes," *The Arab Studies Journal* 23, no. 1 (2015): 128–153.

6 Giorgio Miescher, *Namibia's Red Line: The History of a Veterinary and Settlement Border* (New York: Palgrave Macmillan, 2012).

7 Marion Wallace, *History of Namibia: From the Beginning to 1990* (Oxford: Oxford University Press, 2014).

8 Ricardo Roque, "The Razor's Edge: Portuguese Imperial Vulnerability in Colonial Moxico, Angola," *The International Journal of African Historical Studies* 36, no. 1 (2003): 107.

9   Ricardo Roque, "The Razor's Edge: Portuguese Imperial Vulnerability in Colonial Moxico, Angola," *The International Journal of African Historical Studies* 36, no. 1 (2003): 107.

10 Abolished in 1901, according to Ricardo Roque, "The Razor's Edge: Portuguese Imperial Vulnerability in Colonial Moxico, Angola," *The International Journal of African Historical Studies* 36, no. 1 (2003): 113.

11 John A. Marcum, *The Angolan Revolution: The Anatomy of an Explosion (1950–1962)* No. 15 (Cambridge: MIT Press, 1969).

12 John S. Pustay, *Counterinsurgency Warfare* (New York: Free Press, 1965).

13 Jannie Geldenhuys, *At the Front: A General's Account of South Africa's Border War* (Jeppestown: Jonathan Ball Publishers, 2012), 55.

14 William Minter, ed., *Operation Timber: Pages from the Savimbi Dossier* (Trenton, New Jersey: Africa World Press, 1988).

15 Filipe Ribeiro de Meneses and Robert McNamara, "The Origins of Exercise ALCORA, 1960–1971," *The International History Review* 35, no. 5 (2013): 1113–1134.

16 Jannie Geldenhuys, *At the Front: A General's Account of South Africa's Border War* (Jeppestown: Jonathan Ball Publishers, 2012), 62.

17 Jannie Geldenhuys, *At the Front: A General's Account of South Africa's Border War* (Jeppestown: Jonathan Ball Publishers, 2012), 132.

18 Jannie Geldenhuys, *At the Front: A General's Account of South Africa's Border War* (Jeppestown: Jonathan Ball Publishers, 2012), 132.

19 Jannie Geldenhuys, *At the Front: A General's Account of South Africa's Border War* (Jeppestown: Jonathan Ball Publishers, 2012), 62.

20 Steve Joubert, *Gunship Over Angola: The Story of a Maverick Pilot* (Jeppestown: Jonathan Ball Publishers, 2019).

21 Jannie Geldenhuys, *At the Front: A General's Account of South Africa's Border War* (Jeppestown: Jonathan Ball Publishers, 2012), 55.

22 Jannie Geldenhuys, *At the Front: A General's Account of South Africa's Border War* (Jeppestown: Jonathan Ball Publishers, 2012), 55.

23 Jannie Geldenhuys, *At the Front: A General's Account of South Africa's Border War* (Jeppestown: Jonathan Ball Publishers, 2012), 94–95.

24 Jannie Geldenhuys, *At the Front: A General's Account of South Africa's Border War* (Jeppestown: Jonathan Ball Publishers, 2012), 95.

25 Jannie Geldenhuys, *At the Front: A General's Account of South Africa's Border War* (Jeppestown: Jonathan Ball Publishers, 2012), 41.

26 Jannie Geldenhuys, *At the Front: A General's Account of South Africa's Border War* (Jeppestown: Jonathan Ball Publishers, 2012), 45.

Chapter 16: Is Landscape Haunted?

Edward Eigen

> The truth, the whole truth, and nothing but the truth is what is wanted
> in these days, when nearly everybody is ready to admit that there are,
> doubtless, *more things in heaven and earth than are dreamed of in his
> philosophy* [emphasis added].[1]
>
> —The Boston Globe (1891)

ENTER GHOST *AND* HAMLET [*AND* OLMSTED]

What was Frederick Law Olmsted doing telling ghost stories to William James in the summer of 1891? Spoiler Alert: There is some disturbed and indeed spoiled ground to cover before hearing Olmsted's tale. The most straightforward answer to the above question, though even it includes a bit of historical conjecture, might be that he was responding to James's own notice, placed in journals and newspapers seeking participants for a scientifically conceived "census of hallucinations." Running in the *Boston Globe* under the headline "Do You Hear Voices?," the notice explained the work of the Society for Psychical Research (SPR), dedicated, as it was, to the proposition that "there appears to be, amidst much illusion and deception, an important body of remarkable phenomena, which are *prima facie* inexplicable on any generally recognized hypothesis, and which, if incontestably established, would be of the highest possible value."[2] Per the epigraph, *there are more things in Heaven and Earth, Horatio, / Than are dreamt of in your philosophy* (*Hamlet*, Act 1, Scene 5). Or as James described the SPR's scientific ambitions in more familiar terms, "We are founding here [in Cambridge, MA] a 'Society for Psychical Research,' under which innocent-sounding name ghosts, second sight, spiritualism, and all sorts of hobgoblins are going to be 'investigated' by the most high-toned and 'cultured' members of the community."[3] For all the certain terms James puts in scare (or shudder) quotes,[4] it is the phrase "innocent-sounding" that should place the reader on alert.

The Census of Hallucinations was designed to assemble a "mass of facts" to determine whether "veridical" hallucinations—the apparition of a person at a distance at the time of his or her death—were "accidental coincidences or something more."[5] The question remains—more *what*, exactly? Something more substantial, perhaps more especially if that substance is an ectoplasm (i.e., of an

DOI: 10.4324/9781003148142-17

ideoplastic, non-physical sort).[6] For that matter, what sort of mass of evidence, or evidence of mass, do ghost stories provide, or better yet possess?

Haunted houses and their grounds are part of a larger and unsettling context because of their violently settled, mental topography. This landscape's still-unknown contours and "bourns" were dimly hinted at in another article in the *Boston Globe*, "Going A-Ghost-Hunting," that appeared at the time of the SPR's founding in the winter of 1884. With its initial focus on clairvoyance, telepathy, mental influence, hypnotism, and all their "allied puzzling psychical phenomena," its investigators were preparing to explore the "undiscovered country."[7] By name, this Hamletian place beyond all presently known places was the ultimate scene and setting for the alluring and dreadful, indeed Gothic *and* scientific, imagination of life after death. For Olmsted—who as a landscape architect attended artistically and professionally to the "unity of foreground, middle ground, and background," all with an eye to "scenic effect"; *scenic* being derived from the stage, or "affairs of the drama"[8]—here was an invitation, delivered with the daily news, to engage in the "exact study of this border-land of human experience."[9]

The landscape imagined but not traversed (at least not on foot) replaces and extends those fearful *terrae incognita* reserved for and reassuringly bound by the edges of maps. Instead, in the "undiscovered country," it considers a trans-liminal (ir)reality that lies someplace and sometimes *beyond* experience but that extends well and uncertainly into the equally unfamiliar recesses of human consciousness. It follows a transect drawn between wakefulness and dreaming, scenery and sensibility, rationality and irrationality, spirit and substance, possession and dispossession, all "rounded with a sleep." It takes place on Staten Island. "Is Landscape Haunted?" addresses the future of the past, where and in which history ceases to matter as it was once thought or said to do. A sense of futurity alone guides the untrod path to what lies ahead, where there can be no familiar *and* disconcerting sense of having been there before.[10] This acts as witness, yet again, of that funereal inscription, written in a dead language: *Et in Arcadia Ego*. All else is commentary.

THE UNDISCOVERED COUNTRY, AMERICAN STYLE

"It is the fashion since Shakespeare has had commentators," wrote one reviewer of William Dean Howells's *The Undiscovered Country* (1880), "to discover a purpose in whatever is striking in literature, and it is no doubt the case that the intentions of poets and novelists have been most grossly perverted."[11] A related tendency, apparently free from perversion, is to be found in one curious source of contextualization for Howells's *Hamlet*-ghosted novel about spiritualism in its interpenetrating social, familial, and psychical dimensions.[12] That source appears as a letter to the *New-York Tribune* from Elder Frederick William Evans of the Mount Lebanon Shaker community. Howells's literary reputation might now be a shadow of its former self, but contemporary critics deemed *The Undiscovered*

Country a worthy successor to the tradition of Nathaniel Hawthorne, if not more particularly the latter's essay "The Haunted Mind" (1835), itself a founding text of American hypnagogia. Concerning literary lives and afterlives, Howells's editorial and critical influence as the self-fated "Dean" of American letters spelled invisibility for authors who did not socially and physically embody the genteel version of reality that subtended his own literary realism.[13] However, on the subjects that warranted his regard, Howells deployed "a subtlety of psychological observation and a keenness of penetration wholly unmatched in American fictitious literature outside the pages of Hawthorne."[14]

In "The Haunted Mind," Hawthorne penetrates the disturbed domain of moral reasoning's sleep as he would in depicting the gloomy attics and mossy undergrowth of New England consciousness. "In the depths of every heart," he writes, "there is a tomb and a dungeon, though the light, the music, and revelry above may cause us to forget their existence, and the buried ones, or prisoners whom they hide."[15] A simple but disturbing leitmotif throughout Hawthorne's essay is that which is buried—bodies, histories, artifacts—is seldom left undisturbed, not least of all by landscape architects, and should not be presumed to have a degree of restful peace. And regarding the *leit* ("to lead") of leitmotif, one notable symptom of the unburied past, as noted by James, is the recurrent role of "Indian controls"—i.e., spectralized Indigenous peoples—during episodes of mediumistic possession. Seemingly, they know and occupy a "certain stratum of the *Zeitgeist*" (spirit of the time) of America's subconscious self.[16]

The letter sent by Elder Evans to the *New-York Tribune* represents a somewhat belated response to Howells's investigation of spiritual and domestic realms, but in a sense, one closer to (Evans's) home. Indeed, in addressing the city paper, Evans was not the bringer of news from nowhere but presented the "point of view of an inhabitant of the 'undiscovered country' which Howells has made more undiscovered than it was before."[17] Howells was a literary tourist without a map. Had he misread the place?

Evans was referring to the article "A Shaker Village," published in *The Atlantic Monthly*, of which Howell was the editor. The article served as the empirical fieldwork for his novel on psychic phenomena. Howells's report from this small world was set apart from and yet nested within, the ever-autumnal landscape of transcendentalist New England. It even referenced Evans's "Autobiography of a Shaker," which also appeared in *The Atlantic Monthly*.[18] At a time when the Fox sisters from Rochester had "animated" widespread and avid popular interest in "spirit rapping," it took a semi-skeptical outsider such as Howells to vindicate the status of the Shakers, who were "in full communion with the other world," as the "first spiritual mediums."[19] Although spiritualism is not a local weather system, mid-century America experienced a climate of suggestibility, in which fear, faith, and desire merged into a paid spectacle where the psychic medium, conspiring with séance attendees, became the message. In his book *Modern Spiritualism*, the SPR researcher Frank Podmore discussed the Fox sisters in a chapter titled "In Arcadia," as if evoking the *locus amoenus* (an idealized place) of classical inspiration and that ever-verdant source for built and imaginary landscapes. But he could just as well have taken the haunting place-name from a telephone book; the recipient of the first telephone call, Thomas A. Watson, was keen to investigate this new technology to supplement (or supplant) mediumistic means for communicating with the dead.[20]

"It was in Arcadia that the mysterious rappings were first heard," Podmore writes, referring to the Wayne County, New York, township where the Fox sisters lived and whose house became vibrant with signals from inside its wood frame and from the beyond.[21] Spiritualism itself was said to receive a "death blow" in 1888 when the Fox sisters publicly renounced their claimed abilities of mediumship.[22] In her 1892 obituary, the *New York Times* described Kate Fox Jencken, one of the mediumistic sisters, as "famous years ago in spiritualists' circles."[23] The self-delivered death blow had been given only four years earlier. "Fame is a fickle food," Emily Dickinson ruefully noted.[24] But like the concentrically widening "circles" of Emerson's imagining, the sisters' fame—and what is fame if not the ability to speak from beyond the grave—shaped and reflected the widening compass of spiritualism in American life and death. One correspondent to the SPR, frustrated by its scientifically measured approach to the "spiritistic hypothesis," argued instead for the "erection of a monument to the Fox sisters."[25] Howells was on a far more tentative footing.

Howells spent the summer of 1875 in Shirley, Massachusetts, observing and interacting with the Shakers. He was convinced that he had seen the real thing, or rather the next best thing, which was their humble yet serious striving toward the "realization of a heavenly ideal upon earth."[26] But that season of the year always brings with it the promise of perfection. In the writings of Andrew Jackson Davis—the "Seer of Poughkeepsie," whose mystical tracts contained a pharmaceutical mixture of Shaker and Swedenborgian elements—"Summer Land" (yet another *other* world) was the name for humanity's distant but perhaps attainable "future existence."[27] James notes the frequency with which the term "summer land" is mentioned by those in trance states of hypnosis or mediumistic possession, which is meant to signify a once and final "happy home."[28]

Shirley serves as the model for the town of Vardley in *The Undiscovered Country*, the plot of which detects several discordant chords in the "harmonia" of American spiritualism. Howells's own field research served as the basis for his protagonist, Dr. Boynton's pursuit of a "wholly new field of investigations," that is to say, "communication between embodied and disembodied spirits."[29] As Kermit Vanderbilt notes in one of the first works of literary criticism to take up the broad interpretive implication of Leo Marx's *The Machine in the Garden* (1964), the summer Howells spent in Shirley attuned him to yawning "cultural dislocations" of American life, including rural despair, the lure of the West, urban migration, and the growth of industrialism and unemployment.[30] Howell managed these tensions in his novels with the countervailing narrative settings of city and country, presenting a "sobering," disenchanted "version of American pastoral."[31] But hope, that thing of feathers, never departs the hearts of longing souls. For one, William James, "an insatiable lover of landscape,"[32] invested mind and body in the inhabited reality of a rural ideal. His prized possession was his New Hampshire country house and its scenic acres in the shadow of Mount Chocorua. The place was so named, wrote his brother, Henry James, for a "wild legend, immediately local" of an Indian chief who, pursued by avenging white husbandmen to its topmost peak, "took his leap into the abyss."[33] The novelist, and himself the author of a famous ghost story, imagined the place, not without a sense of disorientation, as a sort of untouched and indeed out-of-place "Arcadia."[34] Seemingly undisturbed by the local legend, William James nonetheless recognized a (natural?) limit to his New England-style *otium liberale*. After inviting his friend, the German psychologist Carl Stumpf, to join him there for the summer, James added, "Why, ah, why is it that '*Freiheit liegt nur in dem Reich' der Träume*?" (Freedom lies only in the realm of dreams).[35] It was from Chocorua that James corresponded with Olmsted regarding his hallucinations. "Pray add the finishing touches <u>quickly</u>," James wrote after receiving a first draft from the former daydreamer-become-doer. After all, Olmsted had made a name for himself by designing public parks that allowed people to escape from controlled urban spaces and access "a sense of enlarged freedom."[36] Work quickly, he urged, "or you may never do so. I find that a thing once postponed is almost impossible to do! I wish other possessors of 'experiences' wrote as easily as you!"[37]

The ever-unstable antipodes of the pastoral schema are marked out in the opening lines of *The Undiscovered Country*, as two of the protagonists wait for a séance to begin:

> Some years ago, at a time when the rapid growth of the city was changing the character of many localities, two young men were sitting, one afternoon in early April, in the parlor of a house on one of those streets which, without having yet accomplished their destiny as business thoroughfares, were no longer the homes of the decorous ease that once inhabited them.[38]

Elder Evans's letter to the *Tribune* includes a similarly unsettling scene from the still-in-the-making "middle landscape," of which Staten Island—particularly the unrealized suburban development plan drafted by Olmsted and Henry Hobson Richardson—could well have served as an example. The Shaker spiritualist refers to an item that appeared under the heading "News from the Suburbs," telling of the ghastly [*sic*] goings-ons in a house on Staten Island reported to be "haunted."[39] For Evans, the story was a

> slight advance toward recognizing the right of Ghosts to live, move, and have their being and a place in mundane affairs. Wherefore should they not come among us and use their powers over men and women, which are very great, to make them better?[40]

Help was needed beyond the expected but expanding competence of architects, landscape architects, and city planners, but not exactly the ambitious mediums

Howells's novel might propose. For Olmsted, the English landscape and Eaton Park, in particular, were potent sources of spiritual inspiration. There, he first wrote of how the artist "directs the shadows of a picture so great that Nature shall be employed upon it for generations, before the work he has arranged for her shall realize his intentions."[41] The statement communicates a profound insight into the futurity of landscape design. It also perpetuates the immemorial trope of subjecting *her* (nature) to *his* (the artist's) intentions.[42]

The landscape architect who knows the land well is haunted not by the past but by the future. In the first anthology of English literary quotations, Edward Bysshe's *The Art of English Poetry* (1702), the "To be, or not to be" soliloquy appears entirely under the heading "Futurity."[43] The unmistakable implication is that death is the future state of things. But still, it remains undiscovered. There's the rub! The past is always coming back, as apparitions or otherwise, without having to be summonsed, often going unnoticed but for that, no less present. There is no cause for uncertainty in what is past; even, as is nearly always the case, it is poorly understood. Rather, for James, the "first practical requisite which a philosophical conception must satisfy" is, in a general way at least, "*banish uncertainty from the future*" (emphasis in the original).[44] James practically addressed the future tense, or rather the "uneasiness" produced by "that vague feeling that it [the future] is impending penetrates all our thought," with reference to custom and habit. It alone ministered to the "haunting sense of futurity" that takes "possession of the mind."[45]

In folklore, some of its woods were said to be haunted, but according to public health experts, Staten Island was endemically plagued by malaria. When asked to define "malaria," the medical statistician Elisha Harris, a member of the Staten Island Improvement Commission, which authored the abovementioned plan, responded that it is "a poison in the atmosphere of certain localities which is recognized only by its effects, of which ague and fever is an example."[46] Disease, like apparitions and manifestations, is first known by its symptoms, be they visual, auditory, or transphenomenal. Long attributed to miasmas, or more particularly to a "granular microphyte" that appears like "spots of oil"[47] on the surface of putrid water, the true culprit would soon enough emerge. A *New York Times* report on the New York Mycological Club's June 1897 semi-disastrous mushroom "hunting" excursion from Manhattan to the more rustic outer borough said that if the club members learned something, it was why, in fact, some of Staten Island's woods were deemed "haunted." And that was because they were haunted by "fierce, gigantic mosquitoes." The mycologists ended the afternoon bearing "scars of conflict."[48] Perceived as a threat to more than just mycological hunting parties, Dr. Alvah H. Doty, Health Officer of the Port of New York, sought to eradicate the mosquito problem on Staten Island by spreading "cheap, practically harmless" crude petroleum oil in the ponds where the larvae hatch.[49]

Returning to the question of hope—and perhaps also redemption, which might be instrumentalized by the landscape architect in the form of far-reaching plans—there are extraliterary indicators (i.e., biographical facts) regarding Howells to suggest that he allowed himself guarded faith in the hereafter. Following the death of his daughter, Winifred, Howells attended a séance in Concord, Massachusetts, with Robertson James, younger brother of William and Henry. He spent the night in the James family house "after the spooking."[50] Howells reported to his father that "the most wonderful things happened. The floor and walls were shaken, and the table lifted and banged with blows as from a hammer and tappings galore." And while he was not entirely convinced the medium was the source of the agitation, he was nonetheless "badly rattled at the time."[51] However, he was evidently disquieted by the séance. Soon after, Howells attended an informal meeting of the SPR, which was still in search of answers. As he described it to his father, the meeting turned into a "ghost-story swap."[52] But this seems to be the mode of research in which the SPR specialized. No special equipment was required. And this is what Olmsted did in his own letter to William James, telling a ghost story (of his own making).

Howells continued to struggle with the hereafter long after *The Undiscovered Country* left the question unanswered. In 1903, he published *Questionable Shapes*, this clause: *according to an advertising copy for a book of "ghost stories"* which "do not seem untrue, as ghost stories usually do."[53] One of its offerings, *Though One Rose from the Dead* (1903)—its title evokes a passage from the Gospel of Luke addressing doubts about prophecy—begins with the narrator substantiating his beliefs with reference to James's *The Will to Believe*.[54] That same year, in one of his "Editor's Easy Chair" columns for *Harper's Monthly Magazine*, Howells once again addressed fitfully but not finally, the "question of the life hereafter" and the "mystery of sleep," which pales before the "mystery of waking." Here, Howells reflected on James's writing in terms that could have been borrowed from his own character, Dr. Boynton's Shakespearian idiom. The Harvard professor's work

> abounds in the substance of things hoped for, the evidence of things not seen. All psychology, which disclaims its putative relationship to the soul, is alive with fresh interest for those who seek to know it through the mind, and a whole order of literature has arisen, calling its psychological, as realism called itself scientific, and dealing with line on its mystical side.[55]

Olmsted narrated his hallucinations to James in the interest of science, if not also a late-in-life effort of self-discovery. Not long after, in a series of letters to the critic Mariana Griswold Van Rensselaer—who was writing a biographical sketch of the landscape architect, by then recognized as an effective "doer" and maker of public places, Olmsted channeled an earlier formative version of himself as a *hallucinari* (of uncertain derivation), one who wanders in mind but perhaps also beyond on foot. "A disposition was born with me, or early became fixed, to vagrancy, to daydreaming, to find my pleasure in an intellectually inactive or not consciously directed contemplation of natural scenery." Natural scenery is very different from the one to which the naturalist or the gardener directs his attention. In a clinical sense, daydreaming includes forms of "mental reproduction" ranging from the hypnagogic state, with complete absence of voluntary control, up to a "distinctly purposive picture of the future with due attention to probable realization."[56] Olmsted acutely diagnosed his own mode of mental self-occupation, which had led him to wander toward a career in landscape architecture. It consisted of a:

> Simple but not unintelligent habit of looking for pleasure in contemplation or rural and picturesque <u>scenery</u>; second, through the effect of being thrown for several years into an unusual intimacy with certain forms of rural scenery and through a habit of daydreaming about rural scenery, a habit favorable to poetic moods and the development of a <u>designing</u> habit[57] (underlining in original).

In the interpenetrating worlds and spheres of the spiritualist and psychical research, there were several fault lines between internal and external experience, phantasm and reality, the living and the dead, and other bordering or borderline states. The lay of the land was perhaps best expressed by James's Irving Street neighbor, a colleague in the philosophy department at Harvard, and fellow SPR researcher Josiah Royce. In his "Report of the Committee on Apparitions and Haunted Houses," Royce explains:

> Quite apart from the objectively verifiable phantasms and presentiments, those which are not veritable may be found to follow types and to show laws whose study shall lead us into yet other parts of that romantic unexplored country. No one who realizes how closely the normal and abnormal are joined in human life, how complex and delicate are their relations, how subtle and significant are their mutual influences, should hesitate to aid in promising research in so profoundly and tragically important a province of the human spirit.[58]

The tragic, comedic, and satiric are just a few of the dramatic genres, each with its own distinct scenery. But the tragedy is most particularly freighted with the responsibility of enacting the law of poetic justice, famously expressed by John Dennis in his essay on Shakespeare as, "the Good must never fail to prosper, and the Bad must always be punish'd: Otherwise the Incidents, and particularly the Catastrophe which is the grand Incident, are liable to be imputed rather to Chance than to the Almighty Conduct and to Sovereign Justice."[59] As for a relevant definition of poetry, it shares traits with Royce's above notion of the exemplary—i.e., illustrative and educative—value of phenomena that have not (yet) been confirmed by rational analysis. According to Aristotle's long-reigning definition, poetry, as recollected in an *Atlantic Monthly* article on the decline of poetic justice, relates what *might be* rather than what *has* been; the latter is the stuff of mere history, which might or might not have a lesson in store.[60]

The data of psychical phenomena consist of the dimly or rather all too vivid sense, the dreaming experience, the fleeting glimpse, the "residual phenomena"[61] of what this *might* be of an undiscovered or "romantic unexplored country." Romances are stories through which the true catastrophe is met and encountered. The catastrophe, the grand incident, that presently matters is not the tragedy that befell Hamlet, though there is still much to learn from his conversation with the ghost and his "immortal soliloquy."[62] In the lay of the land itself, the landscape architect would seek to improve and amend, correct and conceal, and alter when alternation finds where the catastrophe is encrypted and, by chance, riddled out. Hamlet reflects on the fatality of birth in the scene immediately preceding his meeting with the Ghost.

> So, oft it chances in particular men,
> That for some vicious mole of nature in them,
> As in their birth, wherein they are not guilty,
> Since nature cannot choose its origin (*Hamlet*, Act 1, Scene 4).

Helping to put this presentiment in its proper context, one scholar notes that mere months after Olmsted's birth in 1822, the novelist James Fenimore Cooper began publishing his *Leatherstocking Tales* and "peopled the land with a past, a past of imagination."[63] The point here is not to draw on the causal logic of veridical hallucination, in the strict sense of how the dead appear to the living. Instead, it is to engage in a form of critical futurity heralded by Hamlet's soliloquy, specifically by looking forward to D. H. Lawrence's astonishingly clear-sighted assessment of the "lovely half-lies" of Cooper's literary topography. "One day the demons of America must be placated," Lawrence prophesizes, "the ghosts must be appeased, the Spirit of Place atoned for. Then, the true passionate love for American soil will appear. As yet, there is too much menace in the landscape."[64]

NOTES

1 "There are more things in Heaven and Earth, Horatio, than are dreamt of in your philosophy" (Hamlet 1:5). "Searching for Ghosts," *The Boston Globe*, February 11, 1891, reports on the American Psychical Society, founded by Rev. T. Ernest Allen and Rev. Minot J. Savage, and Benjamin Orange Flower, editor of *The Arena*, as a *more* spiritually oriented counterpart to the American branch of the Society for Psychical Research.

2 Council of the American Society for Psychical Research, "Circular No. 1," *Proceedings of the American Society for Psychical Research* 1, no. 1 (July 1885): 2; "Objects of the Society," *Proceedings of the Society for Psychical Research* 1 (1882–1883): 3.

3 William James to Kitty Prince (Cambridge, December 24, 1884), in Julius Seelye Bixler, *Religion in the Philosophy of William James* (Boston, MA: Marshall Jones, 1926), 159.

4 On the (de)familiarizing role of quotation marks, see Marjorie Garber, "' ' (Quotation Marks)," *Critical Inquiry* 25, no. 4 (1999): 653–679.

5 "Do You Hear Voices?" *Boston Daily Globe*, May 12, 1890, 4; William James, "International Congress of Experimental Psychology: Instructions to the Person Undertaking to Collect Answers to the Question on the Other Side (1889)," in *Essays in Psychical Research* (Cambridge: Harvard University Press, 1986), 56. The object of the inquiry is (1) to ascertain approximately the *proportion of persons* who have such experiences, and (2) to obtain details as to the main experiences with a view to examining into their cause and meaning." The questionnaire had Schedule A and Schedule B. A: Have you

ever, when believing yourself to be completely awake, had a vivid impression of seeing, or being touched by a living being or inanimate object, or of hearing a voice; which impression, so far as you could discover, was not due to any external physical cause?" This was accompanied by census-like matrix for age, name, address, sex, occupation. Schedule B, to be answered if Schedule A was answered in the affirmative: "Have you ever, when believing yourself to be completely awake, had a vivid impression of seeing, or being touched by a living being, or inanimate object, or of hearing a voice; which impression, so far as you could discover, was not due to any external physical cause?"

6 See *Charles Richet, Traité de Métaphysique* (Paris: Félix Alcan, 1922), 561; see also Oliver J. Lodge, "Experience of Unusual Physical Phenomena Occurring in the Presence of an Entranced Person (Eusapia Paladino)," *Journal of the Society for Psychical Research* 114, no. 6 (1894): 306–360; Albert Schrenck-Notzing, *Phenomena of Materialisation: A Contribution to the Investigation of Mediumstic Teleplastics*, trans. E. E. Fornier d'Albe (London: Kegan, Paul, Trubner, 1920), 13.

7 "Going A-Ghost-Hunting," *The Boston Globe*, March 1, 1885, 12.

8 Frederick Law Olmsted, "Hints Aidful to Elementary Self-Education in Design in the Common Fields of Landscape Gardening Proper," in *Frederick Law Olmsted: Landscape Architect, 1822–1903*, eds. Frederick Law Olmsted, Jr. and Theodora Kimball (New York: G. P. Putnam's Sons, 1922), 62.

9 Council of the American Society for Psychical Research, "Circular No. 1," *Proceedings of the American Society for Psychical Research* 1, no. 1 (July 1885): 2.

10 See Louis Marin, *Sublime Poussin*, trans. Catherine Porter (Stanford, CA: Stanford University Press, 1999), 109.

11 "The Undiscovered Country," [Atlanta] *Daily Constitution*, 2 July 1880, 2.

12 Michael McGehee, "Religion, Family, and National Belonging in W. D. Howells' The Undiscovered Country," *American Literary Realism* 45, no. 2 (2013): 130.

13 See Gene Andrew Jarrett, *Deans and Truants: Race and Realism in African American Literature* (Philadelphia: University of Pennsylvania Press, 2007; Elizabeth Renker, *Realist Poetics in American Culture, 1866–1900* (New York: Oxford University Press, 2018).

14 From the review in the *New York Evening Post*, cited in the advertisement for the book placed by its publisher Houghton, Mifflin, and Co., in *The Atlantic Monthly* 46, no. 274 (August 1880).

15 Nathaniel Hawthorne, "The Haunted Mind," in *The Token and Atlantic Souvenir*, ed. S. G. Goodrich (Boston, MA: Charles Bowen, 1835), 76.

16 William James, *The Principles of Psychology*, vol. 1 (New York: Henry Holt, 1894), 394.

17 Elder Frederick William Evans, "A Review of the Tribune," *New-York Daily Tribune*, 18 February 1885, 2.

18 F. W. Evans, "Autobiography of a Shaker, Part I," *The Atlantic Monthly* 23 (April 1869): 415–425; "Autobiography of a Shaker, Part II," *The Atlantic Monthly* 23 (May 1869): 594–605.

19 William Dean Howells, "A Shaker Village," *The Atlantic Monthly* 37, no. 224 (June 1876): 700. On the Fox sisters and the rise of séance culture see Simone Natale, *Supernatural Entertainment: Victorian Spiritualism and the Rise of Modern Media Culture* (University Park: Penn State University Press, 2016).

20 Thomas A. Watson, *Exploring Life* (New York: D. Appleton, 1926), 37; Avital Ronell, *The Telephone Book: Technology, Schizophrenia, Electric Speech* (Lincoln: University of Nebraka Press, 1989), 247.

21 Frank Podmore, *Modern Spiritualism: A History and Criticism*, vol. 1 (London: Methuen & Co., 1902), 179.

22 Reuben Briggs Davenport, *The Death-Blow to Spiritualism: Being the True Story of the Fox Sisters, as Was Revealed by Authority of Margaret Fox Kane and Catherine Fox Jencken* (New York: G. W. Dillingham, 1888).

23 "One of the Fox Sisters Dead," *New York Times*, July 3, 1982, 5.

24 "Fame is a fickle food / Upon a shifting plate / Whose table once a / Guest but not /The second time is set." Emily Dickinson, "Fame is a Fickle Food," in *The Poems of Emily Dickinson*, ed. R. W. Franklin, vol. 3 (Cambridge: Harvard University Press, 1998), 1483.

25 "Editorial: Reply to Critics," *The Psychical Review* 1, no. 4 (May 1893): 382.

26 William Dean Howells, "A Shaker Village," *The Atlantic Monthly* 37, no. 224 (June 1876): 699.

27 Andrew Jackson Davis, *A Stellar Key to the Summer Land* (Boston, MA: William White & Co., 1867).

28 William James, *The Principles of Psychology*, vol. 1 (New York: Henry Holt, 1890), 394.

29 William Dean Howells, *The Undiscovered Country* (Boston, MA: Houghton, Mifflin and Company, 1880), 64, 219.

30 Kermit Vanderbilt, "*The Undiscovered Country:* Howells' Version of American Pastoral," *American Quarterly* 17, no. 4 (1965): 651.

31 Kermit Vanderbilt, "The *Undiscovered Country:* Howells' Version of American Pastoral," *American Quarterly* 17, no. 4 (1965): 651.

32 Henry James, *The Letters of William James vol. 1* (Boston, MA: Atlantic Monthly Press, 1920), 272.

33 Henry James, "New England: An Autumn Impression," *North American Review* 180, no. 581 (April 1905): 492.

34 Henry James, "New England: An Autumn Impression," *North American Review* 180, no. 581 (April 1905): 491.

35 William James to Carl Stumpf (Cambridge, MA, February 6, 1887), Riccardo Martinelli, *William James and Carl Stumpf Correspondence: 1882–1910* (Boston, MA: De Gruyter, 2020), 89. The line of poetry is from the final strophe of Friedrich Schiller's poem "Der Antritt des neuen Jahrhunderts ("The Start of the New Century"), see Daniel S. Malachuk, "William James at Chocorua: A Northern Forest Philosopher," in *Nature and Culture in the Northern Forest: Region, Heritage, and Environment in the Rural Northeast*, ed. Pavel Cenkl (Iowa City: University of Iowa Press, 2010), 171–186.

36 Frederick Law Olmsted and Calvert Vaux, "Preliminary Report to the Commission for Laying out a Park in Brooklyn, New York, Being a Consideration of Circumstances of Site and Other Conditions Affecting the Design of Public Grounds" (1866), cited in Andrew Menard, "The Enlarged Freedom of Frederick Law Olmsted," *The New England Quarterly* 83, no. 3 (2010): 508.

37 James to Olmsted (Chocorua, NH, April 2, 1891).

38 William Dean Howells, *The Undiscovered Country* (Boston, MA: Houghton, Mifflin and Company, 1880), 1.

39 "News from the Suburbs: Strange Sights in a Doctor's House" *New York Tribune*, June 14, 1884, 10.

40 Elder Evans, "A Review of the Tribune," *New-York Daily Tribune*, February 18, 1885, 2.

41 Frederick Law Olmsted, *Walks and Talks of an American Farmer in England* (New York: George P. Putnam, 1852), 133.

42 See Gerda Lerner, *The Creation of Patriarchy* (New York: Oxford University Press, 1986).

43 Sayre N. Greenfield, "Quoting *Hamlet* in the Early Seventeenth Century," *Modern Philology* 105, no. 3 (2008): 510–534.

44 William James, "Rationality, Activity, and Faith," *The Princeton Review* 58 (July 1882): 60.

45 William James, "Rationality, Activity, and Faith," *The Princeton Review* 58 (July 1882): 61.

46 Frederick Law Olmsted, Elisha Harris, Joseph Mott Trowbridge, and Henry Hobson Richardson, Staten Island Improvement Commission, *Report of a Preliminary Scheme of Improvements: Presented January 12th, 1871* (New York: James Sutton 1871), 31.

47 Pietro Balestra, "Recherches et expériences sur la nature et l'origine des miasmes paludéens," *Compte Rendu des Séances de l'Académie des Science* 61 (1870): 235–237, cited in Staten Island Improvement Commission, *Report of a Preliminary Scheme of Improvement* (1871), 34.

48 "Hunting for Mushrooms," *New York Times*, June 7, 1897, 7.

49 Alvah H. Doty, "Elimination of the Mosquito," *Journal of the American Medical Association* 45, no. 9 (August 26, 1905): 588; "Fighting Mosquitos with Petroleum," *Harper's Weekly* 45, no. 2330 (August 7, 1901): 829.

50 William Dean Howells to Roberston James (February 10, 1891), in Robert C. Leitz et al., *W. D. Howells, Selected Letters*, vol. 3: 1882–1891 (Boston, MA: Twayne Publishers, 1980), 307, fn. 1.

51 William Dean Howells to William Cooper Howells (Boston, February 15, 1891), in Robert C. Leitz et al., *W. D. Howells, Selected Letters*, vol. 3: 1882–1891 (Boston, MA: Twayne Publishers, 1980), 307.

52 William Dean Howells to William Cooper Howells (Boston, February 15, 1891), in Robert C. Leitz et al., *W. D. Howells, Selected Letters*, vol. 3: 1882–1891 (Boston, MA: Twayne Publishers, 1980), 307.

53 Print advertisement for *Questionable Shapes*, *Harper's Weekly* 47, no. 2427 (June 27, 1903): 1082.

54 William Dean Howells, "Though One Rose from the Dead," in *Questionable Shapes* (New York: Harper, 1903), 159; "And Abraham said unto him, 'If they hear not Moses and the prophets, neither will they be persuaded though one rose from the dead'" (Luke 16:31).

55 William Dean Howells, "Editor's Easy Chair," *Harper's Monthly Magazine* 107, no. 637 (June 1903): 149. See also Charles L. Crow, "Howells and William James: "A Case of Metaphantasmia Solved," *American Quarterly* 27, no. 2 (1975): 169–177.

56 Theodate L. Smith, "The Psychology of Day Dreams," *The American Journal of Psychology*, 15, no. 4 (October 1904): 465.

57 Olmsted letter to Mariana Griswold Van Rensselaer (June 11, 1893).

58 Josiah Royce, "Report of the Committee on Apparitions and Haunted Houses," *Proceedings of the American Society for Psychical Research* 1, no. 3 (December 1887): 229.

59 John Dennis, *An Essay on the Genius and Writings of Shakespear* [*sic*] (London: Bernard Lintott, 1712), 6–7.

60 Raymond Macdonald Alden, "The Decline of Poetic Justice," *The Atlantic Monthly* 105, no. 2 (February 1910): 260.

61 "Objects of the Society," *Proceedings of the Society for Psychical Research* 1 (1882–1883), 3.

62 Mark Twain, *Adventures of Huckleberry Finn (Tom Sawyer's Comrade)* (New York: Charles L. Webster, 1895), 181.

63 Elizabeth Stevenson, *Park Maker: A Life of Frederick Law Olmsted* (New Brunswick, NJ: Transaction Publishers, 2000 [1977]), 7.

64 D. H. Lawrence, *Studies in Classic American Literature* (New York: Thomas Seltzer, 1923), 74.

Chapter 17: Is Landscape Language?

Anne Whiston Spirn

The language of landscape is our native language. Landscape was the original dwelling; humans evolved among plants and animals, under the sky, upon the earth, near water. Everyone carries that legacy in body and mind. Humans touched, saw, heard, smelled, tasted, lived in, and shaped landscapes before the species had words to describe what it did. Landscapes were the first human texts, read before the invention of other signs and symbols. Clouds, wind, and sun were clues to weather, ripples and eddies signs of rocks and life under water, caves and ledges promise of shelter, leaves guides to food; bird calls warnings of predators. Early writing resembled landscape; other languages—verbal, mathematical, graphic—derive from the language of landscape.[1]

The language of landscape can be spoken, written, read, and imagined. Speaking and reading landscape are byproducts of living—of moving, mating, eating—and strategies of survival—creating refuge, providing prospect, growing food. To read and write landscape is to learn and teach: to know the world, to express ideas and to influence others. Landscape, as language, makes thought tangible and imagination possible. Through it humans share experience with future generations, just as ancestors inscribed their values and beliefs in the landscapes they left as a legacy, a "treasure deposited by the practice of speech," a rich lode of literature: natural and cultural histories, landscapes of purpose, poetry, power, and prayer.[2]

Landscape has all the features of language. It contains the equivalent of words and parts of speech—patterns of shape, structure, material, formation, and function. All landscapes are combinations of these. Like the meanings of words, the meanings of landscape elements (water, for example) are only potential until context shapes them. Rules of grammar govern and guide how landscapes are formed, some specific to places and their local dialects, others universal. Landscape is pragmatic, poetic, rhetorical, polemical. Landscape is scene of life, cultivated construction, carrier of meaning. It is language.

Verbal language reflects landscape. Up and down, in and out—the most basic metaphors of verbal language—stem from experience of landscape, like bodily movement through landscape.[3] Verbs, nouns, adverbs, adjectives, and their contexts—parts of speech and the structure of verbal language—mirror landscape processes, products, and their modifiers, material, formal, and spatial. Just as a river combines water, flowing, and eroded banks, sentences combine actions and actors, objects and modifiers. The context of a word or sentence, like that of hill or valley, defines it. Verbal texts and landscapes are nested: word within sentence within paragraph within chapter, leaf within branch within tree within forest. Words reflect observation and experience; dialects are rich in terms specific to landscape of place, like "estuary English," described so vividly by John Stilgoe.[4] Shakespeare, Mark Twain, T. S. Eliot, Anthony Hecht, and Adrienne Rich, like verbal poets of every literature, mine landscape for structure, rhythm, and fresh metaphors of human experience; so do poets of landscape itself, "Capability" Brown, Frederick Law Olmsted, Frank Lloyd Wright, Lawrence Halprin, Martha Schwartz.[5]

Landscape is the material home, the language of landscape is a habitat of mind. Heidegger called language the house of being, but the language of landscape truly is the *house* of being; we dwell within it. To dwell—to make and care for a place—is self-expression. Heidegger traced that verb in High German and Old English; in both, the root for "to dwell" means "to build." In German, the roots for building and dwelling and "I am" are the same. I am because I dwell; I dwell because I build. *Bauen*—building, dwelling, and being—means "to build," "to construct," but also "to cherish and protect, to preserve and care for, specifically to till the soil, to cultivate the mind."[6]

Landscape associates people and place. Danish *landskab*, German *landschaft*, Dutch *landschap*, and Old English *landscipe* combine two roots. *Land* means both a place *and* the people living there. *Skabe* and *schaffen* mean "to shape"; suffixes *-skab* and *-schaft*, as in the English "-ship," also mean association, partnership.[7] Though no longer used in ordinary speech, the Dutch *schappen* conveys a magisterial sense of shaping, as in the biblical Creation. Still strong in Scandinavian and German languages, these original meanings have all but disappeared from English. Webster's Dictionary defines *landscape* as static, "a picture representing a section of natural, inland scenery, as of prairie, woodland, mountains...an expanse of natural scenery seen by the eye in one view"; the Oxford English Dictionary traces the word to a Dutch painting term (*landskip*).[8] But landscape is not a mere visible surface, static composition, or passive backdrop to human theater; therefore, dictionaries must be revised, and the older meanings revived. The words *environment* and *place*, commonly used to replace *landscape* in twentieth-century English, are inadequate substitutes, for they refer to locale or surroundings and omit people. In mid-century, the declining use of *landscape* was in part a reaction to the Nazis' adoption of "blood and soil," a linking of native landscape and racial identity. *Environment* and *place* seem more neutral, but they are abstract, disembodied, sacrificing meaning, concealing tensions and conflicts, ignoring the assumptions *landscape* reveals. *Landscape* connotes a sense of the purposefully shaped, the sensual and aesthetic, the embeddedness in culture. The language of landscape recovers the dynamic connection between place and those who dwell there.

Landscape is loud with dialogues, with story lines that connect a place and its dwellers. The shape and structure of a tree record an evolutionary dialogue between species and environment: eucalypt leaves that turn their edge to bright sun, deciduous leaves that fall off during seasonal heat or cold. And they record dialogues between a tree and its habitat. Tree rings thick and thin tell the water and food of each growing season of the tree's life. Size, shape, and structure— low-branched or high, densely branched or spare—reflect dialogues between a tree and a group of trees in open field or dense forest. Each species has a characteristic form from which individuals deviate, as true of human body shape— muscled or fat, short or tall—as of trees. A coherence of human vernacular landscapes emerges from dialogues between builders and place, fine-tuned over time. They tell of a congruence between snowfall and roof pitch, between seasonal sun angles and roof overhang, wind direction and alignment of hedgerows, cultivation practices and dimensions of fields, family structure and patterns of settlement. Dialogues make up the context of individual, group, and place. The context of life is a woven fabric of dialogues, enduring and ephemeral.

Humans are not the sole authors of landscape. Volcanoes spew lava, remaking land; rain falls, carving valleys. Mountains, gardens, and cities are shaped by volcanoes and rain, plants and animals, human hands and minds. Trees shade ground and shed leaves, produce a more hospitable place for life with similar needs. Beavers cut trees and dam streams to make ponds: a dwelling place. People mold landscape with hands, tools, and machines, through law, public policy, and actions undertaken hundreds, even thousands, of miles away. All living things share the same space, all make landscape, and all landscapes, wild or domesticated, have co-authors, all are phenomena of nature and culture. Others share the language, but only humans (as far as we yet know) reflect, worship, make art, and design landscapes like the gardens of the Villa d'Este that "set the formal strictures" within a natural context "where the tension lectures us on our mortal state."[9]

LANDSCAPE IS MEANINGFUL AND EXPRESSIVE

Landscape has meaning. Rivers reflect, clouds portend. Wilderness, for many now a sacred symbol of undefiled nature, was once a terrifying symbol of chaos. Some meanings are human inventions, and yet significance does not depend on human perception or imagination alone. Significance is there to be discovered, inherent and ascribed, shaped by what senses perceive, what instinct and experience read as significant, what minds know. Any organism with senses has the potential to read and understand landscape. To a deaf man, a rustling bush cannot signal an approaching animal, but moving leaves or vibrating ground may. To a canoer a river is a path, waves and eddy lines are signs to steer by. To a fish a river is a watery world of light and shadow, surface movement is sign of prey. Fly fishermen try to read rivers as fish do in order to trick them, picking then flicking the fly at line's end, mimicking real flies abroad on the stream to convince the fish the fly is real. Norman Maclean describes a master fly-tier who lies under a glass tank filled with water to study the insect he intends to imitate.[10] The best fly fishermen think like fish, become the fish, in an intimate bonding of hunter and hunted.

Landscapes are as small as a garden, as large as a planet. To a person the garden is a landscape, to a people the nation is, to the human species, a planet. A pond is a landscape to a beaver, a tree to a bird, a forest to a tree. Ice floes on a river, lake, or arctic sea, inhabited by birds and seals, are a landscape. Ice crystals on a winter window look like ice floes seen from the air, are uninhabited, yet to a poet a landscape of the imagination. Landscape may be inhabited in imagination alone.

There are landscapes within landscapes within landscapes. Every landscape feature is both a whole and part of one or more larger wholes: leaf and twig, twig and tree, tree and forest; garden and house, house and street, street and town, town and region. Every phenomenon, thing, event, and feeling has a context. A valley is not a valley if it has no ridge or plateau, no up and down. Motion is imperceptible without rest, sound without stillness. Without sense of past and future, there can be no present, without threat no refuge. The same material, form, or action may have different meanings in different settings— water in a desert, water in a sea.

Anomalies are clues to what the wider context is. A "wolf" tree is a tree within a woods, its size and form, large trunk and horizontal branches, anomalous to the environs of slim-trunked trees with upright branches. It is a clue to the open field in which it once grew alone, branches reaching laterally to the light and up. With that field unmowed, unplowed, or ungrazed, younger woodland trees grew thickly together around the older tree, their branches finding light by reaching up. The older tree, engulfed by a dense woodland of younger trees, no longer able to find light horizontally, sends new branches upward. Landscape is dynamic, present context includes the past; the story of the wolf tree is part of the human story.

When valley and river, path and user fail to correspond—when, for example, a valley is vast but a river small, a path broad and well-worn, but those who pass that way infrequent—valley and path may have been shaped by context not now visible or no longer relevant: valley by a great flood, path by an earlier surging crowd. Also, context may be actual or latent—every landscape has both real and potential form—what is, what has been, what will, what might be.

Metaphors grounded in landscape guide how humans think and act. George Lakoff and Mark Johnson demonstrate what Emerson observed: that humans understand and experience one kind of thing in terms of another, projecting bodies and minds onto the surrounding world: trees and clouds seen as bounded, a river seen as having a mouth, a mountain as having a foot, front, back, and side.[11] One might just as easily see things as continuous and undifferentiated; viewing them as separate is more a function of individual consciousness than an inherent quality of landscape. Many metaphors are grounded in fundamental relationships with landscape—moving, making, eating, wasting. The most common refer to space and direction: in and out, up and down. In American culture, high and in are good, down and out are bad; central is important, marginal is not. Landscape imagery conveys feelings and ideas: emotions churning like a stormy sea, rivers of time, clouds where gods live, sacred mountains, Father Sky impregnating Mother Earth with rain as the seed, Zeus and Thor hurling thunderbolts in anger, Siva flashing lightning from his Third Eye, a flare of cosmic intelligence, the god of Jews and Christians dispatching plagues of locusts and disease to punish the wicked. Personification, the attribution of human feelings like intention, anger, love to natural forms and phenomena, is the foundation of myth and religion.

Landscapes are the world itself and may also be metaphors of the world. A tree can be both a tree and The Tree, a path both a path and The Path. A tree in the Garden of Eden represents the Tree of Life, the Tree of Knowledge. It becomes the archetype of Tree. When a path represents the Path of Enlightenment of Buddhism or the Stations of the Cross of Christianity, it is no longer a mere path but The Path. The yellow brick road in *The Wizard of Oz* is both path and Path. The similar is the stuff of metaphor, simile, and personification; contrasts are the stuff of paradox and oxymoron. Landscape actors, objects, and modifiers may enhance meaning without rhetoric: rivers reflect and run, but they do not pun.

Built landscapes may be rhetorical. Landscape features, like hill and street, may be emphasized or embellished for effect, slope steepened to make climb difficult, street broadened and lined with trees to impress the viewer. Gardens of allusion reflect oral and written literature: Shakespearean gardens allude to the bard's plays and poetry, their herbs and blooms references to his works; eighteenth-century English gardens, with their buildings in classical style and pastoral landscape, refer to classical literature. When Mussolini built a monument in 1938 to those who died in a battle of the First World War in Redipuglia, near Italy's northwestern boundary, he used the language of rhetoric. More than 100,000 soldiers are buried there in twenty-two terraces of tombs, arranged from bottom to top in alphabetical order, 60,000 buried at the top of the hill in a common grave surmounted by three crosses, like Calvary. Words engraved in the pavement tell how these soldiers died for the glory of Italy, immortal in memory. Facing the hill of tombs is the grave of their general as if addressing his entombed soldiers. Their inscriptions answer "*Presente.*" "I am here."

The language of landscape can be spoken and read even though never codified, without recourse to rules. People follow paths and make them, plant gardens, are awed by the scale of mountains and cathedrals; great designers use landscape fluently, all without dictionaries or grammars. Thomas Jefferson linked landscape and learning at the University of Virginia, where he sited the library to face the distant mountains and framed the view with flanking buildings. Sigurd Lewerentz and Gunnar Asplund comforted the bereaved in the Hill of Remembrance and Woodland Chapel at Forest Cemetery in Stockholm. Glenn Murcutt associated people, sun, wind, and water in a house at Bingie on the coast of Australia. Even those who exploit landscape cynically may do so masterfully, as Mussolini did, when, at Redipuglia, he fostered feelings of heroic nationalism to promote fascism, or as Disney has exploited it, for profit, at Disneyland and Disney World.

Landscapes are a vast library of literature. The myths of Japan's Fuji and Australia's Uluru, the folksy tales of trolls and pink flamingoes on American lawns, the classical works of earth, water, and wind at Yosemite and the Grand Canyon, the high art of the Alhambra and Manhattan's Central Park, and countless other places, ordinary and extraordinary, record the language of landscape. The library ranges from wild and vernacular landscape, tales shaped by everyday phenomena, to classic landscapes of artful expression, like the relationship of ordinary spoken language to great works of literature. Worship, memory, play, movement, meeting, exchange, power, production, home, and community are pervasive landscape genres. To be fully felt and known, landscape literature must be experienced in situ; words, drawings, paintings, or photographs cannot replace the experience of the place itself, though they may enhance and intensify it.

Landscape literature is a resource to be treasured. Several decades ago, archaeologists in Israel discovered ancient water-gathering systems in the Negev Desert that employed simple channels, check dams, and broad depressions. These techniques, lost for many centuries, have inspired landscape architect Shlomo Aronson and others to reshape whole landscapes to gather water; they planted groves and grassy meadows in the desert, all sustained by dewfall and rain. Water engineer Ken Wright is working with archaeologists to study the water systems of the Incas in Peru, not just to understand, but to use their knowledge. I have studied dozens of community gardens in Philadelphia as landscape expressions. The literature of landscape contains a vast repertoire of similar examples, adaptations to a wide range of circumstances, not just in the diversity of genes and behavior, but in ideas and cultures. Some are cherished and cared for, others are being rediscovered, but entire volumes of landscape literature are being lost and forgotten, whole libraries are being destroyed.

LANGUAGE HAS CONSEQUENCES

The language of landscape is a powerful tool. A person fluent and literate in landscape sees significance where an illiterate person notes nothing. Past and future fires, floods, landslides, welcome or warning are visible to those who can read them in tree and slope, boundary and gate. Knowing how to tell what one wants to express—pragmatics—makes landscape authors more adept; making landscapes appeal to emotion and reason depends on understanding rhetoric. To know landscape poetics is to see, smell, taste, hear, and feel landscape as a symphony of complex harmonies. Natural processes establish the base rhythm that is expressed in the initial form of the land, to which culture, in turn, responds with new and changing themes that weave an intricate pattern, punctuated here and there by high points of nature and art. Landscape symphonies evolve continually in time, in predictable and unpredictable ways, responding to process and to human purpose, and, in landscape symphonies, all dwellers are composers and players.

Humans have always known the language of landscape but now use it piecemeal, with much forgotten. People still read paths and create them, identify boundaries and define territory, delight in a flowering tree, comparing it to a lover, but most people read landscape shallowly or narrowly and tell it stupidly or inadequately. Oblivious to dialogue and storyline, they misread or miss meaning entirely, blind to connections among intimately related phenomena, oblivious to poetry, then fail to act or act wrongly. Absent, false, or partial readings lead to inarticulate expression: landscape silence, gibberish, incoherent rambling, dysfunctional, fragmented dialogues, broken storylines. The consequences are comical, dumb, dire, tragic. Those who admired the yellowwood's excessive, early flowering on the campus in Philadelphia were blind to what the bud scars told, failed to read the flowers' poignant message, could not imagine the tree's connection to soil, plaza, and contractor. When I tried to convince the dean, himself an architect, to find another site for the contractors' trailer and tools, he refused, unconvinced or not caring that the yellowwood would die as a consequence. Those who first built houses over the buried creek in West Philadelphia and those who rebuilt in the same place were illiterate in the language of landscape and so could not read the creek's presence. I tried and failed, at first, to convince planners at the City Planning Commission and engineers at the Philadelphia Water Department that the buried creek was a resource to be exploited and a force to be reckoned with. The yellowwood is dead, but it is not too late to restore Mill Creek—the water, the people, the place.

Ironically, the professionals who specialize, reading certain parts of landscape more deeply than other parts, and shaping them more powerfully, often fail to understand landscape as a continuous whole. Once those who transformed landscapes were generalists: naturalist, humanist, artist, engineer, even priest, all combined. Now pieces of landscape are shaped by those whose narrowness of knowledge, experience, values, and concerns leads them to read and tell only fragments of the story. To an ecologist, landscape is habitat, but not construction or metaphor. To a lawyer, landscape may be property to regulate, to a developer a commodity to exploit, to an architect, a site to build on, to a planner a zone for recreation or residence or commerce or transportation, or "nature preservation." As in the story of the blind men who feel the elephant—each arriving at a false description of the whole animal by touching trunk or tusk or tail alone—so each discipline and each "interest" group reads and tells landscape through its own tunnel vision of perception, value, tool, and action. And as each shouts its own fragment, landscapes of cities, suburbs, and regions are severed, become impoverished, dysfunctional. It is even fashionable now to design buildings, gardens, and cities deliberately as dislocated and unconnected fragments to emphasize the erosion of common ground, a misanthropic view of cultural differences.

Loss of fluency in the language of landscape, in turn, impoverishes verbal language. Words like *bore* and *guzzle* refer to features and processes many no longer perceive—and which can injure or even kill. To know the meaning (and location) of bore and guzzle is to be safe, to survive: a bore is the "noisy rush of the tides against the current in a narrow channel"; a guzzle, the low place in the dunes where water drains and the sea comes crashing through in a "century storm."[12] To know bore is to avoid it at high tide; to know guzzle is to decide not to build and settle there. Such nuances, preserved in specialized, professional language, are now lost to common verbal language. Aboriginal peoples become more "civilized" and less attuned to landscape; young Papua New Guineans, for example, no longer learn to sing with waterfalls and birds. A loss of language and loss of knowledge limits the celebration of landscape as a partnership between people, place, and other life and further reduces the capacity to understand and imagine possible human relationships with non-human nature.

We shape landscape and language; they shape us. To know landscape is to read in Boston's Fens and Riverway artful reconstructions of places laid waste by human occupation. Not to know is to fail to discern what merely grew and what was planted, and thus, for example, to mistake the Fens and Riverway as "preserved" wetland and floodplain forest; they were, rather, the product of human purpose mindful of natural processes of regeneration. Failure to recognize the Fens and Riverway as *designed* and *built*, not happenstance, blinds us to the possibility of designing and building similar transformations elsewhere.

Landscape metaphors modify perceptions, prompt ideas and actions, molding landscape, in turn. To see wilderness as chaos provokes fear and prompts flight, perhaps even the urge to destroy; to believe it sacred fosters appeal, reverence, and the desire to cherish. To know nature as a set of ideas not a place, and landscape as the expression of actions and ideas in place not as an abstraction or as mere scenery promotes an understanding of landscape as a continuum of meaning.[13] Not to know, and to confuse landscape and nature, is to equate landscape with mountain, meadow, farm, and country road, but not highway or town. Yet a designed urban park is no less a landscape than a planted cornfield, the island of Manhattan no less a landscape than its Central Park. Notions of landscape as countryside, but not city, falsely fragment intimate connections and produce such ironies as inner-city schoolchildren bussed out-of-town to study old-field meadows, ignoring the same plants growing on vacant lots next door.[14] To see landscape as mere scenery gives precedence to appearance at the expense of habitability and risks trivializing landscape as decoration—landscaping—concealing the significance of senses other than sight and of parts hidden from view, the deep context underlying the surface. To call some landscapes natural and others artificial or cultural misses the truth that landscapes are never wholly one or the other.

Once most of life was lived outdoors, in constant dialogue with wind, water, soil, plants, and animals; now most livelihoods no longer depend upon literacy in landscape. Or so we imagine. Our most intimate relationships with nature (finding food and water, disposing of waste) are now negotiated by large, distant institutions. Schools and popular media train children to experience the world many steps removed, through textbooks, calculations, and second-hand images and sounds in film, video, and computer. People work in offices with windows that do not open, or with no windows at all. Meanwhile, naturalists, gardeners, and fly fishermen, like modern shamans, preserve and pass on bits of the language—knowledge of bird and insect, soil and seed, water and fish—and write books to reflect on and communicate the meanings of human life and relations with non-human nature, inheritors of Thoreau's *Walden* like Annie Dillard's *Pilgrim at Tinker Creek*, Michael Pollan's *Second Nature*, and Norman Maclean's *A River Runs Through It*.

Even farming and fishing are now high-tech, capital-intensive industries whose owners may never touch soil or sea. In 1990, the category "farmer" was eliminated from the US Census for the first time in American history. "Agricultural laborer" took its place. Laborers, those in most intimate contact with earth and plants, water and fish, now perform work dictated in distant offices by people who may never see the consequences of their policies and plans. At Disney's EPCOT (Experimental Prototypical Community of Tomorrow), an elaborate exhibit called "The Land," sponsored by Kraft Foods, shows a soil-less, factory-like future for American agriculture: plants hang from wires, move as products in an assembly line, roots exposed, sprayed by aerosols of water and fertilizers, divorced from dirt. This is a perilous vision because the sense of control it conveys is false, an illusion. The plants are fragile, dependent upon the sprayers' continued function; lives hang, literally, by a thread. If the sprayers fail, plants shrivel rapidly, for their roots are not held in soil's reservoir of moisture and nutrients.

The power to read, tell, and design landscape is one of the greatest human talents; it enabled our ancestors to spread from warm savannas to cool, shady forests and even to cold, open tundra. But now, the ability to transform landscape beyond the capacity to comprehend it threatens human existence. Having altered virtually every spot on the planet, humans have triggered perturbations that threaten to change it irrevocably and dangerously. Many, as a consequence, feel control slipping, exposed for the illusion it always was. Our lives are like the plants hanging from wires at EPCOT, roots exposed, dependent upon technologies which, should they fail, will spell disaster. Some speak of the "end of nature," but it is nature as *we* know that is threatened, not the planet itself, not the universe.

To recover and renew the language of landscape is to discover and imagine new metaphors, to tell new stories, and to create new landscapes. John Berger describes, and photographer Jean Mohr illustrates, a language of lived experience with which to interpret the common and the particular across the gulf of different cultures.[15] Gregory Bateson says that humans must learn to speak the language "in terms of which living things are organized," in order to read the world not as discrete things, but as dynamic relations, and to practice the art of managing complex, living systems.[16] Aldo Leopold writes of the need for humans to "think like a mountain" to escape the short-sightedness that threatens the larger habitats of which humans are part.[17] Berger, Bateson, Leopold, and others have envisioned or implied the need for such a language; none have elaborated or codified it. The language of landscape is such a language: in terms of it the world is organized and living things behave, humans can think like a mountain, can shape landscapes that sustain human lives and the lives of other creatures as well, can foster identity and celebrate diversity.

The language of landscape prompts us to perceive and shape the landscape *whole*. Reading and speaking it fluently is a way to recognize the dialogues ongoing in a place, to appreciate other speakers' stories, to distinguish enduring dialogues from ephemeral ones, and to join the conversation. The language of landscape reminds us that nothing stays the same, that catastrophic shifts and cumulative changes shape the present. It permits us to perceive pasts we cannot otherwise experience, to anticipate the possible, to envision, choose, and shape the future. We can see what is not immediate, a forest in a meadow, the yellow-wood dying of starved and suffocated roots, and seeing, we can choose to save or snuff its life. Or we can see water underground in the tree along a dry creek bed, in the cracks of a building's foundation, the slumps in pavement in a city; or see the connections between buried, sewered stream, vacant land, and polluted river, and imagine rebuilding communities while purifying water. And we can imagine poetry.

Human survival as a species depends upon adapting ourselves and our landscapes—settlements, buildings, rivers, fields, forests—in new, life-sustaining ways, shaping contexts that acknowledge connections to air, earth, water, life, and to each other, and that help us feel and understand these connections, landscapes that are functional, sustainable, meaningful, and artful. Not everyone will be farmers or fishermen for whom landscape is livelihood, but all can learn to read landscape, to understand those readings, and to speak new wisdom into life in city, suburb, and countryside, to cultivate the power of landscape expression as if our life depends upon it. For it does.

ACKNOWLEDGMENTS

This chapter was originally published in *The Language of Landscape* under the title "Dwelling and Tongue: The Language of Landscape" (pp. 15–26). © Anne Whiston Spirn 1998, reprinted with permission of Yale University Press.

NOTES

1   See, for example, the tablet of Enannatum I, governor of Lagash, which records the delivery of cedar trees to roof a temple, Mesopotamia, ca. 2900 BC, and other examples in Gyorgy Kepes, ed., *Sign, Image, Symbol* (New York: Braziller, 1966). For a review of the literature on evolution of human cognition, see Merlin Donald, *Origins of Mind* (Cambridge, MA: Harvard University Press, 1991).

2   Roland Barthes, *Elements of Semiology* (New York: Farrar, Straus and Giroux, 1968), 16.

3   George Lakoff and Mark Johnson, *Metaphors We Live By* (Chicago, IL: University of Chicago Press, 1980). Each developed these ideas further in subsequent books: George Lakoff, *Women, Fire, and Dangerous Things* (Chicago, IL: University of Chicago Press, 1987); Mark Johnson, *Body in the Mind* (Chicago, IL: University of Chicago Press, 1987).

4   John Stilgoe, *Shallow-Water Dictionary* (Cambridge, MA: Exact Change, 1990) and *Alongshore* (New Haven, CT: Yale University Press, 1994). While I admire Stilgoe's work on the language of the estuary, I reject his definition of landscape as including only land, not water, and countryside, not city.

5   Brown (1716–1783), Olmsted (1822–1903), Wright (1867–1959), Halprin (1916–2009), Schwartz (1950–).

6   Martin Heidegger, "Building Dwelling Thinking," in *Poetry, Language, Thought* (New York: Harper and Row, 1975), 145–147. Some of these implications have been explored by others, including geographer Edward Relph, *Place and Placelessness* (London: Pion Limited, 1976) and architectural theorist Christian Norberg-Schulz, *Concept of Dwelling* (New York: Rizzoli, 1985).

7   Verner Dahlerup, *Ordbog over det Danske Sprog* (Copenhagen: Nordisk, 1931), Jacob Grimm and Wilhelm Grimm, *Deutsches Worterbuch* (Leipzig: Verlag von S. Hirzel, 1885), Arther R. Borden Jr., *A Comprehensive Old English Dictionary* (Washington, DC: University Press of America, 1982). For a review of the histories of the words landscape, nature, land, and country in English, German, and Scandinavian languages, see Kenneth R. Olwig, "Recovering the Substantive Nature of Landscape" *Annals of the Association of American Geographers* 86, no. 4 (1996): 630–653. See also J. B. Jackson, "The Word Itself," in *Discovering the Vernacular Landscape* (New Haven, CT: Yale University Press, 1984), 3–8. I am grateful to Andre Wink for the translation and interpretation of J. Heinsios, *Woordenboek der Nederlandsche Taale* (Martinus Nijhoff, A.W. Sijthoff, 1916).

8   Webster's *New Universal Unabridged Dictionary* (New York: Simon and Schuster, 1983) and *Oxford English Dictionary* (Oxford: Oxford University Press, 1989).

9   Anthony Hecht, "Gardens of the Villa d'Este," in *Hard Hours* (New York: Atheneum, 1967), 95.

10 Norman Maclean, *A River Runs Through It* (Chicago, IL: University of Chicago Press, 1976), 61.

11 George Lakoff and Mark Johnson, *Metaphors We Live By* (Chicago, IL: University of Chicago Press, 1980). See also Ralph Waldo Emerson, "Nature," in *Essays and Lectures* (New York: Library of America, 1983), 7–49.

12 John Stilgoe, *Shallow-Water Dictionary* (Cambridge, MA: Exact Change, 1990), 23, 28.

13 Many cultures have no single name or notion for *nature*. The singular quality of the English word masks a real multiplicity and implies falsely that there is a single definition. See Raymond Williams, "Ideas of Nature," in *Problems in Materialism and Culture* (London: Verso, 1980), 67–85.

14 J. B. Jackson, "The Word Itself," in *Discovering the Vernacular Landscape* (New Haven, CT: Yale University Press, 1984) limits the definition of landscape to deliberately created, "man-made systems"; John Stilgoe, *Common Landscape of America* (New Haven, CT: Yale University Press, 1982), narrows the definition further, to human-made, non-urban land.

15 John Berger and Jean Paul Mohr, *Another Way of Telling* (New York: Random House, 1983).

16 Gregory Bateson, *A Sacred Unity* (New York: Cornelia and Michael Bessie Book, 1991), 310–311 and 253–257.

17 Aldo Leopold, *Sand County Almanac* (New York: Bantam Books, 1966), 129–33.

Chapter 18: Is Landscape Human?

Charles Waldheim

> "Landscape is a medium found in all cultures. . . . Landscape is a
> particular historical formation associated with European imperialism."[1]
> —W.J.T. Mitchell, "Imperial Landscape," *Landscape and Power* (1994)

In his 1994 essay "Imperial Landscape," W.J.T. Mitchell pondered one of landscape's perennial puzzles, which is no closer to being solved today than it was
when he first considered it. The question that Mitchell framed three decades ago
in his edited volume *Landscape and Power* could not be timelier: To what extent
is landscape shared across all cultures? Another way to frame this is to ask: to
what extent is landscape, like language, an essential aspect of human experience? Alternatively, to what extent is landscape a particular cultural formation
developed in Western Europe and exported to certain cultures? At best, this may
be for some a question hardly worth considering. At worst, the implication of the
question might be offensive to some. I think of it as a serious and important
question that examines the very core of what we think landscape is and what it
means.

Multiple authors in this collection address this question in various ways, and
it is among the many motivations for the book project itself. Landscape's implications in colonialism and imperialism are addressed directly by Burcu Yigit Turan
and John E. Crowley in this volume.[2] Turan's "Is Landscape Colonial?" situates
landscape practices in the contexts of nation-state construction, coloniality, and
race. Crowley's "Is Landscape Imperial?" describes the role of the landscape
imaginary, cartographic practices, and representational regimes in the projection
and maintenance of imperial power. The implications of these questions could
not be more significant today as the field struggles to articulate its response to
calls for societal change in relation to structural racism, environmental injustice,
and a history of landscape as a tool of instrumental spatial power over marginalized, disempowered, and dehumanized populations.

W.J.T. Mitchell raised the question of landscape's relative cultural specificity
or universality on the first page of the chapter "Imperial Landscape" under the
subheading "Theses on Landscape." In a bestial ordering of seemingly contradictory and unreconcilable categories recalling Foucault's reading of Borges,[3] Mitchell posited that landscape ought to be understood simultaneously as both

DOI: 10.4324/9781003148142-19

universal to all cultures and yet also culturally specific to the projection of European imperial power:

Theses on Landscape:

5   Landscape is a medium found in all cultures.
6   Landscape is a particular historical formation associated with European imperialism.
7   Theses 5 and 6 do not contradict one another.[4]

LANGUAGE AND ORDER

While we might think of language as a defining human capacity shared across all cultures, language is not exactly "shared," per se. More precisely, languages are culturally constructed over time and hardly universal. The apocryphal source of Borges's Encyclopedia, his 1942 essay "The Analytical Language of John Wilkins," examined the seventeenth-century English natural philosopher's proposal for a universal language based on mathematically derived symbols or ciphers. Wilkins's proposal had little effect, and his entry into the *Encyclopedia Britannica* was removed from the 14th edition. Borges described the "arbitrarities" of Wilkins's classification system with the disclaimer that "it is clear that there is no classification of the universe not being arbitrary and full of conjectures. The reason for this is straightforward: we do not know what thing the universe is."[5] Borges's depiction of the utter lack of correspondence across diverse languages inspired Michel Foucault's *Les Mots et les Choses* (*The Order of Things*). In the Preface to his 1966 volume, Foucault described how the book "first arose out of a passage in Borges, out of the laughter that shattered, as I read the passage, all the familiar landmarks of my thought—our thought, the thought that bears the stamp of our age and our geography."[6] Foucault went on to describe the "wonderment of this taxonomy, the thing that we apprehended in one great leap, the thing that, by means of the fable, is demonstrated as the exotic charm of another system of thought, is the limitation of our own, the stark impossibility of thinking *that*."[7]

If we return to the definition of landscape offered in the Introduction to this volume, we begin to understand what is at stake in this statement: "Landscape is a mode of aesthetic reception between certain human subjects and their environments, imagined or perceived."[8] Should we think of landscape as available to all human subjects across all cultures? What would the implications of that be for the meaning of the field? Note that at this point, the question is not whether it *should* be. While that is another equally important question, this volume considers how and what landscape *means* and how it lands for people in the world. If we accept the dominant narrative and overwhelming consensus that landscape was (and is) a particular product of certain cultures at certain historical moments, then the projection of this cultural product onto other cultures, absent any evidence of its impacts on those cultures, seems a definition of imperialism.

If we posit that this *ought* to be done, as if every culture deserves this wonderous gift of European extraction, we have compounded the problem. The challenge with this line of thought is that it condescends in the very act of generosity that it is putatively motivated by. Any serious consideration of this topic confronts landscape's longstanding *inside/outside* problem and the question of landscape literacy.[9] Must we be educated to see landscape, to "read" it, as they say? If so, how is the expansion of landscape's literacies not simply the ongoing projection of European imperial power through other means?

In the Introduction to *Landscape and Power*, Mitchell asserted that his aspiration for the volume was to change our understanding of landscape "not as an object to be seen or a text to be read, but as a process by which social and subjective identities are formed."[10] As such, Mitchell sought to move our understanding of landscape from a category of art history to a relational medium of subject formation, world-building, and power relations. In so doing, he chronicled the established historical art and geographic literature on the subject, with a particular focus on three characteristics of what he described as a "pure" landscape. First, it is an inherently Western invention. Second, it emerges as a category of cultural production in the early modern era. Third, it is a fundamentally visual medium.[11] From these characteristics, Mitchell constructs the logical premise that landscape is thus an inherently imperial medium conceived in Western Europe and projected, albeit unevenly, onto various cultures and colonies throughout the world. This account has held up surprisingly well over the past three decades as its premise was built upon the literature and claims of multiple fields and disciplines, including art history, cultural geography, and landscape architecture, among others.

ORIGINS AND MEANING

In the second half of the twentieth century, Anglophone art historians on both sides of the Atlantic returned to the origins of landscape in European landscape painting. Mitchell's account returns to British art historian Kenneth Clark's 1949 publication *Landscape Into Art*.[12] Clark's account of landscape origins and meaning grew from his lectures on landscape painting at Oxford. While troubled by Clark's untroubled invocation of "we" in considering landscape and its meanings, Mitchell accepts Clark's preeminent stature on the topic and that his account contributes to the dominant narrative concerning the "pure" form of the field.[13] Another authoritative art historian on the origins and meaning of landscape, Ernst Gombrich, contributed to the topic with his interpretation of the emergence of landscape painting. Gombrich's essay "The Renaissance Theory of Art and the Rise of Landscape" offered an even more enduring, definitive account of landscape as a form of cultural production that emerged as a product of the European Renaissance.[14]

The accounts of landscape's origins in art history in the post-War era were corroborated slightly later by authors working in cultural geography. No less an authority than John Brinkerhoff "J.B." Jackson took up the topic of landscape's origins and meaning through a series of essays in his 1986 publication *Discovering the Vernacular Landscape*. In the first of those, "The Word Itself," Jackson puzzled over the lack of consensus on these questions while acknowledging landscape painting's European origins:

> Why is it, I wonder, that we have trouble agreeing on the meaning of *landscape*? The word is simple enough, and it refers to something which we think we understand; and yet to each of us it seems to mean something different. . . when it was first introduced (or reintroduced) into English it did not mean the view itself, it meant a *picture* of it.[15]

Just two years prior to Jackson's essay, another geographer, Denis Cosgrove, described landscape as an ideological project bound up in the formation of human subjects and world-building:

> Landscape denotes the external world mediated through subjective human experience in a way that neither region nor area immediately suggests. Landscape is not merely the world we see, it is a construction, a composition of that world. Landscape is a way of seeing the world.[16]

Cosgrove's 1984 publication *Social Formation and Symbolic Landscape* presented a robust account of landscape's meaning in relation to certain human subjects and their view of the world. In so doing, he also described the decreasing cultural relevance and persuasiveness of thinking landscape through painting, as the field had been transformed into a form of applied natural science:

> The landscape idea represents a way of seeing—a way in which some Europeans have represented to themselves and to others the world about them and their relationships with it, and through which they have commented on social relations. . . . The landscape idea emerged as a dimension of European elite consciousness at an identifiable period in the evolution of European societies: it was refined and elaborated over a long period during which it expressed and supported a range of political, social, and moral assumptions and became accepted as a significant aspect of taste. That significance declined, again during a period of major social change in the late nineteenth century. Landscape today is pre-eminently the domain either of scientific study and land planning, or of personal and private pleasure.[17]

With this argument, Cosgrove simultaneously articulated the role of landscape as a particular cultural form produced by and for certain European elites while acknowledging the contradiction in which that European cultural imperial project had been transmogrified into a form of applied natural science. While other authors tracked this shift of landscape's meaning from specific cultural production to universally true science, none were as prescient as Cosgrove in understanding the import of this shift on the meaning of landscape for the rest of the world. If landscape was by the time of Cosgrove's account no longer the sole domain of European elites favored by fortune and possessing a landscape literacy, what might it mean as an applied science of environmental planning for populations unburdened by such cultural sophistication? Cosgrove didn't dwell on this question but raised it and productively complicated it. If landscape were no longer the sole provenance of European elites in its newfound universality, wouldn't it still be an ideological construct and implement of power between the wealthy world and the "majority world" that remained unenlightened as to the landscape idea? This shift from culturally specific forms of landscape painting to universally true forms of landscape planning is deserving of its own extended study, well beyond the scope of this modest essay. On the one hand, we might consider the abandonment of the old "pure" account of landscape as a European cultural export as a long overdue cultural corrective, offering an opening to other points of view, as it were. On the other hand, the emergence of landscape as a form of applied natural science in the nineteenth century and the ascendance of landscape in the twentieth century obscured its ideological origins and import. This was Cosgrove's point, in part, that like other forms of universalizing truths proclaimed in the name of science, landscape's reinvention as environmental advocacy through scientific knowledge increased the asymmetrical power relations between elite experts and their citizen subjects. It did so by shifting the form of knowledge from cultural to scientific knowledge. In so doing, it moved the field from a self-conscious elitism founded on literacy and claims of invention toward an equally problematic projection of universal truths alloyed by the moral authority of belief. This shift rendered the human subjects without access to this scientific truth equally impoverished with respect to landscape's agency, as the medium continued to flow from the elites to the masses, yet with its ideology effectively obscured underneath a positivist belief in the scientific basis of its claims.

SCIENCE AND SUBJECTIVITY

Among the myriad implications of this shift of landscape's meaning for human subjects was its effective rendering of all of humanity subject to its truth claims. Consider, for example, Ian McHarg's reformation of the field of landscape architecture in the 1950s, 1960s, and 1970s. Beginning with his appointment as chair of the newly formed Department of Landscape Architecture and Regional Planning at the University of Pennsylvania in 1954, McHarg and his collaborators mobilized a new understanding of landscape architecture as a proto-natural science applied to the planning of the urban region. This new disciplinary and professional formulation was developed through a new curriculum, new faculty expertise, new tools, new literature, and new practices. In 1957, McHarg launched a new course, "Man and Environment," which became one of the most popular courses at the University of Pennsylvania. Six years later, McHarg authored what he considered to mark an important threshold in the development of his ideas under the very same title, "Man and Environment."[18] In 1966, he joined other leading landscape architects in signing the "Declaration of Concern," advocating for a central role for landscape architects in the environmental movement. In 1969, his canonical publication *Design with Nature* consolidated his theoretical and methodological contributions to the new form of landscape architecture as an applied natural science. McHarg changed the field's trajectory internationally through public lectures, television programs, and an array of conferences, exhibitions, and public engagements. He also revealed a new, or perhaps latent, definition of landscape and what it might mean. Throughout these programs and publications, McHarg consistently implied an understanding of landscape as a universal medium attendant to all human beings. This understanding was evident in McHarg's *Design with Nature,* as well as his reference to cities as a form of cancer on the planet. In 1971, he summed this up succinctly in titling a public lecture: "Man: Planetary Disease."[19] While McHarg was not an early adherent to the Voluntary Human Extinction Movement, he was an extraordinary rhetorician with an often-proselytizing manner and preacher's stamina.[20] While the gendered language of "Man" in his formulation of landscape subjects has not aged well, it would be a mistake to dismiss this as a generational anachronism. McHarg's formulation of "Man" as the subject of landscape is not simply dated, gendered language (although it is). Beyond the generational shift in gendered language, McHarg clearly refers to a *landscape* that exists a priori across all cultures. We might read this in a couple of ways. On the one hand, McHarg could be understood as simply a modernist technocrat, exporting his knowledge from Europe to benefit less fortunate populations. On the other hand, we might take McHarg as understanding landscape as something shared by all people, a priori, and prior to cultural construction. This sounds a lot like an unreconstructed and uncritical understanding of landscape as a kind of applied natural science. Given McHarg's predisposition to reconstruct landscape architecture as a form of regional planning through the application of scientific knowledge, this account seems the simplest and most elegant explanation.

Perhaps more relevant to this argument is that McHarg's student and protégée Anne Whiston Spirn has long argued that landscape is analogous to language as a fundamental human capacity. While different cultures express landscape and language in different modalities, it is essentially a shared capability that, in some ways, defines us as humans. The excerpt from Spirn's "Is Landscape Language?" reprinted in this volume touches on this sentiment.[21] Spirn productively contrasts that sense of landscape as a potential shared human medium with an equally persuasive call for the urgency of maintaining and developing "landscape literacy" as a more specialized form of cultural knowledge. Through this dialectic, Spirn has productively addressed the question of landscape's putative universality across all cultures while maintaining a sense of it as a highly developed intellectual, cultural, and social art.

IMPLICATIONS AND ASPIRATIONS

The shift of landscape from an elite form of cultural production and reception specific to certain subjects in certain cultures in favor of universally true natural science has effectively continued the colonial power relations between empire and colony through different means. Perhaps in this reading, Mitchell found no contradiction between landscape as a historical formation of European imperialism and landscape as a medium found in all cultures. Perhaps both forms of landscape, landscape painting as cultural production and landscape planning as applied natural science, have been equally culpable in maintaining power relations between elite knowledge producers and the subjects of their worldbuilding exercises. Rather than relitigating the field's identity as pure art or science, reconsidering its history could be more useful and might offer a third term. This new landscape would be a decolonized relational condition in which we learn to read the various culturally specific and regionally diverse forms of mediation between human subjects and their worlds via a radical decentering of origins, agendas, and authorities. This would suggest that we focus our disciplinary attention on the languages and literacies that attend to landscape through the innumerable instances of human beings shaping their relations to the world through various forms of mediation. This third iteration of landscape might be worthy of being considered human, even if the two previous versions might not be worthy of the privilege.

NOTES

1 W.J.T. Mitchell, "Imperial Landscape," in *Landscape and Power*, ed. W.J.T. Mitchell (Chicago, IL: University of Chicago Press, 1994), 5.

2 See Burcu Yiğit-Turan, "Is Landscape Colonial?," chapter 3; and John E. Crowley, "Is Landscape Imperial?," chapter 6 in this volume.

3 Jose Luis Borges's 1942 essay, "The Analytical Language of John Wilkins," describes a 'Chinese Encyclopedia,' the *Celestial Emporium of Benevolent Knowledge* where it is claimed that "animals are divided into: (a) belonging to the Emperor, (b) embalmed, (c) tame, (d) sucking pigs, (e) sirens, (f) fabulous," etc. See Jose Luis Borges, "The Analytical Language of John Wilkins,"*Other Inquisitions, 1937–1952*, trans. Ruth L. C. Simms (Austin: University of Texas Press, 1964), 101–105.

4 W.J.T. Mitchell, "Imperial Landscape," in *Landscape and Power*, ed. W.J.T. Mitchell (Chicago, IL: University of Chicago Press, 1994), 5.

5 Jose Luis Borges, "The Analytical Language of John Wilkins,"*Other Inquisitions, 1937–1952*, trans. Ruth L. C. Simms (Austin: University of Texas Press, 1964), 103.

6 Michel Foucault, "Preface," in *The Order of Things: An Archaeology of the Human Sciences* (New York: Vintage, 1994), xv.

7 Michel Foucault, "Preface," *The Order of Things: An Archaeology of the Human Sciences* (New York: Vintage, 1994), xv.

8 Doherty and Waldheim, "Introduction," in this volume. This triad of reception recalls something of Félix Guattari's "Three Ecologies," the mental, the social, and the environmental. See Félix Guattari, *The Three Ecologies,* trans., Ian Pindar and Paul Sutton (London and New York: Continuum, 2000).

9 The concept of a "landscape literacy" has been articulated most productively by Anne Whiston Spirn as a form of reading beyond superficial appearances to understand a landscape's underlying processes, both biophysical and societal. See Anne Whiston Spirn, "Landscape Literacy and Design for Ecological Democracy," *Grounding Urban Natures: Histories and Futures of Urban Ecologies*, eds. Henrik Enstson and Sverker Sörlin (Cambridge: MIT Press, 2019), 109–136.

10 W.J.T. Mitchell, "Introduction," in *Landscape and Power* (Chicago, IL: University of Chicago Press, 1994), 1.

11 W.J.T. Mitchell, "Imperial Landscape," in *Landscape and Power*, ed. W.J.T. Mitchell (Chicago, IL: University of Chicago Press, 1994), 7.

12 W.J.T. Mitchell, "Imperial Landscape," in *Landscape and Power*, ed. W.J.T. Mitchell (Chicago, IL: University of Chicago Press, 1994), 6–13; Kenneth Clark, *Landscape Into Art* (London: J. Murray, 1949).

13 W.J.T. Mitchell, "Imperial Landscape," in *Landscape and Power*, ed. W.J.T. Mitchell (Chicago, IL: University of Chicago Press, 1994), 6–7.

14 Ernst Gombrich, "The Renaissance Theory of Art and the Rise of Landscape," in *Norm and Form* (London: Phaidon, 1966), 107–121.

15 J. B. Jackson, "The Word Itself," in *Discovering the Vernacular Landscape* (New Haven, CT: Yale University Press, 1986), 1.

16 Denis Cosgrove, "The Idea of Landscape," in *Social Formation and Symbolic Landscape* (Kent, UK: Croom Helm, 1984), 13.

17 Denis Cosgrove, "Introduction," in *Social Formation and Symbolic Landscape* (Kent, UK: Croom Helm, 1984), 1–2.

18 Ian McHarg, "Man and Environment," in *The Urban Condition: People and Policy in the Metropolis*, ed. Leonard J. Duhl (New York: Basic Books, 1963), 44–58.

19 Ian McHarg, "Man: Planetary Disease," North American Wildlife and Natural Resources Conference, Portland, Oregon, 1971.

20 The Voluntary Human Extinction Movement (VHEMT) proposes the voluntary end of Homo sapiens to end human suffering and allow nature to proceed unfettered by humans. See https://www.vhemt.org (accessed January 10, 2023).

21 See Anne Whiston Spirn, "Is Landscape Language?," chapter 17 in this volume; and Spirn, *The Language of Landscape* (New Haven, CT: Yale University Press, 1998).

Index